WEARABLE TECHNOLOGIES

Edited by **Jesús Hamilton Ortiz**

Wearable Technologies
http://dx.doi.org/10.5772/intechopen.71472
Edited by Jesús Hamilton Ortiz

Contributors

Farouq Aliyu, Senem Kurşun Bahadır, Umut Kivanc Sahin, Venere Ferraro, Mila Stepanovic, Silvia Ferraris, Masahiko Suzuki, Hiroo Terashi, Hiroshi Mitoma, Makiko Yogo, Masayo Morita, Mutsumi Iijima, Yasuyuki Okuma, Satoshi Orimo, Akito Hayashi, Hiroya Utsumi, Mitsuru Yoneyama, Masato Takada, Paulo Fernandes Silva Júnior, Alexandre Jean René Serres, Georgina Karla Freitas Serres, Edmar Candeia, Joabson Nogueira, Raimundo Freire, Ewaldo Santana, Soochan Kim, Kanitthika Kaewkannate, Han Jin, Nikolay Bazaev, Nikita Zhilo, Viktor Grinval'd, Sergey Selishchev, Brian Leaker, Arsam Shiraz, Andreas Demosthenous, Jung-Sim ROh, Pandiyarasan Veluswamy, Suhasini Sathiyamoorthy, Ikeda Hiroya, Manikandan Elayaperumal, Malek Maaza, Noushin Nasiri, Antonio Tricoli

First published in London, United Kingdom, 2018 by IntechOpen
IntechOpen is the global imprint of INTECHOPEN LIMITED, registered in England and Wales, registration number: 11086078, The Shard, 25th floor, 32 London Bridge Street
London, SE19SG – United Kingdom
Printed in Croatia

British Library Cataloguing-in-Publication Data
A catalogue record for this book is available from the British Library

Additional hard copies can be obtained from orders@intechopen.com

Wearable Technologies, Edited by Jesús Hamilton Ortiz
p. cm.
Print ISBN 978-1-78984-003-2
Online ISBN 978-1-78984-004-9

For EU product safety concerns:
IN TECH d.o.o., Prolaz Marije Krucifikse Kozulić 3, 51000 Rijeka, Croatia,
info@intechopen.com or visit our website at intechopen.com.

Meet the editor

Jesus Hamilton Ortiz obtained his Bachelor's Degree in Mathematics at the Universidad Santiago de Cali, his DEA in Telecommunications Engineering at the Technical University of Madrid, his PhD in Computer Engineering from the Universidad de Castilla La Mancha, and his PhD(c) in Telecommunication Engineering from the Autonomous University of Madrid. He was previously an assistant professor and researcher at the Universidad de Castilla La Mancha and is currently a full-time professor in the School of Computer Science and Electronic Engineering at the Universidad Santiago de Cali. He is also an editor associate in *IEEE Access*. He is a founding partner of SME close mobile R&D, focused on R&D projects around wireless networks and RPAS.

Contents

Preface

This book includes recent developments in key areas of interest in health, new materials, and technology TI. It is also suitable as a text for graduate students, engineers, physicians, etc. The book is organized into three sections:

1. Health
2. New Materials
3. Technology TI

"Wearables technology" is a blanket term for electronics that can be worn on the body, either as an accessory or as part of material used in clothing. One of the major features of wearable technology is its ability to connect to the internet, enabling data to be exchanged between a network and the device. This ability to both send and receive data has pushed wearable technology to the forefront of the Internet of Things. This book supplies the foundation and insight you will need to make your own contributions to the field. It includes coverage of tools or devices to improve health and use in sports activities such as fitness, new materials such as conductors made of organic polymers, embroidered circuits made of smart materials, bio-inspired antennas used in the textile area, and middleware for the wearables industry.

Dr. Jesús Hamilton Ortiz
Closemobile R&D startup Spain
Closemobile SAS Colombia
COMBA group
Universidad Santiago de Cali, Columbia

Section 1

Health

Chapter 1

Advances in Wearable Sensing Technologies and Their Impact for Personalized and Preventive Medicine

Noushin Nasiri and Antonio Tricoli

Additional information is available at the end of the chapter

http://dx.doi.org/10.5772/intechopen.76916

Abstract

Recent advances in miniaturized electronics, as well as mobile access to computational power, are fostering a rapid growth of wearable technologies. In particular, the application of such wearable technologies to health care enables to access more information from the patient than standard episodically testing conducted in health provider centres. Clinical, behavioural and self-monitored data collected by wearable devices provide a means for improving the early-stage detection and management of diseases as well as reducing the overall costs over more invasive standard diagnostics approaches. In this chapter, we will discuss some of the ongoing key innovations in materials science and micro/nanofabrication technologies that are setting the basis for future personalized and preventive medicine devices and approaches. The design of wire- and power-less ultra-thin sensors fabricated on wearable biocompatible materials that can be placed in direct contact with the body tissues such as the skin will be reviewed, focusing on emerging solutions and bottlenecks. The application of nanotechnology for the fabrication of sophisticated miniaturized sensors will be presented. Exemplary sensor designs for the non-invasive measurement of ultra-low concentrations of important biomarkers will be discussed as case studies for the application of these emerging technologies.

Keywords: non-invasive health care, biosensors, wearable electronics

1. Electrochemical sensors

Electrochemical sensors have since long being integrated into microfluidic systems for the measurement of very low concentration of biomarkers [1, 2]. The first electrochemical sensors were developed for oxygen monitoring in the second half of the twentieth century [3]. Miniaturized electrochemical sensors for many toxic gases were commercialized in the 1980s showing good sensitivity to the permissible exposure limit ranges and selectivity [4]. Compared to optical, mass and thermal sensors, electrochemical sensors are especially attractive because of their remarkable sensitivity, experimental simplicity and low cost [5, 6]. Furthermore, they benefit small power requirements facilitating their integration in small sensing systems. Currently, a variety of electrochemical sensors are being used as detectors in stationary and portable devices for personal safety, clinical, industrial, environmental and agricultural applications [7–9].

Electrochemical detection is usually performed by reducing or oxidizing an electro-active analyte and monitoring the current or potential between the working (WE) and the counter electrodes (CE) as a function of time [5, 6]. As many compounds present relatively easy oxidable or reducible sites, the electrochemical detection mechanism is applicable to a wide number of relevant analytes. There are three main types of electrochemical sensors: potentiometric, amperometric and conductometric [10]. For potentiometric sensors, a local equilibrium is established at the sensor interface, where either the electrode or the membrane potential is measured. The information about the composition of a sample is obtained from the potential difference between two electrodes [11, 12]. Amperometric sensors exploit the use of a potential applied between a reference and a working electrode, to cause the oxidation or the reduction of an electro-active species, which result in currents in a proportional current [13, 14]. Conductometric sensors are based on the measurement of the conductivity at a series of frequencies [15].

Figure 1A shows a schematic of the operation principle of an electrochemical sensor [16]. Commonly, a constant voltage, sufficiently high to initiate the reduction or the oxidation of the analyte, is applied between the working and the counter electrode (**Figure 1A**). The working electrode is usually nanostructured and functionalized with a molecule-specific layer of enzymes or antibodies, providing a selective surface for the redox of the targeted analyte [16, 17]. This results in the rise of a catalytic current between the working and the counter electrode, which is proportional to the analyte concentration [2, 16].

While electrochemistry methods have since long been applied to the characterization of molecules dispersed in liquids, and electrochemical biosensors are commercially available, their integration in wearable sensors is relatively new. In particular, the fabrication of miniaturized electrochemical cells and circuitry that can be placed in close contact with body tissues and related fluids has been challenging. The development of flexible electronic micro-fabrication approaches has enabled the engineering of skin-like planar electrochemical cells that require very small amount of liquids and can be placed on the human skin [18] (**Figure 1B**), iris [6, 19] (**Figure 1C**) and saliva [20] (**Figure 1D**). For instance, the design of functional micro-electrodes on flexible polymer lenses has enabled the online measurement of glucose and other key biomarkers directly from human tears [6, 19] (**Figure 1C**). Typical layouts consist of working (WE), counter (CE) and reference (RE) electrodes [21] deposited on flexible insulating and biocompatible substrates such as 2-methacryloyloxyethyl phosphorylcholine (MPC) polymer [6, 22], polydimethyl siloxane (PDMS) [6, 22] and polyethylene terephthalate (PET) [19]. A current between the working and the counter electrode is generated by targeted redox reactions between the biomarker and the modified working electrode while the opposite reaction takes place at the counter electrode [21, 23]. This results in the generation of an electrochemical

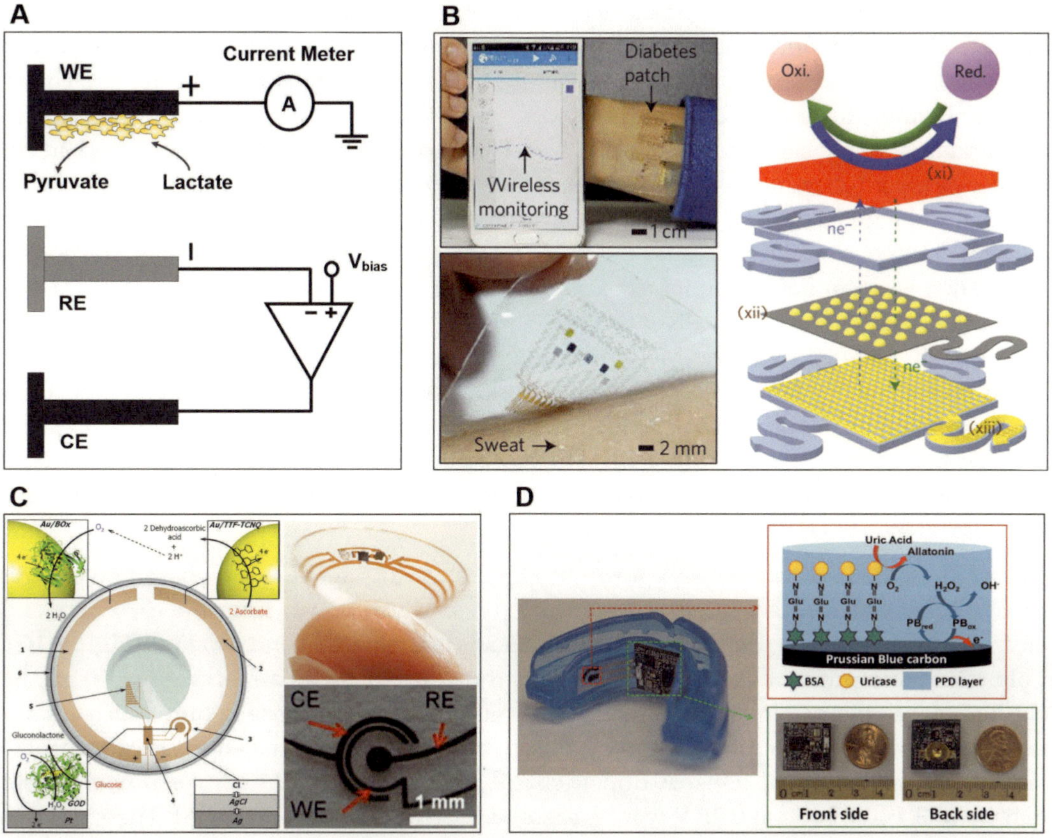

Figure 1. (A) Typical operation principles and sensing mechanisms for electrochemical lactate sensors for other metabolite the functional layer of the working electrode is designed to provide specific bonding and reaction sited to the targeted molecules [73]. Reproduced with permission [73]. Copyright 2017, Wiley Online Library. (B) In-vivo demonstration of a wearable diabetes monitoring and therapeutic system. The electrochemical analyser as an integrated Bluetooth circuit to communicate with external devices (top left). Optical image of the electrochemical sensor array on the human skin collecting perspiration for glucose monitoring (bottom left). Schematic of the GP-hybrid electrochemical unit, which consists of an electrochemically active layers (xi), gold-doped graphene (xii) and a serpentine Au mesh (xiii), from top to bottom (right) [18]. Reproduced with permission [18]. Copyright 2016, Nature Publishing Group. (C) Schematic layout and characterization of a tear glucose electrochemical sensor with an integrated biofuel cell (BFC) on a contact lens [6, 19]. Reproduced with permission [6, 19]. Copyright 2013, ACS Publications, Copyright 2012, Elsevier. (D) Demonstration of a mouthguard biosensor with integrated wireless amperometric circuit board, reagent layer made of a chemically-modified working electrode and the wireless amperometric circuit board: front side and back side [20]. Reproduced with permission [20]. Copyright 2015, Elsevier.

current which, in diluted conditions, is proportional to the concentration of the biomarker [17]. A major advantage is that the selectivity can be enhanced both by coating the nanotexture with highly selective functional groups such as enzymes, antibodies and carefully engineered peptides, and by selecting the appropriate voltage potential for activation of the selected reaction.

Lee et al. [18] reported soft materials, device designs and system integration strategies for a new class of diabetes monitoring and therapy devices based on functionalized chemical vapour deposition (CVD) graphene. Graphene biochemical sensors with solid-state Ag/AgCl counter electrodes have shown enhanced electrochemical activity, sensitivity and selectivity for the detection of important biomarkers contained in human sweat (**Figure 1B**). Such hybrid interconnections and physical sensors efficiently transmit the signal through the stretchable array and supplement electrochemical sensors, respectively. The orchestrated monitoring of biomarkers and physiological cues with sweat control and transcutaneous drug delivery achieves a closed-loop, point-of-care treatment for diabetes (**Figure 1B**). This demonstrated a stretchable device featuring a serpentine bilayer of gold mesh and gold-doped graphene as an efficient electrochemical interface for the stable transfer of electrical signals. The patch consists of a heater, temperature, humidity, glucose and pH sensors and polymeric microneedles that can be thermally activated to deliver drugs transcutaneously. They showed that the patch can be thermally actuated to deliver Metformin and reduce blood glucose levels in diabetic mice. The diabetes patch is laminated on the human skin and is electrically coupled to a portable electrochemical analyser, which in turn wirelessly transfers data to remote mobile devices and supplies power to the patch (**Figure 1B**). The humidity sensor monitors the increase in RH. It takes about 15 min for the sweat-uptake layer of the patch to collect sweat for the measurement and corresponds to an RH of over 80% [18].

Falk et al. [6] demonstrated a self-powered glucose measurement system based on a miniature biofuel cells (BFCs). It generates ca. 3.1 μW cm^{-2} of power from the ascorbate and oxygen available in basal tears without influencing the glucose concentration (**Figure 1C**). Calibration and proof of concept were demonstrated ex-situ with a macrocell of 30 mL by chronopotentiometry. Notably, the device showed no sensor response to pure glucose solutions, while a strong electrochemical response was achieved with the addition of ascorbate and ascorbate-dopamine fuels. An open circuit voltage of 0.54 V and a maximal power density of 3.1 μW cm^{-2} at 0.25 V were measured with human basal tears [6]. These devices showcased a stable current density output of 0.55 μA cm^{-2} at 0.4 V over 6 h of continuous operation.

Kim et al. [20] presented electrochemical sensors integrated on mouthguards featuring a Bluetooth low-energy communication system for the real-time amperometric monitoring of uric acid (**Figure 1D**). An electrochemical three-electrode layout was screen-printed on the mouthguard and functionalized with a uricase enzyme-modified layer (**Figure 1D**). A 2.45 GHz chip antenna was employed for wireless transmission. These sensors were validated by in vivo measurements of salivary uric acid (SUA) levels in hyperuricaemia patients and healthy individuals. The device was powered with two 1.55 V, 33 mAh watch batteries that were connected in series and consumed in average 21 mW in an active mode and 0.6 mW in a sleep mode. Using a measurement frequency of 60 s enabled battery charge to last up to ca. 5 days. A clear distinction in the SUA levels of the healthy control (178.5 ± 20.7 mM) and hyperuricaemia patients (822.6 ± 26.25 mM) was observed, indicating a good potential of this approach for the diagnostics of hyperuricaemia. A sensitivity of 2.45 mA per mM of uric acid was determined with uric acid solutions with 100–600 mM with an R^2-correlation coefficient of 0.998. Further decreasing the devices size, while reducing power consumption and requirement of integrated batteries may, has the potential to lead to significant commercial applications.

2. Capacitive sensors

Non-contact capacitive sensors work by measuring the changes in capacitance induced by the interaction of a sensing layer with an analyte [24, 25]. When a voltage is applied across two electrodes in an open circuit configuration, an electric field is created between them causing positive and negative charges to collect on the electrode (**Figure 2A**). If the polarity of the voltage is reversed, the charges will also reverse. Usually, capacitive sensors use alternating voltage which causes the charges to continually reverse their positions. The moving of the charges creates an alternating electric current which is strongly influenced by the interaction with the target analyte (**Figure 2B**). More in specific, the current is a function of the capacitance, and the capacitance is determined by the area, proximity of the electrodes and the properties of the space between the electrodes. The larger and closer the electrodes, the higher the current.

Capacitive sensing is becoming increasing popular as an alternative to optical and mechanical sensors for multiple applications such as proximity/gesture detection, material analysis and liquid level sensing. Amongst the advantages of capacitive sensing are its applicability to different kinds of materials such as skin, plastics, metals and liquids, and its contactless and wear-free sensing mechanism, high miniaturization potential, low cost and low-power requirements. Amongst other configurations, capacitive biosensing can be based on capacitive coupling with the human body. This results in a reliable solution for measuring liquid levels (**Figure 2C**) and material composition and creating human-to-machine interfaces such as digital buttons. **Figure 2C–E** displays three basic implementations for capacitive sensing, such as liquid-level sensing (**Figure 2C**), proximity/gesture recognition (**Figure 2D**) and material analysis (**Figure 2E**).

Chen et al. [24] demonstrated low-force capacitive contact lens sensors (**Figure 2F**) for the continuous monitoring of the intraocular pressure (IOP). This is a primary indicative factor in the diagnosis and treatment progress monitoring of glaucoma. The innovative design utilized capacitive sensors for the measurement of these ultra-low pressure variations. To facilitate the wireless transmission of the pressure sensor readings, a curvature-sensing capacitor (C) was coupled with an inductive coil (L) having a fixed inductance resulting in an LC resonant circuit. Changes in the curvature of the soft cornea, and thus in the contact lens electrode spacing, result in variation of the capacitance and resonance frequency of the LC circuit. The resonance frequency was measured wirelessly via a reader integrated in the glass frame worn by the user [26]. The sensing element consisted of a variable gap-sensing capacitor. The electrode with a soft gap was embedded in a soft silicone rubber layer on the corneal side of the lens, while a reference electrode and an inductive coil were fabricated in a hard silicon rubber layer on the outer layer of the lens (**Figure 2F**). The flexible-sensing capacitor was electrically coupled with the hard inductive coil. The sensor lens layout was designed for human eye. For an average person with a cornea radius curvature of 8 mm, the curvature changes due to the typical IOP variation between 5 and 40 mmHg is 0.12 mm for a typical cornea. To maximize the linearity of the sensor response to the pressure change, the difference

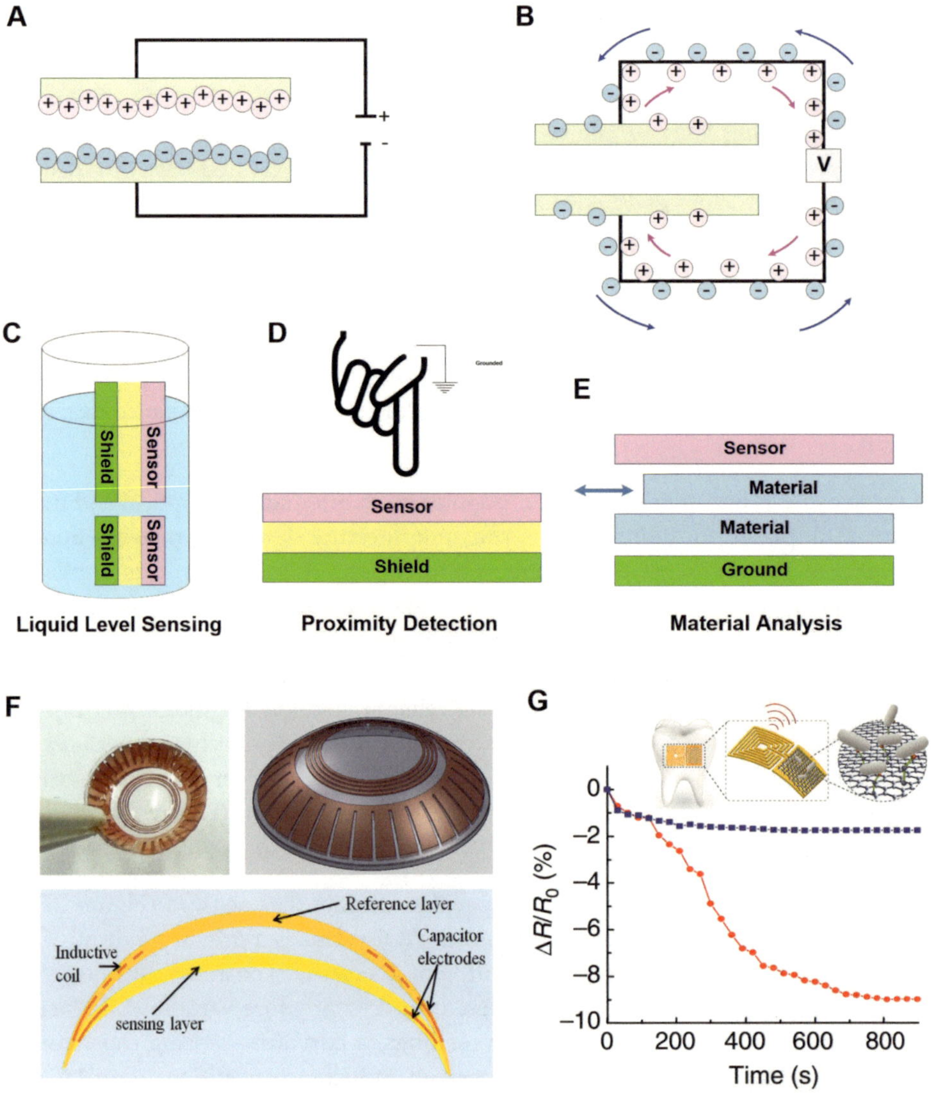

Figure 2. (A) Simplified schematic of conductive sensors: applying a voltage to two electrically insulated electrodes causes positive and negative charges to collect on each side. (B) Applying an alternative voltage causes the charges to move back and forth between the objects, creating an alternating current which is influenced by the space between the electrodes such as the concentration of a target analyte. Basic implements for capacitive sensing: (C) liquid level sensing (parallel fingers), (D) proximity detection (isolated sensor) and (E) material analysis (parallel plate). (F) Capacitive contact lens sensor for IOP monitoring. The sensing element (LC circuit) is embedded into silicone rubber in a double-layer contact lens sensor [24]. Reproduced with permission [24]. Copyright 2013, Elsevier. (G) Biotransfer of the nanosensing architecture onto the surface of a tooth with magnified schematic of the sensing element illustrating wireless readout [25]. Percentage change in graphene resistance versus time following exposure to ~100 H. pylori cells in human saliva (red line). The response to 'blank' saliva solution is shown as blue line [25]. Reproduced with permission [25]. Copyright 2012, Nature Publishing Group.

in radius between the sensing and the reference layers had to be maximized. However, an overly large difference would reduce the IOP sensitivity. A linear, but sufficient range of sensitivity was obtained with a 0.5 mm curvature change for the capacitor. In addition, a circular spiral multi-turn inductive coil designed to have a high quality factor (Q) was used to maximize the reading resolution of the LC resonance circuit.

Mannoor et al. [25] presented a novel approach for passive, wireless, graphene nanosensors to be placed in contact with biomaterials. A biointerfaced-sensing platform, which can be tuned to detect target analytes, was fabricated via silk bioresorption. In particular, they demonstrated their integration onto a tooth for remote monitoring of respiration and bacteria detection in saliva (**Figure 2G**). The architecture consisted of a parallel resonant circuit with a gold inductive coil for wireless transmission and interdigitated capacitive electrodes in contact with graphene-resistive sensors. The resulting system is a passive wireless telemetry device, without the need of on-board power sources and external connections. More in detail, a single-layer, thin-film, inductor-capacitor (LC) resonant circuit integrated in parallel with the resistive graphene monolayer enabled the wireless readout and battery-free operation (**Figure 2G**). The binding of pathogenic bacteria on the graphene nanosensor resulted in variation of its conductance, which induces changes in the characteristic frequencies and bandwidth of sensor resonance (**Figure 2G**). Upon recognition and binding of the specific bacterial targets by the immobilized peptides, the variation of the electrical conductivity of the graphene film was wirelessly monitored using an inductively coupled radio frequency (RF) reader device. The exposure to a 1 μl sample of human saliva containing ~100 *Helicobacter pylori* cells resulted in a real-time variation of the sensor signal as presented in **Figure 2G**. By contrast, 1 μL of "blank" saliva solution without any bacterial cells, used as a control, showed good selectivity and application potential.

3. Chemiluminescence sensors

Luminescence is the emission of light from an electronically excited compound returning to the ground state. The source of excitation energy serves as a basis for a classification of the various types of luminescence and includes electromagnetic waves, heat, friction, electric field or chemical reaction (**Figure 3A**). When the energy source for the excitation is a chemical reaction, this phenomenon is called chemiluminescence (CL). Chemiluminescence measurements usually consist of monitoring the rate of production of photons, which in turn depends on the rate of the luminescent reaction [27, 28]. The light intensity is directly proportional to the concentration of the limiting reactant involved in the luminescence reactions. Thanks to the achievement in miniaturized and integrated optical-sensing systems, very low levels of light intensity can be measured allowing the development of very sensitive analytical methods and systems. Chemiluminescence-based sensors have been then developed with the aim of combining the selective of light-emitting reactions with the convenience of electronic sensors. More recently, chemi- and electrochemiluminescence (ECL) detections have also been used instead of fluorescence for the development of biochips and microarrays.

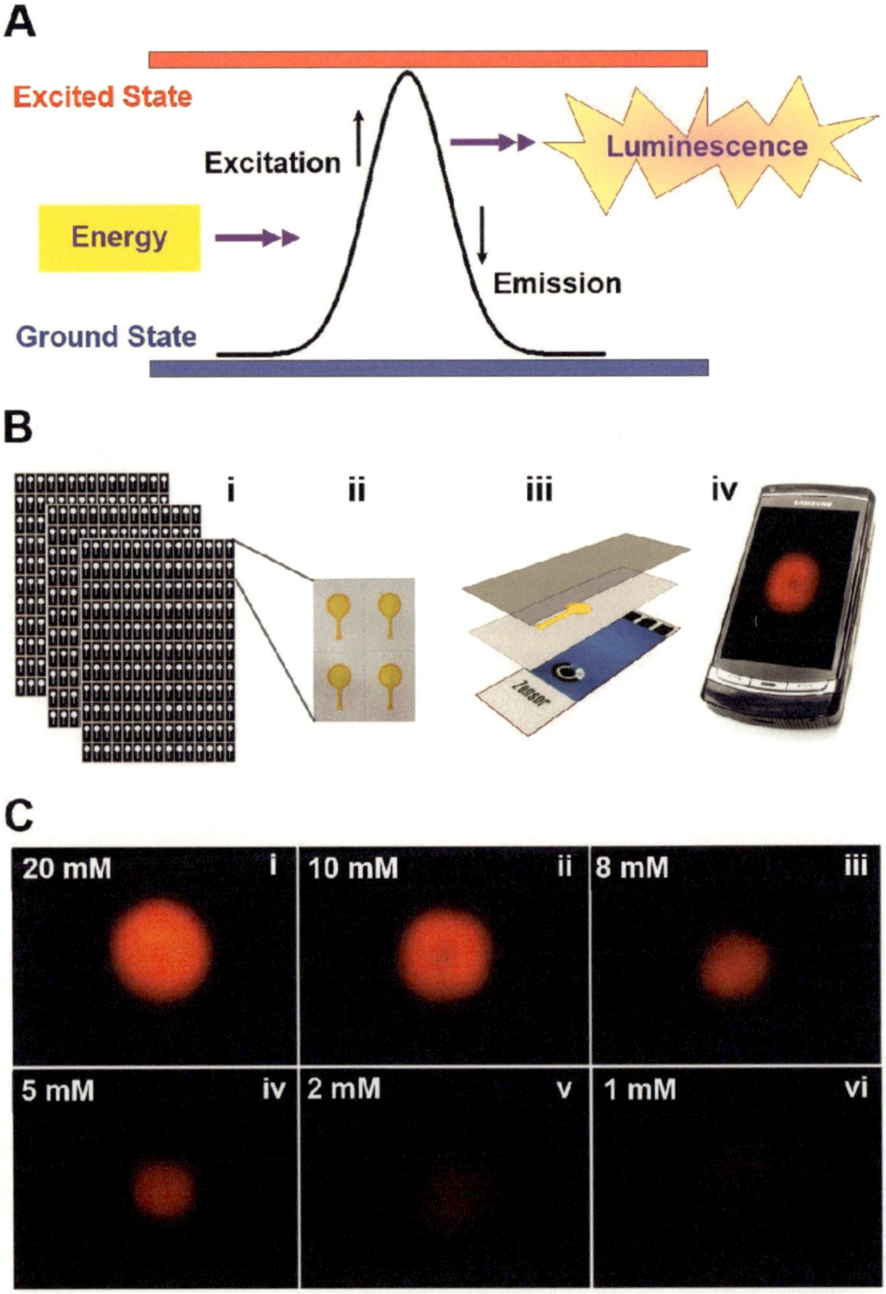

Figure 3. (A) Schematic of luminescence phenomenon consisting in the emission of light from an electronically excited compound returning to the ground state. (B) Fabrication and operation of a paper-based microfluidic ECL sensor [31]. Upon full wetting of the sensing area, the sensor is placed in proximity of the lens of the camera phone and a potential of 1.25 V is applied [31]. The resulting emission is captured and analysed (C) Digital photographic images of ECL emission obtained with various DBAE concentrations (1–20 mM) using a camera phone [31]. The ECL reactions were initiated by stepping the potential from 0 to 1.25 V upon application of a sample droplet [31]. Reproduced with permission [31]. Copyright 2011, ACS Publications.

Freeman and Seitz reported in 1978 the first CL sensor for hydrogen peroxide measurement [29]. Since then, various types of CL-based sensors have been commercially implemented and are extensively used for the analysis of inorganic, organic and biological/pharmaceutical compounds [30]. In recent years, a higher sensitivity of CL-based sensors over other photoluminescence-based ones has attracted significant research focus. The high sensitivity of CL-based sensing is attributed to the suppression of the noise introduced light scattering. Furthermore, CL-based devices feature a simpler set-up with lower background emissions than other photoluminescence detection systems. One of the major remaining deficiencies in the application of CL sensors to routine analysis systems is the short lifetime and signal drift, caused by the irreversible consumption of the CL reagents. While many CL sensors have designs based on the recycling of the CL reagents, which are usually bound to polymers and have a decreased consumption of the luminescent reagent, the stability of this type of sensors is compromised by the bleaching of the reagents or they partition into hydrophobic regions of the film over time.

Recently, Delaney et al. [31] presented an ECL sensor based on paper-based microfluidic devices. Low cost, disposable ECL sensors were fabricated with inkjet-printed paper fluidic substrates and screen-printed electrodes (**Figure 3B**). Such devices can be read with a conventional photodetector or a mobile camera phone. Importantly, because ECL is performed in the dark, unlike the case of colorimetric detection, it is independent of ambient light. The paper fluidic element was placed in direct contact with a screen-printed electrode and tightly enclosed in clear plastic (**Figure 3B**) [31]. A limit of detection of 0.9 μM was achieved with a linear response between 3 μM and 5 mM using a photomultiplier tube as the photodetector. Quantitative results could also be collected using a mobile camera phone by analysing the red intensity from digital images of the ECL emission, using a simple program that can be run on the mobile phone itself. **Figure 3C** shows the photographic images of the ECL emission from the sensors for a range of concentrations of 2-(Dibutylamino)ethanol (DBAE) between 1 and 20 mM, produced by stepping the potential from 0 to 1.25 V. The images reveal an unambiguous relationship between concentration and the intensity of the colour. The red pixel intensity was analysed for each image and a numerical value associated with each intensity was obtained [31].

4. Chemiresistive sensors

Semiconductor-based chemiresistive sensors are amongst the most investigated and widely used devices for the detection of combustible and toxic gases owing to their low cost, miniaturization potential and circuit simplicity [32, 33]. Such chemiresistive gas sensors are able to detect a wide variety of reactive reducing or oxidizing gases at very low ppb concentrations via a strong variation of their resistivity, which are not detectable with catalytic combustion and electrochemical-based gas sensors [34, 35]. A major challenge is their inherent non-specificity, which makes it challenging to use for the identification of unknown and complex gas mixtures. Chemiresistor gas sensors mainly operate on the basis of surface reactions, which cause change in the electrical resistance due to the modulation of the surface charge layer and semiconductor band bending [36]. These types of sensors feature excellent sensitivity, short response time, low cost, and very good suitability for the integration in miniaturized portable instruments making them appealing for numerous applications including breath analysis, alarm systems and electronic nose [34–36]. In spite of their numerous benefits, chemiresistive sensors have shown different challenges with respect to their commercialization such as poor reproducibility, long-time instability due to aging of the sensing materials and surfaces, cross-sensitivity to other gases and also to water vapour [37, 38].

Chemiresistive-type sensors using various semiconducting metal oxides, such as WO_3, MoO_3, SnO_2 and NiO, have been demonstrated for use as exhaled breath sensors, due to their superior sensitivity to volatile organic compounds (VOCs) and integration in portable devices [38–42]. A major requirement is the development of selective semiconductor material and surface that can provide fast response times [43]. Here, the sensing mechanism of chemiresistive devices is schematically described in **Figure 4A** and **B**. Oxygen molecules physisorb on the oxide surface, with electrons being transferred from the metal oxide to the oxygen which is ionosorbed [44, 45]. Several oxygen species are adsorbed including molecular (O_2) and atomic (O, O_2) ions with their fraction depending from the operating temperature (**Figure 4A** and **C**). These adsorbed ions scatter electrons within the Debye length (d) of the metal oxide forming a depleted region (**Figure 4B** and **C**), and band bending, with a reduced electron mobility near the oxide surface [44, 46].

The reception function of a common chemiresistive semiconductor, such as SnO_2 (in air) for a reducing or oxidizing analyte, consists of the variation of the ionosorbed oxygen concentration [44, 47]. If a reducing analyte is injected in the sensing chamber, such as EtOH, H_2 or CO, the surface concentration of oxygen ions is decreased (**Figure 4B**), releasing some of the trapped electrons back to the semiconductor and reducing the concentration of the scattering centres. As a result, the electron mobility and carrier concentration of the semiconductor are increased [38, 44]. In more detail, the ionosorbed oxygen scatters electrons within the Debye length of SnO_2, reducing its electron mobility. For large grains ($d_p >> \delta$), the sensing mechanism is controlled by the grain boundaries (**Figure 4C**). For ultra-fine nanoparticles, there are two possible mechanisms as a function of the grain size (d_p) [44, 46, 48]. If the grain size is larger than twice the Debye length, a conduction channel with bulk mobility exists within a diameter ($L_C = d_p - 2\delta$) from the grain centre (**Figure 4C**) [46, 49, 50]. A change in the scattering centre concentration (O_2, O, O_2) will then only result in a change in the conduction channel width (L_C). By contrast, if the grain is smaller or equal to twice d, then the whole grain is depleted (**Figure 4C**); therefore, a reduction of the ionosorbed oxygen may 'open' a conduction channel through the oxide (**Figure 4C**) [49, 51].

Recently, Moon et al. [41] present a high-performance chemiresistive electronic nose (CEN) based on an array of metal oxide semiconductor thin films, metal-catalysed thin films and nanostructured thin films. It consisted of differently nanostructured thin films based on metal oxides, including tungsten trioxide (WO_3), tin dioxide (SnO_2) and indium oxide (In_2O_3) (**Figure 4D**). These nanostructured thin films were synthesized by e-beam deposition in a glancing angle deposition. The chemiresistive-sensing properties of these nine sensing element arrays were explored for the diagnostics of asthma, halitosis and kidney disorders. The CEN operating at 168°C demonstrated a sensitive and selective detection of NO, H_2S and NH_3 with an 80% relative humidity (RH) atmosphere (**Figure 4E**). The estimated detection limits of this sensor array for these gas molecules were in the sub-ppb range, suggesting their potential use

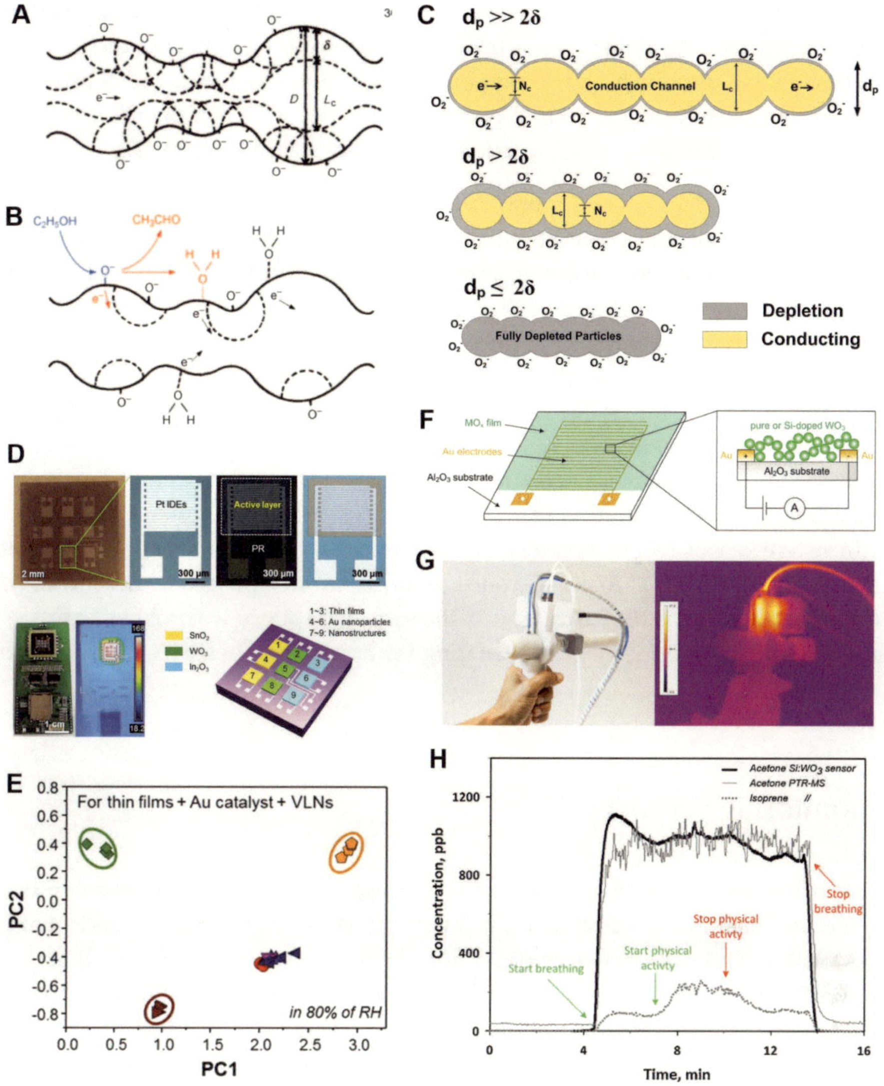

Figure 4. Schematic model of the chemiresistive sensing mechanism of semiconductor metal oxides based on the oxygen ionosorption on the semiconductor surface in (A) pure dry air and (B) with ethanol [44]. Reproduced with permission [44]. Copyright 1982, AIP Publishing. (C) As a function of the ratio between the particle diameter (d_p) and the Debye length (δ), three sensing mechanisms are expected [46]. (Reproduced with permission [46]. Copyright 2015, Wiley Online Library. (D) Optical microscope images of the sequential fabrication process of the CEN on a single chip (1 × 1 cm) containing an active layer (1 × 1 mm) on Pt interdigitated electrodes [41]. Reproduced with permission [41]. Copyright 2016, ACS Publications. (E) PCA plot showing the responses to 8 gases with 80% RH [41]. (F) Detector schematic: a metal oxide (MO_x) film is deposited onto a sensor substrate consisting of an Al_2O_3 support with interdigitated Au electrodes [55]. Reproduced with permission [55]. Copyright 2010, ACS Publications. (G) Picture of the operating hand-held sampler and sensing unit with a disposable mouthpiece (left). IR camera image (right) confirmed the absence of any hot-spots close to the mouth piece or holder, the temperatures are always below 40°C there [52]. Reproduced with permission [52]. Copyright 2015, IOP Publishing. (H) Estimated acetone concentration with the $Si:WO_3$ sensor (thick solid line) and acetone (thin solid line) and isoprene (dotted line) concentrations measured by PTR-MS during breath analysis of a healthy subject during physical activity [39]. Reproduced with permission [39]. Copyright 2012, Elsevier.

for the diagnosis of asthma, halitosis and kidney disease via breath analysis. The enhanced sensitivity of these sensing materials is attributed to the spill-over effect of the noble metal Au nanoparticle catalysts and the high porosity of villi-like nanostructures, which provides a large surface-to-volume ratio. Using principal component analysis (PCA), the CEN can detect and distinguish the H_2S, NH_3 and NO biomarkers in exhaled breath (**Figure 4E**).

Righettoni et al. [39, 52] developed a portable acetone sensor (**Figure 4F** and **G**) consisting of flame-deposited [53, 54] and in situ-annealed, Si-doped epsilon-WO_3 nanostructured films, capable of measuring ultra-low acetone concentrations (down to 20 ppb) with a high signal-to-noise ratio in ideal (dry air) and realistic (up to 90% RH) conditions. The detector films consist of highly sensitive and selective pure and Si-doped WO_3 nanoparticle films (10–13 nm in diameter) made of the gas phase and aerosol-deposited onto interdigitated electrodes [55] as previously demonstrated for SnO_2, ZnO and Si-SnO_2 nanoparticles [43, 56–58]. The breath acetone content of test persons was monitored continuously and compared to that measured via proton transfer reaction mass spectrometry (PTR-MS). Notably, in addition to providing similar breath acetone concentration readings to the PTR-MS and a better signal-to-noise ratio, the chemiresistive sensor response times were below 15 s, making these devices attractive for breath analysis (**Figure 4H**). Acetone concentrations of 20 ppb were measured with high signal-to-noise ratios above 10. Furthermore, the sensor response was robust against variation in gas flow rates down to 0.2 $L.min^{-1}$, facilitating the application of such sensors for real breath measurement devices.

5. pH-monitoring sensors

Monitoring of chronic wounds is a relatively unexplored area that presents significant challenges to modern health-care providers. Chronic wounds are defined as wounds that either do not heal or heal very slowly or may reoccur after healing. Generally, wounds that do not heal within 3 months are termed chronic [59]. They affect more than nine million people across the United States and Europe with an annual cost exceeding US$ 39 billion [60, 61]. Usually, diabetic, obese and elderly people tend to suffer from chronic wounds [62, 63]. Untimely, treatment may lead rapidly to infections and complications. The diagnosis and the treatment of chronic wounds are quite complex, providing a major challenge to the health-care staff [64]. One of the major causes of disruption of the healing process, which may lead to a chronic wound, is bacterial infection which may result in the overgrowth of a newly formed capillary-rich granulation tissue over the wound [65]. This condition, termed as over-granulation, can hamper the healing process and may result in a protruding, friable flesh that is very sensitive and bleeds easily.

There is currently no commercially available wireless device to continuously monitor the wound-healing process. Patients rely on medical staff for physical inspection of the wound, which requires repeated trips to clinics or prolonged hospitalization. Few devices have been reported in the study that monitor parameters related to the wound-healing process. These include a bandage in a solution form that can be painted onto the skin to form a thin film [66].

The film emits oxygen-dependent phosphorescence that can be used to map the oxygen levels of the underlying skin tissue. In another example, a flexible electrode array has been developed through the inkjet printing of gold nanoparticle on flexible polyethylene naphthalate to measure the impedance spectrum of the tissues for early detection of pressure ulcers [67]. Electrodes have been demonstrated to measure moisture levels [68] as well as bacteria [69] in wound dressings. Changes in pH values have also been related to the presence of infection. An infected wound shows slightly basic pH due to certain enzyme activities, bacterial colonization and formation of protein structures [70]. However, these approaches rely on the sensing of a single parameter to monitor the wound-healing process, while the healing is a complex process that may require information about a number of factors affecting the wound healing. An attractive solution would be the integration of low cost, wearable, compact, wireless, real-time wound-monitoring system on the adhesive bandages that are commonly used to protect wounds from the external environment. Such a system could be worn by patients everyday and be able to issue early warnings to the patients regarding any abnormality in the healing process, as well as wirelessly send the data recordings of multiple parameters related to the wound-healing process to the remote medical staff.

Recently, Farooqui et al. [71] reported a wearable system to wirelessly monitor chronic wounds using simple bandage strips. The system comprised a set of inkjet-printed sensors on a disposable bandage to monitor bleeding, pH levels and external pressure on the wound site. The sensor and wireless circuitry electronics were integrated on the bandage in a reusable fashion, thus maintaining the disposability of the bandage strip in contact with the wound (**Figure 5A–C**). This wearable bandage could alert the patient and the health-care providers regarding any abnormality in the wound-healing process through the integrated wireless module [71]. The pH levels on the wound site were detected by changes in resistance of one of the electrodes (carbon-based) placed on the bandage [71]. The wireless tests were performed while the bandage is worn on the body as shown in **Figure 5C**. A fluid was injected underneath the bandage using a narrow tube, attached to a syringe containing the fluid, imitating a wound-bleeding process. A Zigbee wireless receiver was placed to receive an information signal from the bandage. When the fluid was pumped from the syringe and reaches the bottom side of the bandage, the transmitter on the bandage was activated and sent information to the receiver [71].

Guinovart et al. [72] demonstrated a wearable potentiometric pH cell embedded into an adhesive bandage for real-time monitoring of pH variations in a wound. Screen-printed silver-silver chloride electrodes were deposited on a commercial adhesive bandage forming a potentiometric cell. The fabrication process is illustrated in **Figure 5D**. A transparent dielectric insulator was first printed onto the cellulose pad to cover the area where the potentiometric electrodes would be printed [72]. Then, conductive traces that defined the electrode underlayer were made by screen-printing of an Ag/AgCl conductive ink. Thereafter, a carbon layer was overlaid on the traces to fabricate the working electrode. Finally, an insulator ink was used to shield the conductive tracks and expose the electro-active areas and contact pads for micro-wiring [72]. To assess the ability of the sensor to dynamically monitor pH changes in wounds, a poly(ethylene glycol) (PEG) hydrogel, cured on top of both electrodes, was used to emulate a healing segment of tissue. **Figure 5E** shows a typical pH time response. First, a sharp decrease in the pH is observed. This corresponds to the diffusion of the supplied solution to the

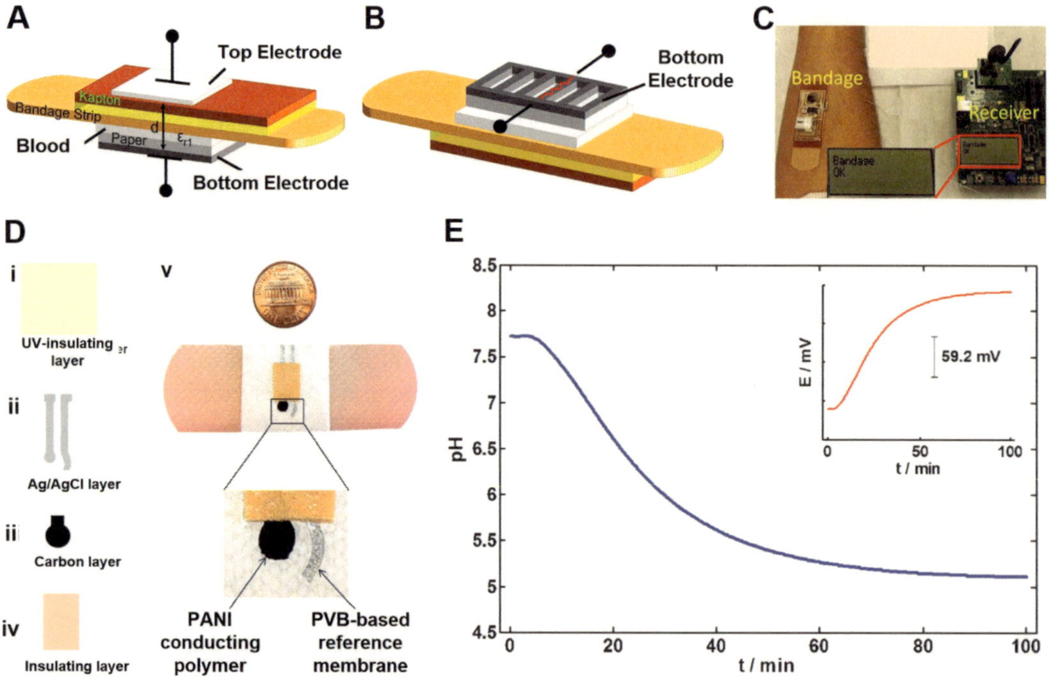

Figure 5. (A) Two sensor electrodes are mounted on the top and the bottom side of a bandage forming a capacitor. (B) The pH levels are sensed when the resistance of the carbon based bottom electrode changes in response to the pH levels. (C) The bandage (with exposed electronics) is worn on the forearm [71]. Reproduced with permission [71]. Copyright 2016, Nature Publishing Group. (D) Fabrication process for a pH-sensing bandage device. (i–iv) An insulating layer is printed, followed by an Ag/AgCl and a carbon layer, and finally an insulating layer with skin-like colour. (v) Images displaying the printed potentiometric sensor on an adhesive bandage [72]. Reproduced with permission [72]. Copyright 2014, Wiley Online Library. (E) Real-time recording of pH changes over a 100-min interval using a PEG hydrogel simulating the composition of a wound [72].

electrode surface. After 40 min, the pH reaches a similar value to the added solution pH level. The corresponding electromotive force (EMF) profile is shown in the inset of **Figure 5E**. These findings demonstrate that the new pH-sensitive bandage is able to detect pH fluctuations at a wound site with relatively long time intervals (up to 100 min).

6. Conclusions

Advanced wearable-sensing technologies have the potential to offer a minimally intrusive telemedicine platform for individual health service and dramatically alter the landscape of health-care delivery as well as our understanding of many diseases. In this chapter, we have reviewed some recent achievements in miniaturised sensor technologies for the non-invasive detection of disease biomarkers for medical diagnostics. We have classified these technologies according to their working principles and detecting mechanisms. Notably, the development of portable devices is rapidly revolutionising the device layout, resulting in the engineering of wearable sensor systems on tattoos, bandages and contact lenses. Highly selective body fluid sensors for readily available sweat, saliva and tears have been developed leveraging on existing electrochemical, capacitive and pH-monitoring technologies, which can benefit from an extensive library of enzymes for biomolecule detection. Similarly, non-invasive breath analysis approaches based on solid-state devices made of unique nanomaterial compositions can sense important biomarkers such as acetone, one the primary breath markers for diabetes, down to a few particles per billion. By enhancing reducing health-care costs and providing novel health data with unprecedented frequency, these smart wearable sensors can play a critical role in revolutionising the future health-care system and improving our quality of life.

Acknowledgements

A.T. gratefully acknowledges the support of Australian Research Council DP150101939, Australian Research Council DE160100569 and Westpac2016 Research Fellowship.

Conflict of interest

The authors have no conflict of interest to declare.

Author details

Noushin Nasiri[1] and Antonio Tricoli[2]*

*Address all correspondence to: antonio.tricoli@anu.edu.au

1 Faculty of Science, Institute for Biomedical Materials and Devices, University of Technology Sydney, Sydney, NSW, Australia

2 Nanotechnology Research Laboratory, Research School of Engineering, Australian National University, Canberra, Australia

References

[1] Xu S, Zhang Y, Jia L, Mathewson KE, Jang K-I, Kim J, et al. Soft microfluidic assemblies of sensors, circuits, and radios for the skin. Science. 2014;**344**(6179):70-74. DOI: 10.1126/science.1250169

[2] Jia W, Bandodkar AJ, Valdés-Ramírez G, Windmiller JR, Yang Z, Ramírez J, et al. Electrochemical tattoo biosensors for real-time noninvasive lactate monitoring in human perspiration. Analytical Chemistry. 2013;**85**(14):6553-6560. DOI: 10.1021/ac401573r

[3] Zhu W, Bao C, Li F, Zhou X, Yang J, Yu T, et al. An efficient planar-heterojunction solar cell based on wide-bandgap $CH_3NH_3Pb_{I2}.1Br_{0.9}$ perovskite film for tandem cell application. Chemical Communications. 2016;**52**(2):304-307. DOI: 10.1039/C5CC07673K

[4] Stetter JR, Zaromb S, Penrose WR. Sensor Array for Toxic Gas Detection. Google Patents; 1987

[5] Kim J, Jeerapan I, Imani S, Cho TN, Bandodkar A, Cinti S, et al. Noninvasive alcohol monitoring using a wearable tattoo-based Iontophoretic-biosensing system. ACS Sensors. 2016;**1**(8):1011-1019. DOI: 10.1021/acssensors.6b00356

[6] Falk M, Andoralov V, Silow M, Toscano MD, Shleev S. Miniature biofuel cell as a potential power source for glucose-sensing contact lenses. Analytical Chemistry. 2013;**85**(13):6342-6348. DOI: 10.1021/ac4006793

[7] Li H, Mu X, Yang Y, Mason AJ. Low power multimode electrochemical gas sensor Array system for wearable health and safety monitoring. IEEE Sensors Journal. 2014;**14**(10):3391-3399. DOI: 10.1109/JSEN.2014.2332278

[8] Tang L, Zeng G-M, Shen G-L, Li Y-P, Zhang Y, Huang D-L. Rapid detection of Picloram in agricultural field samples using a disposable Immunomembrane-based electrochemical sensor. Environmental Science & Technology. 2008;**42**(4):1207-1212. DOI: 10.1021/es7024593

[9] Hayat A, Marty J. Disposable screen printed electrochemical sensors: Tools for environmental monitoring. Sensors. 2014;**14**(6):10432. DOI: 10.3390/s140610432

[10] Bakker E, Telting-Diaz M. Electrochemical sensors. Analytical Chemistry. 2002;**74**(12): 2781-2800. DOI: 10.1021/ac0202278

[11] Schazmann B, Morris D, Slater C, Beirne S, Fay C, Reuveny R, et al. A wearable electrochemical sensor for the real-time measurement of sweat sodium concentration. Analytical Methods. 2010;**2**(4):342-348. DOI: 10.1039/B9AY00184K

[12] Guinovart T, Parrilla M, Crespo GA, Rius FX, Andrade FJ. Potentiometric sensors using cotton yarns, carbon nanotubes and polymeric membranes. The Analyst. 2013;**138**(18): 5208-5215. DOI: 10.1039/C3AN00710C

[13] Yin Z, Wu J, Yang Z. Amperometric sensors based on Ni/Al and co/Al layered double hydroxides modified electrode and their application for hydrogen peroxide detection. Biosensors & Bioelectronics. 2011;**26**(5):1970-1974. DOI: 10.1016/j.bios.2010.08.049

[14] Yang Y-L, Chuang M-C, Lou S-L, Wang J. Thick-film textile-based amperometric sensors and biosensors. The Analyst. 2010;**135**(6):1230-1234. DOI: 10.1039/B926339J

[15] Li D, Hu J, Wu R, Lu JG. Conductometric chemical sensor based on individual CuO nanowires. Nanotechnology. 2010;**21**(48):485502. DOI: 10.1088/0957-4484/21/48/485502

[16] Kharitonov SA, Barnes PJ. Biomarkers of some pulmonary diseases in exhaled breath. Biomarkers. 2002;**7**(1):1-32. DOI: 10.1080/13547500110104233

[17] Bollella P, Fusco G, Tortolini C, Sanzò G, Favero G, Gorton L, et al. Beyond graphene: Electrochemical sensors and biosensors for biomarkers detection. Biosensors & Bioelectronics. 2016;**89**:152-166. DOI: 10.1016/j.bios.2016.03.068

[18] Lee H, Choi TK, Lee YB, Cho HR, Ghaffari R, Wang L, et al. A graphene-based electrochemical device with thermoresponsive microneedles for diabetes monitoring and therapy. Nature Nanotechnology. 2016;**11**:566. DOI: 10.1038/nnano.2016.38

[19] Thomas N, Lähdesmäki I, Parviz BA. A contact lens with an integrated lactate sensor. Sensors and Actuators B: Chemical. 2012;**162**(1):128-134. DOI: 10.1016/j.snb.2011.12.049

[20] Kim J, Imani S, de Araujo WR, Warchall J, Valdés-Ramírez G, Paixão TRLC, et al. Wearable salivary uric acid mouthguard biosensor with integrated wireless electronics. Biosensors & Bioelectronics. 2015;**74**:1061-1068. DOI: 10.1016/j.bios.2015.07.039

[21] Wang W-S, Kuo W-T, Huang H-Y, Luo C-H. Wide dynamic range CMOS Potentiostat for Amperometric chemical sensor. Sensors. 2010;**10**(3):1782. DOI: 10.3390/s100301782

[22] Chu MX, Miyajima K, Takahashi D, Arakawa T, Sano K, Sawada S-I, et al. Soft contact lens biosensor for in situ monitoring of tear glucose as non-invasive blood sugar assessment. Talanta. 2011;**83**(3):960-965. DOI: 10.1016/j.talanta.2010.10.055

[23] Azzouzi S, Rotariu L, Benito AM, Maser WK, Ali MB, Bala C. A novel amperometric biosensor based on gold nanoparticles anchored on reduced graphene oxide for sensitive detection of L-lactate tumor biomarker. Biosensors & Bioelectronics. 2015;**69**:280-286. DOI: 10.1016/j.bios.2015.03.012

[24] Chen G-Z, Chan I-S, Lam DCC. Capacitive contact lens sensor for continuous non-invasive intraocular pressure monitoring. Sensors & Actuators, A: Physical. 2013;**203**:112-118. DOI: 10.1016/j.sna.2013.08.029

[25] Mannoor MS, Tao H, Clayton JD, Sengupta A, Kaplan DL, Naik RR, et al. Graphene-based wireless bacteria detection on tooth enamel. Nature Communications. 2012;**3**:763. DOI: 10.1038/ncomms1767

[26] Schnakenberg U, Walter P, vom Bögel G, Krüger C, Lüdtke-Handjery HC, Richter HA, et al. Initial investigations on systems for measuring intraocular pressure. Sensors and Actuators, A: Physical. 2000;**85**(1):287-291. DOI: 10.1016/S0924-4247(00)00426-X

[27] Liu X, Freeman R, Golub E, Willner I. Chemiluminescence and chemiluminescence resonance energy transfer (CRET) aptamer sensors using catalytic hemin/G-quadruplexes. ACS Nano. 2011;**5**(9):7648-7655. DOI: 10.1021/nn202799d

[28] Li LL, Liu KP, Yang GH, Wang CM, Zhang JR, Zhu JJ. Fabrication of graphene–quantum dots composites for sensitive electrogenerated chemiluminescence immunosensing. Advanced Functional Materials. 2011;**21**(5):869-878. DOI: 10.1002/adfm.201001550

[29] Freeman TM, Seitz WR. Chemiluminescence fiber optic probe for hydrogen peroxide based on the luminol reaction. Analytical Chemistry. 1978;**50**(9):1242-1246. DOI: 10.1021/ac50031a012

[30] Zhang Z, Zhang S, Zhang X. Recent developments and applications of chemiluminescence sensors. Analytica Chimica Acta. 2005;**541**(1–2):37-46. DOI: 10.1016/j.aca.2004.11.069

[31] Delaney JL, Hogan CF, Tian J, Shen W. Electrogenerated chemiluminescence detection in paper-based microfluidic sensors. Analytical Chemistry. 2011;**83**(4):1300-1306. DOI: 10.1021/ac102392t

[32] Whitfield MD, McKeag RD, Pang LY, Chan SS, Jackman RB. Thin film diamond UV photodetectors: Photodiodes compared with photoconductive devices for highly selective wavelength response. Diamond and Related Materials. 1996;**5**(6–8):829-834. DOI: 10.1016/0925-9635(95)00419-X

[33] Pacholski C, Kornowski A, Weller H. Self-assembly of ZnO: From nanodots to nanorods. Angewandte Chemie, International Edition. 2002;**41**(7):1188-1191. DOI: 10.1002/1521-3773(20020402)41:7<1188::AID-ANIE1188>3.0.CO;2-5

[34] Mirica KA, Azzarelli JM, Weis JG, Schnorr JM, Swager TM. Rapid prototyping of carbon-based chemiresistive gas sensors on paper. Proceedings of the National Academy of Sciences of the United States of America. 2013;**110**(35):E3265-E3E70. DOI: 10.1073/pnas.1307251110

[35] Ramgir N, Datta N, Kaur M, Kailasaganapathi S, Debnath AK, Aswal DK, et al. Metal oxide nanowires for chemiresistive gas sensors: Issues, challenges and prospects. Colloids and Surfaces, A: Physicochemical and Engineering Aspects. 2013;**439**:101-116. DOI: 10.1016/j.colsurfa.2013.02.029

[36] Meng F-L, Guo Z, Huang X-J. Graphene-based hybrids for chemiresistive gas sensors. Trends in Analytical Chemistry. 2015;**68**:37-47. DOI: 10.1016/j.trac.2015.02.008

[37] Fratoddi I, Venditti I, Cametti C, Russo MV. Chemiresistive polyaniline-based gas sensors: A mini review. Sensors and Actuators, B: Chemical. 2015;**220**:534-548. DOI: 10.1016/j.snb.2015.05.107

[38] Tricoli A, Righettoni M, Teleki A. Semiconductor gas sensors: Dry synthesis and application. Angewandte Chemie, International Edition. 2010;**49**(42):7632-7659. DOI: 10.1002/anie.200903801

[39] Righettoni M, Tricoli A, Gass S, Schmid A, Amann A, Pratsinis SE. Breath acetone monitoring by portable Si:WO_3 gas sensors. Analytica Chimica Acta. 2012;**738**:69-75. DOI: 10.1016/j.aca.2012.06.002

[40] Righettoni M, Tricoli A, Pratsinis SE. Thermally stable, silica-doped ε-WO_3 for sensing of acetone in the human breath. Chemistry of Materials. 2010;**22**(10):3152-3157. DOI: 10.1021/cm1001576

[41] Moon HG, Jung Y, Han SD, Shim Y-S, Shin B, Lee T, et al. Chemiresistive electronic nose toward detection of biomarkers in exhaled breath. ACS Applied Materials & Interfaces. 2016;**8**(32):20969-20976. DOI: 10.1021/acsami.6b03256

[42] Moon HG, Shim Y-S, Kim DH, Jeong HY, Jeong M, Jung JY, et al. Self-activated ultrahigh chemosensitivity of oxide thin film nanostructures for transparent sensors. Scientific Reports. 2012;**2**:588. DOI: 10.1038/srep00588

[43] Zayasu K, Sekizawa K, Okinaga S, Yamaya M, Ohrui T, Sasaki H. Increased carbon monoxide in exhaled air of asthmatic patients. American Journal of Respiratory and Critical Care Medicine. 1997;**156**(4):1140-1143. DOI: 10.1164/ajrccm.156.4.96-08056

[44] Ogawa H, Nishikawa M, Abe A. Hall measurement studies and an electrical conduction model of tin oxide ultrafine particle films. Journal of Applied Physics. 1982;**53**(6):4448-4455. DOI: 10.1063/1.331230

[45] Harrison PG, Willett MJ. Tin oxide surfaces. Part 20—Electrical properties of tin (IV) oxide gel: Nature of the surface species controlling the electrical conductance in air as a function of temperature. Journal of the Chemical Society, Faraday Transactions. 1989;**85**(8):1921-1932. DOI: 10.1039/F19898501921

[46] Nasiri N, Bo R, Wang F, Fu L, Tricoli A. Ultraporous Electron-depleted ZnO nanoparticle networks for highly sensitive portable visible-blind UV Photodetectors. Advanced Materials. 2015;**27**(29):4336-4343. DOI: 10.1002/adma.201501517

[47] Barsan N, Weimar U. Understanding the fundamental principles of metal oxide based gas sensors; The example of CO sensing with SnO_2 sensors in the presence of humidity. Journal of Physics Condensed Matter. 2003;**15**(20):R813. DOI: 10.1088/0953-8984/15/20/201

[48] Nasiri N, Bo R, Hung TF, Roy VAL, Fu L, Tricoli A. Tunable band-selective UV-Photodetectors by 3D self-assembly of heterogeneous nanoparticle networks. Advanced Functional Materials. 2016;**26**(40):7359-7366. DOI: 10.1002/adfm.201602195

[49] Nasiri N, Bo R, Fu L, Tricoli A. Three-dimensional nano-heterojunction networks: A highly performing structure for fast visible-blind UV photodetectors. Nanoscale. 2017; **9**(5):2059-2067. DOI: 10.1039/C6NR08425G

[50] Jesenak M, Banovcin P, Havlicekova Z, Dobrota D, Babusikova E. Factors influencing the levels of exhaled carbon monoxide in asthmatic children. The Journal of Asthma. 2014; **51**(9):900-906. DOI: 10.3109/02770903.2014.936448

[51] Nasiri N, Bo R, Chen H, White TP, Fu L, Tricoli A. Structural engineering of Nano-grain boundaries for low-voltage UV-Photodetectors with gigantic photo- to dark-current ratios. Advanced Optical Materials. 2016;**4**(11):1787-1795. DOI: 10.1002/adom.201600273

[52] Righettoni M, Ragnoni A, Güntner AT, Loccioni C, Pratsinis SE, Risby TH. Monitoring breath markers under controlled conditions. Journal of Breath Research. 2015;**9**(4):047101. DOI: 10.1088/1752-7155/9/4/047101

[53] Nasiri N, Elmoe TD, Liu Y, Qin QH, Tricoli A. Self-assembly dynamics and accumulation mechanisms of ultra-fine nanoparticles. Nanoscale. 2015;**7**(21):9859-9867. DOI: 10.1039/C5NR00877H

[54] Yahuitl Osorio M, The D, Noushin N, Thomas PW, Antonio T, Kylie RC. Flame-made ultra-porous TiO_2 layers for perovskite solar cells. Nanotechnology. 2016;**27**(50):505403. DOI: 10.1088/0957-4484/27/50/505403

[55] Righettoni M, Tricoli A, Pratsinis SE. Si:WO_3 sensors for highly selective detection of acetone for easy diagnosis of diabetes by breath analysis. Analytical Chemistry. 2010; **82**(9):3581-3587. DOI: 10.1021/ac902695n

[56] Tricoli A, Graf M, Mayer F, Kuühne S, Hierlemann A, Pratsinis SE. Micropatterning layers by flame aerosol deposition-annealing. Advanced Materials. 2008;**20**(16):3005-3010. DOI: 10.1002/adma.200701844

[57] Tricoli A, Graf M, Pratsinis SE. Optimal doping for enhanced SnO_2 sensitivity and thermal stability. Advanced Functional Materials. 2008;**18**(13):1969-1976. DOI: 10.1002/adfm.200700784

[58] Tricoli A, Nasiri N, Chen H, Wallerand AS, Righettoni M. Ultra-rapid synthesis of highly porous and robust hierarchical ZnO films for dye sensitized solar cells. Solar Energy. 2016; **136**:553-559. DOI: 10.1016/j.solener.2016.07.024

[59] Nunan R, Harding KG, Martin P. Clinical challenges of chronic wounds: Searching for an optimal animal model to recapitulate their complexity. Disease Models & Mechanisms. 2014;**7**(11):1205-1213. DOI: 10.1242/dmm.016782

[60] Sen CK, Gordillo GM, Roy S, Kirsner R, Lambert L, Hunt TK, et al. Human skin wounds: A major and snowballing threat to public health and the economy. Wound Repair and Regeneration. 2009;**17**(6):763-771. DOI: 10.1111/j.1524-475X.2009.00543.x

[61] Posnett J, Gottrup F, Lundgren H, Saal G. The resource impact of wounds on health-care providers in Europe. Journal of Wound Care. 2009;**18**(4):154. DOI: 10.12968/jowc.2009.18.4.41607

[62] Gist S, Tio-Matos I, Falzgraf S, Cameron S, Beebe M. Wound care in the geriatric client. Clinical Interventions in Aging. 2009;**4**:269. DOI: 10.2147/CIA.S4726

[63] Blakytny R, Jude E. The molecular biology of chronic wounds and delayed healing in diabetes. Diabetic Medicine. 2006;**23**(6):594-608. DOI: 10.1111/j.1464-5491.2006.01773.x

[64] Hampton S. Understanding overgranulation in tissue viability practice. British Journal of Community Nursing. 2007;**12**(Sup4):S24-S30. DOI: 10.12968/bjcn.2007.12.Sup4.43000

[65] Gardner SE, Frantz RA, Doebbeling BN. The validity of the clinical signs and symptoms used to identify localized chronic wound infection. Wound Repair and Regeneration. 2001;**9**(3):178-186. DOI: 10.1046/j.1524-475x.2001.00178.x

[66] Li Z, Roussakis E, Koolen PG, Ibrahim AM, Kim K, Rose LF, et al. Non-invasive transdermal two-dimensional mapping of cutaneous oxygenation with a rapid-drying liquid bandage. Biomedical Optics Express. 2014;**5**(11):3748-3764. DOI: 10.1364/BOE.5.003748

[67] Swisher SL, Lin MC, Liao A, Leeflang EJ, Khan Y, Pavinatto FJ, et al. Impedance sensing device enables early detection of pressure ulcers in vivo. Nature Communications. 2015;**6**: 6575. DOI: 10.1038/ncomms7575

[68] Kim I-Y, Suh S-H, Lee I-K, Wolfe RR. Applications of stable, nonradioactive isotope tracers in in vivo human metabolic research. Experimental & Molecular Medicine. 2016;**48**(1): e203. DOI: 10.1038/emm.2015.97

[69] McColl D, Cartlidge B, Connolly P. Real-time monitoring of moisture levels in wound dressings in vitro: An experimental study. International Journal of Surgery. 2007;**5**(5):316-322. DOI: 10.1016/j.ijsu.2007.02.008

[70] Schreml S, Szeimies RM, Karrer S, Heinlin J, Landthaler M, Babilas P. The impact of the pH value on skin integrity and cutaneous wound healing. Journal of the European Academy of Dermatology and Venereology. 2010;**24**(4):373-378. DOI: 10.1111/j.1468-3083.2009.03413.x

[71] Farooqui MF, Shamim A. Low cost inkjet printed smart bandage for wireless monitoring of chronic wounds. Scientific Reports. 2016;**6**:28949. DOI: 10.1038/srep28949

[72] Guinovart T, Valdés-Ramírez G, Windmiller JR, Andrade FJ, Wang J. Bandage-based wearable potentiometric sensor for monitoring wound pH. Electroanalysis. 2014;**26**(6): 1345-1353. DOI: 10.1002/elan.201300558

[73] Tricoli A, Nasiri N, De S. Wearable and miniaturized sensor technologies for personalized and preventive medicine. 2017;**27**(15):1605271. DOI: 10.1002/adfm.201605271

Chapter 2

A Proposal for New Algorithm that Defines Gait-Induced Acceleration and Gait Cycle in Daily Parkinsonian Gait Disorders

Masahiko Suzuki, Makiko Yogo, Masayo Morita, Hiroo Terashi, Mutsumi Iijima, Mitsuru Yoneyama, Masato Takada, Hiroya Utsumi, Yasuyuki Okuma, Akito Hayashi, Satoshi Orimo and Hiroshi Mitoma

Additional information is available at the end of the chapter

http://dx.doi.org/10.5772/intechopen.75483

Abstract

We developed a new device, the portable gait rhythmogram (PGR), to record up to 70 hrs of movement-induced accelerations. Acceleration values induced by various movements, averaged every 10 min, showed gamma distribution, and the mean value of this distribution was used as an index of the amount of overall movements. Furthermore, the PGR algorithm can specify gait-induced accelerations using the pattern-matching method. Analysis of the relationship between gait-induced accelerations and gait cycle duration makes it possible to quantify Parkinson's disease (PD)-specific pathophysiological mechanisms underlying gait disorders. Patients with PD showed the following disease-specific patterns: (1) reduced amount of overall movements and (2) low amplitude of gait-induced accelerations in the early stages of the disease, which was compensated by fast stepping. Loss of compensation was associated with slow stepping gait, (3) narrow range of gait-induced acceleration amplitude and gait cycle duration, suggesting monotony, and (4) evident motor fluctuations during the day by tracing changes in the above two parameters. Prominent motor fluctuation was associated with frequent switching between slow stepping mode and active mode. These findings suggest that monitoring various movement- and gait-induced accelerations allows the detection of specific changes in PD. We conclude that continuous long-term monitoring of these parameters can provide accurate quantitative assessment of parkinsonian clinical motor signs.

Keywords: portable gait rhythmogram, gait disorder, gait analysis, wearable device, Parkinson's disease

1. Introduction

1.1. Why are wearable monitors necessary in clinical management?

Patients with various neurological and musculoskeletal conditions present with a variety of gait abnormalities. Parkinson's disease (PD), parkinsonism-related diseases, and cerebellar ataxias show clinically specific features of gait disorders [1–10]. PD is characterized by poverty of movements (akinesia) and slowness/smallness of executed movement (bradykinesia) [1–10]. Parkinsonian gait disorders are assessed by using clinical score (Unified Parkinson's Disease Rating Scale (UPDRS)) [11], Timed Up and Go testing [12], and gait analysis on short-distance walking [10, 13]. However, these methods have limitations regarding exact assessment. First, there are dispersions in estimation on UPDRS depending on the examiners. Second, information outside the hospital examination rooms is lacking. Finally, gait analysis on short-distance walking can be easily influenced by emotional stress caused by increased attention [10, 13].

Proper assessment of the characteristic features of each gait disorder should satisfy two conditions: (1) measure specific kinematic and temporal parameters that reflect deficits in particular neural circuitry [13] and (2) measure voluntary control of gait movements continuously over a long period of time during activities of daily living (ADL) [1, 5–8]. For these reasons, many types of wearable devices that monitor activities including gait have been proposed (see **Table 1**) [14–24]. Maetzler et al. [14] proposed the following clinical features as targets for wearable

Types of sensors	
Devices that mainly monitor overall activities	
TricTrac RT3 [9, 20]	Accelerometer- and gyroscope-mounted sensor
Devices that mainly monitor gait disorders	
Step watch [19, 20]	Specific ankle-worn microprocessor-based step counter
DynaPort [19, 20]	Accelerometer-mounted sensor attached on the trunk
Devices that monitor changes in overall activities and gait disorders	
Mobility lab [21, 22]	Accelerometer- and gyroscope-mounted sensor attached on the limbs and trunk
Physiolog [23]	Accelerometer-, gyroscope-, and barometric pressure-mounted sensors attached on the limbs and trunk
AX3 [24]	Accelerometer-mounted sensor attached on the limbs or trunk

Table 1. Features of currently available wearable devices.

devices: (1) motor disabilities, including axial disability (gait and transfer deficits, freezing of gait, imbalance, and frequent falling), bradykinesia in the distal limb, dyskinesia, resting tremor, dysarthria, and secondary low activity, and (2) non-motor clinical features, which include sleep disturbance and autonomic dysfunction [14]. So far, gait-induced accelerations and angular velocities have been used to assess gait akinesia and bradykinesia in PD [14, 25].

1.2. Advantages of accelerometer-mounted wearing device

Accelerometer-mounted devices were originally developed to record all movement-induced accelerations [26–29]. Since acceleration-mounted sensors have advantages in quantitative assessment of PD-specific changes underlying PD gait disorders, among other types of sensors [14, 15], acceleration signals have recently been analyzed [30–36].

Especially, a series of studies by Weiss and colleagues have clarified the clinical validity of quantification of gait-induced accelerations [12, 37–40]. They calculated the gait cycle time and the step-to-step variability in PD patients [37–40]. Importantly, the step-to-step variability was higher in fallers than non-fallers [37, 38] and in freezers than nonfreezers [39]. They also showed that the amplitudes of gait-induced accelerations can be good indices for gait disorders [37–40]. These studies suggest that quantification of gait-induced accelerations, cycles, and amplitudes can identify PD-specific changes in gait control. Thus, further development of improved algorithm is desirable.

1.3. Proposal of a new algorithm: relationship between gait-induced acceleration and gait cycle duration

In order to assess physiologically gait akinesia/bradykinesia in PD, we have developed a new system (the portable gait rhythmogram (PGR)) for long-term monitoring of gait-induced accelerations during walking and daily routine activity. This system consists of two components: the first component is a wearable device with an acceleration sensor, and the second one is an automatic gait detection algorithm. Gait-induced accelerations are deduced from limb and trunk movements using a mathematical algorithm known as the "pattern-matching method" [41–54]. Thus, the PGR can quantitatively identify various movement-induced accelerations and gait-induced accelerations, respectively.

In a series of publications, we described the motor signs of patients with PD recorded by the PGR [41–54]. Mathematical analysis of various movement-induced accelerations allowed us to quantify poverty of movements. Furthermore, examination of gait-induced accelerations clarified two major features of parkinsonian gait: (1) disabilities in force generation and rhythm setting and (2) loss of dynamic modulation of the force and rhythm depending on the context. By tracing gait fluctuations, it was possible to trace the time and degree of motor fluctuation.

The present article is a review of our published data on the PGR and gait disorders [41–54] and proposes a new algorithm that defines the relationship between gait-induced acceleration and gait cycle duration so as to quantify comprehensively PD-specific pathophysiological mechanisms underlying PD gait disorders.

2. Identification of gait-induced accelerations using pattern-matching method and analysis of gait force-rhythm relationship

2.1. Monitoring of motion-induced acceleration

Acceleration signals were measured by a portable PGR equipped with a triaxial accelerometer (Mimamori-gait system, LSI Medience Corporation: size, 75 mm × 50 m × 20 mm; weight, 120 g) (**Figure 1A**) [41–54]. The sensor is sensitive not only to dynamic acceleration associated with rapid body movements but also to those with static acceleration due to gravity. The PGR was secured with a fixation belt to the front center of the subject's waist (predefined position) at the start of the study. The subject was instructed to go about her/his daily activities in free-living environment for a period of 24 hrs with the device attached at all times except when changing clothes or taking a bath. When standing in the anatomical position, the orientation of the three acceleration axes—x, y, and z—was medial/lateral, vertical, and anterior/posterior, respectively. The rhythmogram measured three-dimensional accelerations (a_x, a_y, a_z) associated with limb and trunk movements, as well as those induced by step-in and kickoff during gait. Data were collected at a sampling rate (Δt) of 10 ms and stored on a Secure Digital card inserted into the device for later analysis using a custom-made software program. A fully charged PGR can achieve 70 hrs of continuous recording.

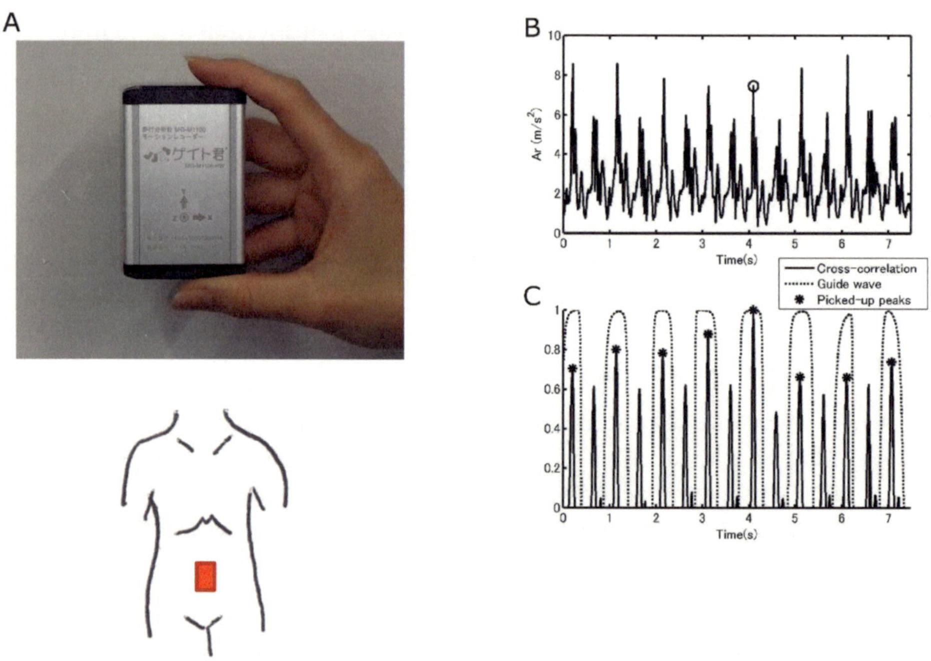

Figure 1. The portable gait rhythmogram device and illustration of the sensor placements (A), Examples of total acceleration signal during walking (B), and corresponding cross correlation (solid line), guide wave (dotted line), and picked-up peaks (asterisk) (C).

http://dx.doi.org/10.5772/intechopen.75483

2.2. Identification of acceleration induced by gait motion

2.2.1. Overview of analytical tools for gait detection

For the analysis of gait characteristics from the three-dimensional acceleration signal, it is important to correctly extract the timing of every stride event during walking (hereafter referred to as stride peak or gait peak). Several basic gait features can be utilized for this purpose. First, walking is a repeated body movement and therefore is accompanied by rhythmic acceleration patterns. This rhythmicity can be captured by the conventional template-matching method [41–44]. Details of the mathematical calculations were reported in our previous methodological articles [42–44, 50]. **Figure 2** shows a summary diagram of the algorithm.

In the present study, the recorded acceleration signal is filtered by a high-pass filter $sT_F/(1 + sT_F)$ to remove slow trends caused by body inclination. The time constant T_F is set at 0.7 sec. From the filtered signal, we choose a three-dimensional (3D) template wave composed of p consecutive points:

$$[Ax(t+i\Delta t)\ \ Ay(t+i\Delta t)\ \ Az(t+i\Delta t)] \tag{1}$$

where $i = 0, 1, \ldots p-1$.

Then, the normalized cross correlation between this wave and an arbitrary signal segment of the same length as the template with a time shift of T is obtained by the following equation:

$$\frac{\frac{1}{p}\sum_{i=0}^{p-1}[Ax(t+i\Delta t)Ax(t+i\Delta t+T)+Ay(t+i\Delta t)Ay(t+i\Delta t+T)+Az(t+i\Delta t)Az(t+i\Delta t+T)]}{\left\{\frac{1}{p}\sum_{i=0}^{p-1}[Ax\,(t+i\Delta t)^2+Ay\,(t+i\Delta t)^2+Az\,(t+i\Delta t)^2]\right\}^{\frac{1}{2}}\left\{\frac{1}{p}\sum_{i=1}^{p}[Ax\,(t+i\Delta t+T)^2+Ay\,(t+i\Delta t+T)^2+Az\,(t+i\Delta t+T)^2]\right\}^{\frac{1}{2}}}. \tag{2}$$

Here, the average value is subtracted from each acceleration component beforehand. The normalized cross correlation characterizes the similarity between two signal segments (therefore, the rhythmicity of the signal), offering the basis of self-adaptive algorithm of automatic gait peak detection [41–44]. Moreover, this value is a robust measure because it is rather insensitive to the manner of setting the coordinate system. This feature is especially useful in the analysis of long-term acceleration data, since the position or orientation of the device might be altered from the predefined one during the day; for example, it sometimes occurs that the device is worn upside down and/or back to front after the patient changes clothes.

We also paid attention to another important feature of gait: biphasicity vs. monophasicity. This means that, during one gait cycle, both vertical and anterior/posterior accelerations change in a biphasic manner, that is, a similar pattern is repeated twice per cycle, while medial/lateral acceleration shows a monophasic pattern. The biphasicity/monophasicity inherent in human gait helps to identify correct stride peaks with high precision as described below.

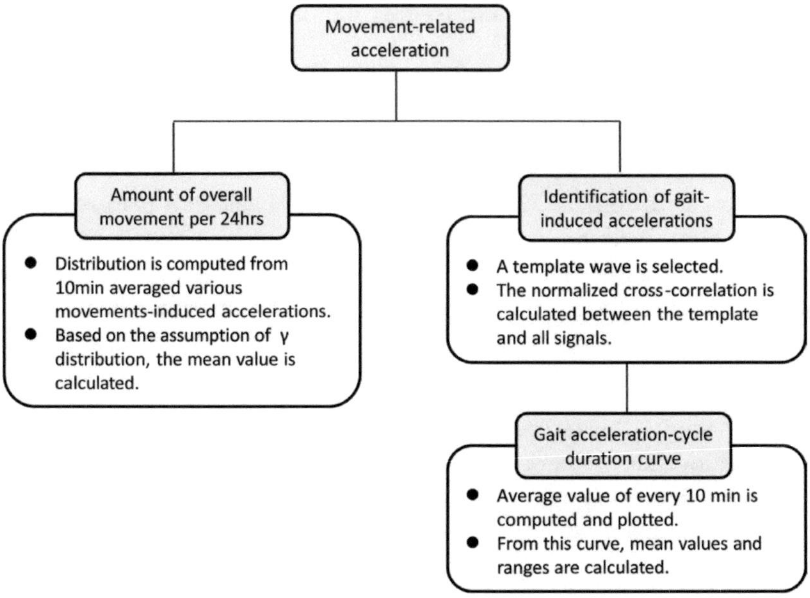

Figure 2. Diagram of the algorithm using the pattern-matching method and analysis of the relationship between gait force and rhythm.

2.2.2. *Estimation of continuously walking region*

Possible walking regions with stride cycle <2 sec are estimated from the whole time series by the following procedure:

- Total acceleration $A_r(t)$ is calculated from the filtered acceleration components by the equation:
- $Ar(t) = \sqrt{Ax(t)^2 + Ay(t)^2 + Az(t)^2}$.
- From this time series, local maximum points above a certain threshold are picked up. The time indices of these points are hereafter referred to as LMP. The threshold value is typically set to 1.5 m/s^2 to ensure that other cyclic events with low intensity, such as tremor, can be excluded.
- Selection of a template wave is composed of p consecutive points centered at one of those LMP, $[A_x(t_0)\ A_y(t_0)\ A_z(t_0)]$. Typically, p is chosen as 50, which corresponds to a time interval of 0.5 sec when $\Delta t = 10$ msec.
- Calculation of the normalized cross correlation between the template and all signal segments centered at other LMP, $[A_x(t)\ A_y(t)\ A_z(t)]$ for $0 \leq t\text{-}t_0 \leq 2$ (sec).
- Among the segments with correlation coefficient > 0.5, a segment that yields the highest correlation coefficient is selected. The template wave is replaced with this segment wave. The existence of a segment with large correlation coefficient (>0.5) strongly suggests that the signal within the current time range is highly periodic.
- Procedures (3) and (4) are repeated until no segment waves with correlation coefficient > 0.5 are found. The central time of the final template thus obtained is denoted as T_1.

- From the initial t_0 and the template selected in procedure 2, procedures (3), (4), and (5) are repeated backward in time, that is, $0 \le t_0 - t \le 2$. The central time of the final template is denoted as T_2.
- The region between T_2 and T_1 is extracted from the acceleration signal. This is an active rhythm block characterized by highly rhythmic behavior of moderate or large intensity with a cycle duration of <2 sec.
- Procedures (2)–(7) are repeated. In this way, the entire time series is divided into active rhythm blocks and inactive and/or nonrhythmic blocks. The former can be regarded as continuously walking regions.

2.2.3. *Estimation of stride peak candidates*

From each of the continuously walking regions, stride peak candidates are estimated by the following template-matching algorithm:

- Selection of a point t_0 from LMP and generation of a template wave with p consecutive points centered at this point. The template size p can be typically set to 50, but more precisely, it should be optimized based on the characteristic timescale of the original acceleration signal [42].
- Calculation of the normalized cross correlation between the template and all signal segments centered at other LMP.
- Selection of the segments with correlation coefficient > 0.5. Calculation of the average value of those correlation coefficients.
- Procedures (1)–(3) are repeated for all LMP. The template wave that yields the largest average correlation coefficient is chosen as the optimized template.
- Calculation of the normalized cross correlation time series by sliding the optimized template over the current continuously walking region.
- Extraction of the local maximum points from the normalized cross correlation time series. Among these points, the ones with positive correlation coefficient are the desired stride peak candidates.

Hereafter, the time index of the optimized template center is indicated by t_c.

2.2.4. *Detection of stride peaks*

The candidate peaks obtained so far are not fiducial points due to the possible presence of extra peaks related to the contact of both feet (step peaks) and other spurious points. Therefore, stride peaks are determined with the help of a "guide wave" by the following procedure:

- The three-dimensional acceleration signal $[A_x(t)\ A_y(t)\ A_z(t)]$ is integrated twice in combination with high-pass filtering to avoid integration drift. The integrated signal $[D_x(t)\ D_y(t)\ D_z(t)]$ corresponds to the relative displacement of the body where the device was attached.

This operation works to smooth the original signal and enhances the monophasic nature of medial/lateral movements during walking.

- (2) Consider a fixed unit vector pointing from $[D_x(t_c\text{-}w)\ D_y(t_c\text{-}w)\ D_z(t_c\text{-}w)]$ to $[D_x(t_c+w)\ D_y(t_c+w)\ D_z(t_c+w)]$ and an arbitrary unit vector pointing from $[D_x(t\text{-}w)\ D_y(t\text{-}w)\ D_z(t\text{-}w)]$ to $[D_x(t+w)\ D_y(t+w)\ D_z(t+w)]$ in 3D space. The inner product of these two vectors is calculated. The inner product is time-dependent, giving a new time series, which is called here "guide wave." Here, the time range of the unit vector w is decided so that the obtained guide wave behaves like a square wave with the same monophasic pattern as the medial/lateral (x-axis) signal. Usually, it is almost equal to the time range of the optimized template wave, that is, $2w = p$.
- Selection of the candidate stride peak with the largest correlation coefficient per single guide wave cycle. These are fiducial stride peaks.

The method of making a guide wave is not limited to the one presented here. The key is to find a suitable monophasic signal, that is, a signal composed of a wavelet that is repeated once per single gait cycle. In [42], we designed a different type of guide wave by utilizing the anisotropic character of the x-axis acceleration signal.

Figure 1B and **C** illustrates the validity of this procedure. **Figure 1B** shows an example of the total acceleration wave. The center point of the optimized template is marked by an open circle. The detected peaks using the guide wave are indicated by asterisks in **Figure 1C**. It is clear that only the stride peaks, that is, gait peaks associated with the same leg, are selected, while peaks due to the other leg are not counted. In our earlier work, we found that this type of peak detection algorithm performs fairly well for the gait of PD patients under supervised but non-laboratory conditions, with more than 95% accuracy [43].

2.2.5. Estimation of "gait cycle duration" and "gait-induced acceleration"

After identifying correct stride peaks, the duration of a single gait cycle and the amplitude of gait-induced acceleration per cycle are calculated as follows. Let two adjacent stride peaks be indexed as $i = 1$ and $i = N$. Then, the gait cycle duration is given by $(N-1)\Delta t$, and gait acceleration is defined by $\frac{\sum_{i=1}^{N-1} A_r(i\Delta t)}{N-1}$. These measures are calculated from all peak-to-peak intervals. Since gait accelerations correlate with floor reaction forces, the amplitude of gait acceleration is selected as an index of floor reaction force. Thus, the obtained gait cycle duration and the amplitude of gait-related acceleration are averaged for every 10-min period of recording.

2.2.6. Accuracy of analysis using pattern-matching method in detecting gait-induced signals

To further assess the sensitivity of the PGR, we compared the results of PGR with those of a standard system using floor reaction forces [unpublished data]. We asked the subject to step in response to a particular gait cycle determined by a metronome (0.87–1.30 sec) and compared the reappearance of the cycle recorded from the two devices.

Using the PGR, we recorded metronome-guided 10 m walking in six normal control subjects (age, 43.3 ± 7.7 years, mean ± SD, four men, and two women). The metronome was set at 92, 100, 112,

http://dx.doi.org/10.5772/intechopen.75483

120, 132, and 138 beats per minute. The subjects were asked to contact the right or left foot corresponding to the beat. Thus, the expected gait cycle (the duration from the ground contact of one foot to the next contact of the same foot) was 1.30, 1.20, 1.07, 1.00, 0.91, and 0.87 sec, respectively.

In this control study, subjects equipped with the PGR device walked on the force plate (WalkWay MW-1000, Anima, Tokyo). Since the force plate was 2.4 m long, only part of the metronome-guided 10 m walking was recorded simultaneously by the two devices. The gait cycles recorded by the two equipments were compared across all subjects and all metronomic paces by using the Bland–Altman plot and intraclass correlation (ICC (2,1)) methods. The subjects were asked to walk in sync with the beats of the metronome. The cycle of the metronome-guided walking was set at five stages: 1.30, 1.20, 1.07, 1.00, 0.91, and 0.87 sec. **Figure 3** shows an example of the recorded accelerations by PGR during the metronome recording. **Figure 3A** shows an example where a difference was observed between the expected and recorded gait cycles. The expected cycle was 0.87 sec, whereas the recorded cycle was 0.94 sec. On the other hand, **Figure 3B** shows an example where the recorded gait cycle matched the expected gait cycle.

Table 2 shows a summary of the gait cycles recorded by the PGR and the force plate, for each subject. We could not identify the gait cycle from the force plate data in four cases out of the total of 36 measurements. Therefore, we used 32 pairs of data for comparison of the two methods. **Figure 3C** shows the Bland–Altman plot, demonstrating the difference in two gait cycles as a function of their mean values. The 95% confidence interval of the mean difference was [−0.0191 to 0.0003], which includes zero, and no significant regression was observed ($p = 0.49$),

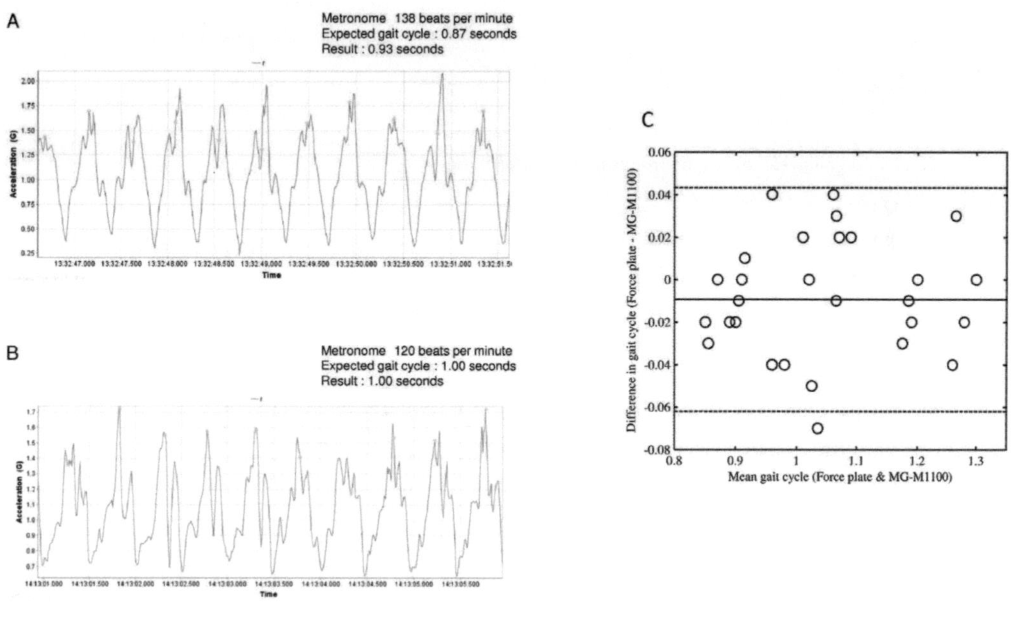

Figure 3. (A, B) Examples of gait-induced accelerations during metronome-guided walking. The expected rate defined by the metronome was 0.87 sec in (A) and 1.00 sec in (B). (C) Bland–Altman plot of two gait cycles estimated by force plate and PGR. The differences between the two sets of gait cycles are plotted against their mean values. The solid line represents the mean difference, while the dashed lines are the 95% confidence limit lines (bias ± 1.96 SD).

Metronome (beats/min)	92	100	112	120	132	138
Expected gait cycle (sec)	1.30	1.20	1.07	1.00	0.91	0.87
Subject 1						
Force plate (sec)		1.10	1.08	1.08		0.98
MG-M1100 (sec)	1.18	1.08	1.05	1.04	1.01	0.94
Subject 2						
Force plate (sec)	1.27	1.18	1.00	0.94	0.88	0.84
MG-M1100 (sec)	1.29	1.19	1.05	0.98	0.90	0.86
Subject 3						
Force plate (sec)	1.24	1.18	1.06	1.02	0.92	
MG-M1100 (sec)	1.28	1.20	1.07	1.00	0.91	0.87
Subject 4						
Force plate (sec)	1.30	1.20	1.06	0.96	0.89	0.87
MG-M1100 (sec)	1.30	1.20	1.07	1.00	0.91	0.87
Subject 5						
Force plate (sec)	1.28	1.16	1.00	0.96	0.91	0.84
MG-M1100 (sec)	1.25	1.19	1.07	1.00	0.91	0.87
Subject 6						
Force plate (sec)		1.20	1.08	1.02	0.90	0.84
MG-M1100 (sec)	1.26	1.20	1.06	1.02	0.91	0.87

Table 2. Recorded gait cycle during metronome-guided walking.

suggesting that there was no systematic error in the Bland-Altman plot. Moreover, the two gait cycles yielded a fairly high ICC value of 0.98 ($p < 0.0001$), with 95% confidence interval value of [0.96-0.99]. These results indicate that the recorded gait cycle by the PGR was statistically in good agreement with that of the force plate. The results showed no significant differences between the two methods, indicating that PGR accurately monitors gait movements with sensitivity comparable to that of the force plates. Since more steps were analyzed in the recordings using PGR than those using the force plate, fluctuation appears to be small. The results of the control experiments indicate that PGR can quantitatively estimate gait disorders in daily life with good sensitivity.

2.2.7. Merits of proposed gait measures

One notable merit of the developed method is that both gait cycle duration and gait-induced acceleration can serve as quantitative indices for the assessment of daily gait performance. Other basic parameters that are widely used in gait analysis are walking speed and step length. While these two parameters are not measured directly by our method, it is possible to estimate them with good accuracy from the recorded gait measures, since there is a close correlation between (logarithm of) gait-induced acceleration and walking speed, when these two parameters are expressed by body height [50].

However, it should be noted that any gait parameter, when used separately, is more than often insufficient to allow complete understanding of ambulatory gait behavior. For example, a person usually changes gait cycle duration (or cadence) over the course of a day both voluntarily and involuntarily. Real life comprises a diversity of environmental constraints, making it almost impossible for us to keep on walking freely. Therefore, even if the gait parameter appears to deteriorate in a certain time of the day, one cannot decide whether it is caused by disorders in internal factors (e.g., the subject's physical conditions) or by changes in walking pace due to external perturbations. In most cases, such "forced" changes are transient, and their effects could be reduced by taking the long-time average of the parameter. However, the combination of gait cycle duration and gait acceleration provides a more powerful solution to this problem, as will be demonstrated in this paper and as was reported previously [43, 44]. The two measures, when plotted on a plane, will fall on a rather well-ordered structure that is individual-specific and robust against external constraints. The robustness of the gait cycle-acceleration relationship is considered to result from the human tendency to automatically select preprogrammed preferred walking behavior in response to environmental changes [43]. Based on this property, we can obtain insightful information on intrinsic gait problems from ambulatory monitoring, which is the greatest merit of using the proposed gait measures.

3. Quantification of "poverty of movements" and "slowness/smallness in executed movements": "amount of overall movements per 24 hrs"

Poverty in movements and slowness/smallness in executed movements are the core motor clinical signs of PD. Overall activities should be examined to quantify these two clinical signs. In order to quantify overall activities, we analyzed the distribution of 10-min averaged various movement-induced accelerations (**Figure 4A** and **B**) and calculated the mean value of the distribution with the assumption that the curve fits with gamma distribution [45, 53]. The gamma distribution is defined by the following formula:

$$f(x) = x^{k-1} e^{-x/\theta}/\Gamma(k)\, \theta^k \text{ for } x > 0$$

The mean value was defined as the *amount of overall movements per 24 hrs*, representing an index of poverty of movements and slowness/smallness of executed movements.

A total of 50 patients with PD and 17 normal control subjects participated in this analysis [45]. The distribution curve shifted to the left, suggesting a decrease in activities (**Figure 4A** and **B**). The *amount of overall movements per 24 hrs* was smaller in PD patients compared with the mean value of normal control subjects (0.69 ± 0.19 m/sec^2). Furthermore, 38 of the 50 patients (76%) showed a decrease below 0.50 m/sec^2, the value of mean (1sd). The *amount of overall movements per 24 hrs* decreased in proportion with the severity of PD. The *amount of overall movements per 24 hrs* for patients with score ≥ 30 (0.31 ± 0.09 m/sec^2) was significantly lower than that of

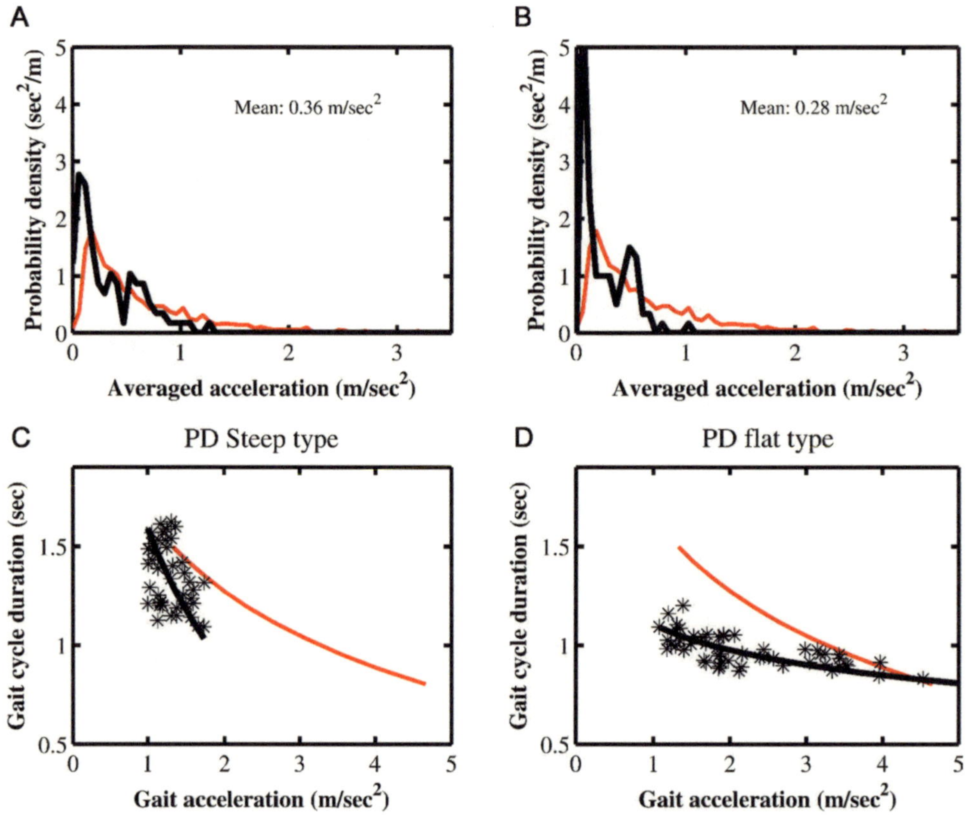

Figure 4. (A, B) Probability distribution of the averaged acceleration per 10 min, obtained from 24 hrs of continuous recording. (C, D) examples of gait acceleration-cycle duration relationship in a patient with Parkinson's disease (PD) based on 24 hrs of gait recording. Black thick line: Corresponding regression line. In (C) and (D), red thin line: Regression line obtained from the summed average of 17 normal control subjects. The regression line of the PD patient is steep in (C), whereas it is flat in (D).

patients with scores 1–15 (0.40 ± 0.09 m/sec^2) ($p < 0.05$) and that of patients with scores 16–30 (0.48 ± 0.09 m/sec^2) ($p < 0.01$). Furthermore, in 50 patients with untreated PD, the *amount of overall movements per 24 hrs* was significantly associated with the UPDS part II score and part III score [53]. These results suggest that the *amount of overall movements per 24 hrs* can be a good index for poverty of movements and slowness/smallness of executed movements in ADL.

4. Quantitative estimation of parkinsonian gait using two parameters: "gait-induced acceleration" and "gait cycle duration"

4.1. Relationship between gait-induced acceleration and gait cycle duration

The relationship between gait-induced acceleration and gait cycle duration (*gait acceleration-cycle duration curves*) [45, 46] is shown in **Figure 4C** and **D**. Both parameters were recorded continuously for more than 24 hrs, and then the average value of every 10 min was computed and plotted. Interestingly, a regression line can be determined for the plotted data. In **Figure 4C** and **D**, the thick black line represents the regression line of data of a PD patient, while the thin red line represents the regression line for the data of 17 normal control subjects. The data clearly show a correlation between accelerations and cycle; for example, walking with a slow cycle is associated with low amplitude of gait accelerations. The slope of the linear regression line for data of the normal control subjects was 1.20 ± 0.29.

The *gait acceleration-cycle duration curves* of PD patients showed two distinct patterns compared to those of normal control subjects [45, 46]. First, the position of the curve in PD deviated from that of the normal control curve; the curve of PD patients was shifted to the left on the acceleration axis (**Figure 4C** and **D**), suggesting a decrease in the mean amplitude of gait-induced acceleration. Furthermore, the curve of PD patients was shifted upward (**Figure 4C**) or downward (**Figure 4D**) on the gait cycle duration axis, suggesting prolongation or shortening of the mean gait cycle duration, respectively. Second, the range value around the regression line was narrow in PD patients [45, 46]. A narrow range of acceleration was associated with a steep slope of the regression line (**Figure 4C**), whereas a narrow range of gait cycle duration was associated with a flat slope of the regression line (**Figure 4D**).

Based on these findings, we next quantified these two kinds of changes in t*he gait acceleration-cycle duration curve*, that is, changes in the mean value and range, as described below.

4.2. Changes in mean values of gait-induced acceleration and gait cycle duration: Possible index of PD deficits in gait control neural circuits

We first examined changes in the mean amplitude of gait-induced acceleration and gait cycle duration during daily walking in 64 PD patients and 17 normal control subjects [48]. The mean values of the gait-induced acceleration and gait cycle duration were computed (**Figure 5**). PD patients walked with significantly lower gait accelerations compared with normal control subjects, but there were no significant differences among the disease stage groups. In contrast, PD patients with modified Hoehn and Yahr (mH&Y) stage 1.5 walked with significantly faster gait cycle duration (faster rhythm) than those with stages 2.5–3.0 and normal control subjects, although there was no significant difference in gait cycle duration between patients with stages 2.5–3.0 PD and normal control subjects. The lower gait accelerations (i.e., inappropriate force production) can be the cause of narrow stride [42, 43]. Thus, the above results suggest that PD patients cannot walk with a sufficient stride throughout the clinical course, which can be compensated by increasing stepping [48, 55–57] at the early stage, but not at the advanced stage [48]. Our data suggest that computation of acceleration of gait movements and gait cycle duration from daily walking can be useful in the assessment of the clinical stage of the gait disorder.

4.3. Narrow range of gait-induced acceleration and gait cycle duration: Possible index of monotony in PD

Another characteristic change in the shape of the *gait acceleration-cycle duration curve* is the narrowness of the range of force and/or rhythm [45, 46]. When PD patients walked with a narrow range of change in gait acceleration (i.e., floor reaction forces), the slope of the regression line was steep (*steep type*) (**Figure 4C**). In contrast, when PD patients walked with a narrow

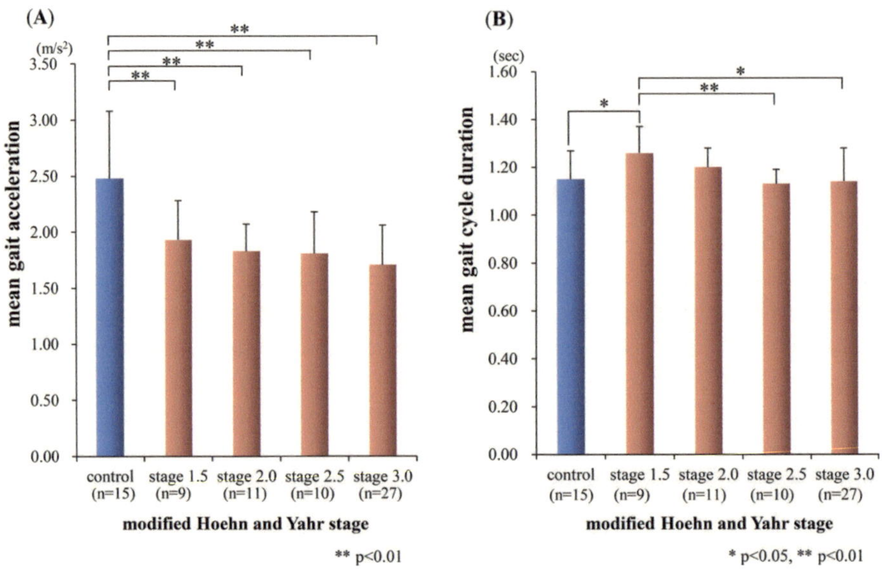

Figure 5. The mean gait acceleration (A) and mean gait cycle (B) in PD patients according to the modified Hoehn and Yahr stages. Differences between each stage group and normal control subjects, and among the stage groups were analyzed for statistical significance (*$p < 0.05$, **$p < 0.01$, by *t*-test) [40].

range of change in the gait cycle duration (i.e., rhythm), the regression line was flat (*flat type*) (**Figure 4D**). Other patients walked with a narrow range of both acceleration and cycle, and their acceleration and cycle duration data were scattered within a narrow range (*lump type*). Under these conditions, the patient always walks with the preferred floor reaction forces or cycle, that is, monotone walking.

For quantitative assessment of these changes, we defined *%alteration range* as an index of monotone walking [46]. The mean and standard deviation (SD) of gait acceleration and cycle duration were calculated from the *gait acceleration-step cycle duration curve*, which was constructed from a single long-term recording from each patient. Here, we used the SD as an index of alteration range. The interindividual mean of SD was calculated from data of 17 control subjects. Thus, the ratio of the SD of each patient to the interindividual mean of SD of the normal control subjects was defined as *%alteration range*. The cutoff level was set at 75% in analysis of the *%alteration range*.

Changes in *%alteration range* were examined in 40 patients with PD. In this analysis, the regression in 16 of the 40 (40%) patients was similar to that of the control (*%alteration range* > 75%). Furthermore, the *steep type* (*%alteration range* of gait acceleration of <75%) was identified in 12 of the 40 patients (30%), whereas the *flat type* (*%alteration range* of gait cycle of <75%) was noted in 8 of the 20 (20%) patients, and the *lump pattern* was noted in 10% of patients (4/40) [46].

It has been reported that the maximum activities that can be generated in a muscle burst is low in PD patients [58] and that PD patients can only execute movements of different amplitudes at a single, slow velocity without dynamic changes of the movement velocity [59]. These studies suggest a possible mechanism for the narrow range of force and rhythm in gait, that is, PD patients cannot dynamically vary the force and rhythm in their daily walking, resulting in a narrow range of floor reaction forces and step cycles. Our data suggest that quantification of both the range of gait-induced acceleration and gait cycle duration provides a useful index of monotony, a fundamental pathophysiology in PD.

5. Quantitative estimation of motor fluctuation

Although dopamine replacement therapy is effective at the early stage, patients with PD develop motor fluctuation as the disease progresses [13, 47, 60–62]. By detecting changes in gait parameters, we tried to determine motor fluctuations. We review here the following two methods.

5.1. Method I: identification of "off" by tracing daily changes in gait-induced accelerations and gait cycle duration

We traced changes in gait-induced accelerations and cadences (steps/min, a parameter identical to gait cycle duration). These changes in gait force and rhythm parameters were compared simultaneously with patients' diaries. A total of 44 patients with PD, who showed simultaneous changes between accelerations and cadences, and 17 normal control subjects participated in this study [47].

No subjective fluctuations were noticed by 23 of these 44 patients. However, the PGR detected pathological shift of accelerations and cadences in 19 of the 23 patients (83%). These patients showed fast stepping or slow stepping with low amplitude of accelerations. On the other hand, the other 21 patients complained of subjective off in their diaries. Long-term tracing of gait-induced accelerations and cadences identified no changes in 11 of these 21 patients (52%). Surprisingly, no synchronization was noted between the subjective off and gait off determined by the PGR in 30 of the 44 patients (68%). This discrepancy could be in part due to deficits in attention to motor deficits [63]. Our data suggest that tracing simultaneous changes in acceleration and cadence can raise an interesting question: "How do PD patients subjectively notice the off time?"

Clinically, improvements in gait fluctuations of PD patients after the addition or increase in dose of an anti-parkinsonism medicine were detected by examining changes in gait force and rhythm parameters [54]. The present method had a high sensitivity for detecting improvements than the UPDRS scores.

5.2. Method II: quantitation of degree of motor fluctuation by calculating frequency mode change in gait

5.2.1. *Two strategies*

We reported previously [42, 43] that the relationship between gait-induced acceleration and gait cycle duration can be described by the following function derived from an inverted pendulum model:

$$a_v = \frac{1}{T_c}\left[\alpha\left(\frac{1}{T_c} - \frac{1}{T_0}\right) + \beta\right]^2,$$

where *Tc* is the gait cycle, *av*. is the vertical acceleration magnitude averaged over *Tc*, and α, β, and *T0* are the three separate parameters. Parameters α and β are subject-specific, representing a particular walking strategy of an individual, α quantifies how to regulate gait acceleration according to different gait paces, and β is something like a basic potential force that drives natural walking. In contrast, T_0 is assumed to be a constant across subjects. Specifically, we set T_0 = 1.4 sec in order to uncorrelate α and β with each other [42, 43]. The above mathematical method was used to build better fitting regression curves from the data. In other words, this mathematical method for regression curves represents improvement compared to that shown in **Figure 4C** and **D**.

Figure 6A shows that two regression lines are necessary for better fitting of plots obtained from long-term daily walking, suggesting that a subject usually walks with two types of gait mode [49]. The steep regression line with small accelerations for a particular cycle means quiet walking, for example, walking within the house, whereas the flat regression line with large accelerations for a particular cycle reflects active walking, for example, walking in the street. These results suggest that the subject changes the walking strategy during daily walking between "quiet mode" and "active mode." To identify the involvement of deficits in higher functions, we quantitatively examined the "switching" from one mode to another [49].

5.2.2. Frequency of mode change

A total of 26 patients with PD and 13 normal control subjects participated in this study [49]. We introduced the parameter of *frequency of mode change* (FC), which describes the frequency of changes in walking strategies and is calculated mathematically by the following equation:

$$\mathrm{FC} = \frac{\sum_{t=1}^{n-1} |I(t+1) - I(t)|}{n-1},$$

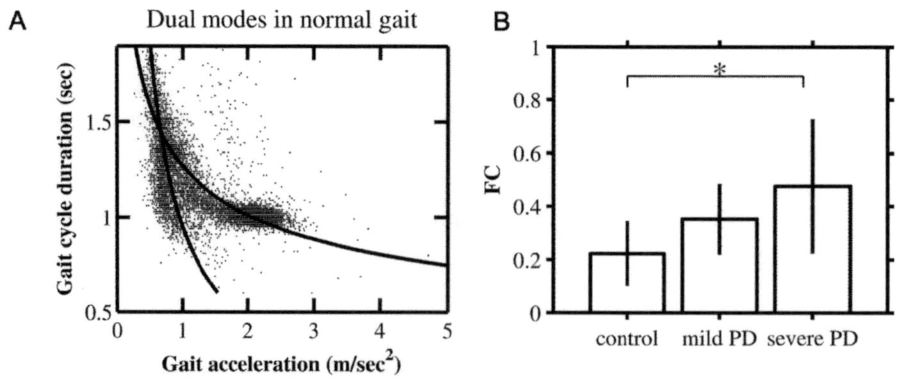

Figure 6. (A) Examples of gait acceleration-cycle duration relationship in a normal subject based on 24 hrs of gait recording. The amplitude of gait-related accelerations and duration of gait cycles were plotted for all stride events. Two regression lines were required for better fitting. (B) Frequency of mode change (FC) that describes the frequency of changes into different walking strategies between quiet mode and active mode.

where $I(t)$ reflects the walking mode [i.e., active (0) or quiet (1)] in each non-overlapping 1 hr. interval during the day and t signifies the time index [$t = 1, 2, \ldots, n$ ($n \leq 24$)]. FC was 0.22 ± 0.12 in normal subjects compared with 0.35 ± 0.13 in patients with mild PD (mH&Y stage <2.5) and 0.48 ± 0.25 in patients with severe PD (mH&Y stage ≥2.5) ($p < 0.05$) (**Figure 6B**) [49]. In PD, the quiet mode coincides approximately with slow stepping gait mode. These results show that estimation of frequency of change between quiet and active modes provides an estimate of the degree of motor fluctuation.

6. Conclusion

Our newly developed wearable device and algorithm, the PGR, allows continuous monitoring of gait for up to 70 hrs and can differentiate precisely gait-induced accelerations from other movements. The PGR can monitor both the overall motor activities and gait movements in daily activities. The findings obtained from long-term monitoring using PGR are classified into three categories.

6.1. Poverty of movements and slowness/smallness of executed movements

Acceleration values induced by various movements, averaged every 10 min, showed gamma distribution. The mean value in this distribution was used as an index of the *amount of overall movements per 24 hrs*. The *amount of overall movements per 24 hrs* was lower in PD patients, especially in patients with UPDRS-III ≥ 30, suggesting that this parameter reflects the range from poverty of movements to slowness/smallness of executed movements.

6.2. Gait disorders

Analysis of *the gait acceleration-cycle duration curve* provided the mean values and ranges of gait-induced acceleration and gait cycle duration. Quantitative analysis showed force generation deficits, disabilities of gait rhythm setting, and loss of dynamic modulation of the force and rhythm in each patient with PD. The quantitative measures can be used for assessment of the degree of parkinsonian gait disorders. These changes in gait-induced acceleration reflect PD-specific deficits in neural circuits involved in gait control [1, 2, 10]. Thus, it is hoped that long-term monitoring of these parameters can identify the pathophysiological changes in gait disorders in PD.

6.3. Motor fluctuation

Furthermore, the device also provides assessment of motor fluctuation from the data of gait fluctuation in any particular day. By tracing changes in gait-induced accelerations and gait cycle durations, we identified the time of motor off, which was not necessarily noticed by patients. Measurement of the frequency of changes in quiet mode (i.e., slow steeping mode) and active mode provides information on the degree of severity of motor fluctuation.

6.4. Open questions and possible clinical applications

Taken together, measurement of overall movement- and gait-induced accelerations using a wearable device could be useful for clinical assessment of parkinsonian status. Especially, the

Quantification of "poverty of movements" and "slowness/ smallness of executed movements": • Acceleration values induced by various movements are averaged every 10 min from the long-term monitoring data • The mean value in this gamma distribution can be an index of the *amount of overall movements per 24 hrs*
Gait disorders: • Gait-induced accelerations are averaged every 10 min from the long-term monitoring data • The *gait acceleration-cycle duration curve* is calculated from these plots • Changes in the mean values of acceleration and cycle duration can reflect deficits in force production and rhythm setting mechanisms, respectively • Narrowness of accelerations and cycle duration can reflect loss of dynamic modulation of the force and rhythm control
Motor fluctuation: • The time of motor off can be determined by tracing changes in gait-induced accelerations and gait cycle durations • Measurement of frequency of changes between quiet and active modes provides an index of the severity of motor fluctuation

Table 3. Protocol for quantitative assessment of poverty of movements and slowness/smallness of executed movements, gait disorders, and motor fluctuation.

new algorithm that defines the relationship between gait-induced acceleration and gait cycle duration could be useful in clinical applications. However, there are still the following open questions:

- Our studies were conducted in a small number of normal subjects and patients with PD. Thus, the outcome from our data has not been validated.
- Methodologically, our analysis does not include quantification of stride length. For example, analysis using an inverted pendulum model should be useful [64].

Here, we provisionally propose a comprehensive and quantitative examination protocol for poverty of movements and slowness/smallness of executed movements, gait disorders, and motor fluctuation in patients with PD using an accelerator-mounted wearable device (see **Table 3**). This physiology-based protocol would make it possible to quantify the extent of parkinsonian motor signs and/or the effect of drugs. We anticipate that further analysis of daily walking using the PGR would contribute to the development of better therapies for PD.

Conflict of interest

The authors declare no conflict of interest.

http://dx.doi.org/10.5772/intechopen.75483

Author details

Masahiko Suzuki[1]*, Makiko Yogo[1], Masayo Morita[1], Hiroo Terashi[2], Mutsumi Iijima[3], Mitsuru Yoneyama[4], Masato Takada[5], Hiroya Utsumi[6], Yasuyuki Okuma[7], Akito Hayashi[8], Satoshi Orimo[9] and Hiroshi Mitoma[10]

*Address all correspondence to: suzukimd@jikei.ac.jp

1 Department of Neurology, Katsushika Medical Center, The Jikei University School of Medicine, Tokyo, Japan

2 Department of Neurology, Tokyo Medical University, Tokyo, Japan

3 Department of Neurology, Tokyo Women's Medical University, Tokyo, Japan

4 Mitsubishi Chemical Corporation, Yokohama R&D Center, Yokohama, Japan

5 LSI Medience Corporation, Tokyo, Japan

6 Department of Neurology, Shioya Hospital, International University of Health and Welfare, Tochigi, Japan

7 Department of Neurology, Shizuoka Hospital, Juntendo University, Shizuoka, Japan

8 Department of Rehabilitation, Urayasu Hospital, Juntendo University, Chiba, Japan

9 Department of Neurology, Kanto Chuo Hospital, Tokyo, Japan

10 Medical Education Promotion Center, Tokyo Medical University, Tokyo, Japan

References

[1] Dietz V. Human neuronal control of automatic functional movements: Interaction between central programs and afferent input. Physiological Reviews. 1992 Jan;**72**(1):33-69

[2] Dietz V. Neurophysiology of gait disorders: Present and future application. Electroencephalography and Clinical Neurophysiology. 1997 Sep;**103**(3):333-355

[3] Schaafsma JD, Balash Y, Gurevich T, Bartels AL, Hausdorff JM, Giladi N. Characterization of freezing of gait subtypes and the response of each to levodopa in Parkinson's disease. European Journal of Neurology. 2003 Jul;**10**(4):391-398

[4] Critchley M. Arteriosclerotic parkinsonism. Brain. 1929;**52**:23-83

[5] Nutt JG, Marsden CD, Thompson PD. Human walking and higher-level gait disorders, particularly in the elderly. Neurology. 1993 Feb;**43**(2):268-279

[6] Thompson PD, Marsden CD. Gait disorder of subcortical arteriosclerotic encephalopathy: Binswanger's disease. Movement Disorders. 1987;**2**(1):1-8

[7] Thompson PD. Gait Disorders Accompanying Diseases of the Frontal Lobes. In: Růžička E, Hallet M, Jankovic J, editors. Advances in Neurology. Gait Disorders. Philadelphia, USA: Lippincott Williams & Wilkins; 2001. pp. 235-241

[8] Nutt JG. Gait disorders. In: Jankovic J, Tolosa E, editors. Parkinson's disease and movement disorders. Baltimore, USA: Urban & Schwarzenberg; 1988. pp. 377-383

[9] FitzGerald PM, Jankovic J. Lower body parkinsonism: Evidence for vascular etiology. Movement Disorders. 1989;**4**(3):249-260

[10] Mitoma H, Hayashi R, Yanagisawa N, Tsukagoshi H. Characteristics of parkinsonian and ataxic gaits: A study using surface electromyograms, angular displacements and floor reaction forces. Journal of the Neurological Sciences. 2000 Mar;**174**(1):22-39

[11] Fahn S, Elton RL. UPDRS Development Committee. Unified Parkinson's Disease Rating Scale. In: Fahn S, Marsden CD, Calne D, Goldstein M, editors. Recent Development in Parkinson's Disease. Florham Park, NJ: Macmillan; 1987. pp.153-164

[12] Weiss A, Mirelman A, Buchman AS, Bennet DA, Hausdorff JM. Using a body-fixed sensor to identify subclinical gait difficulties in older adults with IADL disability: Maximizing the output of the time up and go. PLoS One. 2013 Jul 29;**8**(7):e6885

[13] Knutsson E. An analysis of parkinsonian gait. Brain. 1972;**95**:475-486

[14] Maetzler W, Domingos J, Srulijes K, Ferreira JJ, Bloem BR. Quantitative wearable sensors for objective assessment of Parkinson's disease. Movement Disorders. 2013 Oct;**28**(12):1628-1637

[15] Kubota KJ, Chen JA, Little MA. Machine learning for large-scale wearable sensor data in Parkinson's disease: Concepts, promises, pitfalls, and futures. Movement Disorders. 2016 Sep;**31**(9):1314-1326

[16] Hobert MA, Maetzler W, Aminian K, Chiari L. Technical and clinical view on ambulatory assessment in Parkinson's disease. Acta Neurologica Scandinavia. 2014 Sep; **130**(3):139-147

[17] Pasluosta CF, Gassner H, Winkler J, Klucken J, Eskofier BM. An emerging era in the management of Parkinson's disease: Wearable technologies and the internet of things. IEEE Journal of Biomedical and Health Informatics. 2015 Nov;**19**(6):1873-1881

[18] Godinho C, Domingos J, Cunha G, Santos AT, Fernandes RM, Abreu D, Gonçalves N, Matthews H, Isaacs T, Duffen J, Al-Jawad A, Larsen F, Serrano A, Weber P, Thoms A, Sollinger S, Graessner H, Maetzler W, Ferreira JJ. A systematic review of the characteristics and validity of monitoring technologies to assess Parkinson's disease. Journal of NeuroEngineering and Rehabilitation. 2016;**13**:24

[19] Dijkstra B, Kamsma YP. Zijlstra,W. Detection of gait and postures using a miniaturized triaxial accelerometer-based system: Accuracy in patients with mild to moderate Parkinson's disease. Archives of Physical Medicine and Rehabilitation. 2010;**91**(8):1272-1277

[20] de Groot S, Nieuwenhuizen MG. Validity and reliability of measuring activities, movement intensity and energy expenditure with the DynaPort MoveMonitor. Medical Engineering & Physics. 2013;**35**(10):1499-1505

[21] Mancini M, Horak FB. Potential of APDM mobility lab for the monitoring of the progression of Parkinson's disease. Expert Review of Medical Devices. 2016;**13**(5):452-466

[22] Dewey DC, Miocinovic S, Bernstein I, Khemani P, Dewey RB III, Querry R, Chitnis S, Dewey RB Jr. Automated gait and balance parameters diagnose and correlate with severity in Parkinson disease. Journal of the Neurological Sciences. 2014;**345**(1-2):131-138

[23] Mariani B, Jiménez MC, Vingerhoets FJG, Aminian K. On-shoe wearable sensors for gait and turning assessment of patients with Parkinson's disease. IEEE Transactions on Biomedical Engineering. 2013;**60**(1):155-158

[24] Fisher JM, Hammerla NY, Ploetz T, Andras P, Rochester L, Walker RW. Unsupervised home monitoring of Parkinson's disease motor symptoms using body-worn accelerometers. Parkinsonism and related disorders. 2016. DOI: doi.org/10.1016/j.parkreldis.2016.09.009

[25] Suzuki M, Mitoma H, Yoneyama M. Quantitative analysis of motor status in Parkinson's disease using wearable devices. From methodological considerations to problems in clinical applications. Parkinson's Disease. 2017. Article ID 6139716

[26] van Hilten JJ, Middelkoop HA, Kerkhof GA, Roos RA. A new approach in the assessment of motor activity in Parkinson's disease. Journal of Neurology, Neurosurgery, and Psychiatry. 1991 Nov;**54**(11):976-979

[27] Saito N, Yamamoto T, Sugiura Y, Shimizu S, Shimizu M. Lifecorder: A new device for the long-term monitoring of motor activities for Parkinson's disease. Internal Medicine. 2004 Aug;**43**(8):685-692

[28] Hoff JI, van den Plas AA, Wagemans EA, van Hilten JJ. Accelerometric assessment of levodopa-induced dyskinesias in Parkinson's disease. Movement Disorderes 2001 Jan;**16**(1):58-61

[29] Hoff JI, van den Meer V, van Hilten JJ. Accuracy of objective ambulatory accelerometry in detecting motor complications in patients with Parkinson's disease. Clinical Neuropharmacology 2004 Mar-Apr;**27**(2):53-57

[30] Steins D, Sheret I, Dawes H, Esser P, Collett J. A smart device inertial-sensing method for gait analysis. Journal of Biomechanics. 2014 Nov 28;**47**(15):3780-3785

[31] Cancela J, Pastorino M, Arredondo MT, Nikita KS, Villagra F, Pastor MA. Feasibility study of a wearable system based on a wireless body area network for gait assessment in Parkinson's disease patients. Sensors (Basel, Switzerland). 2014 Mar;**14**(3):4618-4633

[32] Godfrey A, Del Din S, Barry G, Mathers JC, Rochester L. Instrumenting gait with an accelerometer: a system and algorithm examination. Medical Engineering & Physics. 2015 Apr;**37**(4):400-407

[33] Rispens SM, van Schooten KS, Pijnappels M, Daffertshofer A, Beek PJ, van Dieen JH. Identification of fall risk predictors in daily life measurements: Gait characteristics' reliability and association with self-reported fall history. Neurorehabilitation and Neural Repair, 2015 Jan; **29**(1):54-61

[34] van Schooten KS, Pijnappels M, Rispens SM, Elders PJ, Lips P, van Dieen JH. Ambulatory fall-risk assessment: Amount and quality of daily-life gait predict falls in in daily life measurements: Gait characteristics' reliability and association with self-reported fall history. Neurorehabilitation and Neural Repair. 2015 Jan;**29**(1):54-61

[35] Del Din S, Hickey A, Hurwitz N, Mathers JC, Rochester L, Godfrey A. Measuring gait with an accelerometer-based wearable: Influence of device location, testing protocol and age. Physiological Measurement. 2016 Sep 21;**37**(10):1785-1797

[36] Del Din S, Godfrey A, Mazzà C, Lord S, Rochester L. Monitoring of Parkinson's disease: Lesson from the field. Movement Disorders. 2016 Sep;**31**(9):1293-1313

[37] Weiss A, Brozgol M, Dorfman M, Herman T, Shema S, Giladi N, Hausdorff JM. Does the evaluation of gait quality daily life provide insight into fall risk ? A novel approach using 3-day accelerometer recordings. Neurorehabilitation and Neural Repair. 2013 Oct;**27**(8):742-752

[38] Weiss A, Herman T, Giladi N, Hausdorff JM. Objective assessment of fall risk in Parkinson's disease using a body-fixed sensory worn for 3 days. PLoS One. 2014 May 6;**9**(5): e96675

[39] Weiss A, Herman T, Giladi N, Hausdorff JM. New evidence for gait abnormalities among Parkinson's disease patients who suffer from freezing of gait: Insights using a body-fixed sensor worn for 3 days. Journal of Neural. Transmission (Vienna). 2015 Mar;**122**(3):403-410

[40] Weiss A, Brozgol M, Giladi N, Hausdorff JM. Can a single lower trunk body-fixed sensor differentiate between level-walking and stair descent and ascent in older adults? Preliminary findings. Medical Engineering & Physics. 2016 Oct;**38**(10):1146-1151

[41] Mitoma H, Yoneyama M, Orimo S. 24-hour recording of parkinsonian gait using a portable gait rhythmogram. Internal Medicine. 2010;**49**(22):2401-2408

[42] Yoneyama M, Mitoma H, Watanabe K, Kurihara Y. Accelerometry-based gait analysis and its application to Parkinson's disease assessment. Part 1: Detection of stride event. IEEE Transactions on Neural Systems and Rehabilitation Engineering. 2014 May;**22**(3):613-622

[43] Yoneyama M, Mitoma H, Watanabe K, Kurihara Y. Accelerometry-based gait analysis and its application to Parkinson's disease assessment. Part 2: A new measure for quantifying walking behavior. IEEE Transactions on Neural Systems and Rehabilitation Engineering. 2013 Nov;**21**(6):999-1005

[44] Yoneyama M, Mitoma H, Okuma Y. Accelerometry-based long-term monitoring of movement disorders: From diurnal gait behavior to nocturnal bed mobility. Journal of Mechanics in Medicine and Biology. 2013;**13**:1350041

[45] Utsumi H, Terashi H, Ishimura Y, Takazawa T, Hayashi A, Mochizuki H, Okuma Y, Orimo S, Takahashi K, Yoneyama M, Mitoma H. Quantitative assessment of gait bradykinesia in Parkinson's disease using a portable gait rhythmogram. Acta Medica Okayama. 2012;**66**(1):31-40

[46] Terashi H, Utsumi H, Ishimura Y, Takazawa T, Okuma Y, Yoneyama M, Mitoma H. Deficits in scaling of gait force and cycle in parkinsonian gait identified by long-term monitoring of acceleration with the portable gait rhythmogram. ISRN Neurology. 2012:e306816

[47] Utsumi H, Terashi H, Ishimura Y, Takazawa T, Okuma Y, Yoneyama M, Mitoma H. How far do the complaints of patients with Parkinson's disease reflect motor fluctuation? Quantitative analysis using a portable gait rhythmogram. ISRN Neurology. 2012:e372030

[48] Terashi H, Utsumi H, Ishimura Y, Mitoma H. Independent regulation of the cycle and acceleration in parkinsonian gait analyzed by a long-term daily monitoring system. European Neurology. 2013;**69**(3):134-141

[49] Yoneyama M, Mitoma H, Higuma M, Sanjo N, Yokota T, Terashi H. Ambulatory gait behavior in patients with dementia: A comparison with Parkinson's disease. IEEE Transactions on Neural Systems and Rehabilitation Engineering. 2016 Aug;**24**(8):817-826

[50] Yoneyama M, Mitoma H, Hayashi A. Effect of age, gender, and walkway length on accelerometry-based gait parameters for healthy adult subjects. Journal of Mechanics in Medicine and Biology. 2015;**16**:1650029

[51] Yoneyema M, Okuma Y, Utsumi H, Terashi H, Mitoma H. Human turnover dynamics during sleep: Statistical behavior and its modeling. Physical Review E. 2014. DOI: DOI 10.1103/PhysRevE00.002700

[52] Higuma M, Sanjo N, Mitoma H, Yoneyama M, Yokota T. Whole-day gait monitoring in patients with Alzheimer's disease: Relationship between attention and gait cycle. JAD Reports. 2017. DOI: 10.3233/ADR-170001

[53] Terashi H, Mitoma H, Yoneyama M, Aizawa H. Relationship between amount of daily movement measured by a triaxial accelerometer and motor symptoms in patients with Parkinson's disease. Applied Sciences. 2017;**7**:486

[54] Iijima M, Mitoma H, Uchiyama S, Kitagawa K. Long-term monitoring gait analysis using a wearable device in daily lives of patients with Parkinson's disease: The efficacy of selegiline hydrochloride for gait disturbance. Frontiers in Neurology. 2017;**8**:542

[55] Morris ME, Iansek R, Matyas TA, Summers JJ. The pathogenesis of gait hypokinesia in Parkinson's disease. Brain. 1994 Oct;**117**(Pt5):1169-1181

[56] Morris ME, Iansek R, Matyas TA, Summers JJ. Ability to modulate walking cadence remains intact in Parkinson's disease. Journal of Neurology, Neurosurgery, and Psychiatry. 1994 Dec;**57**(12):1532-1534

[57] Morris ME, Iansek R, Matyas TA, Summers JJ. Stride length regulation in Parkinson's disease. Normalization strategies and underlying mechanisms. Brain. 1996 Apr; **119**(Pt2):551-568

[58] Hallett M, Khoshbin S. A physiological mechanism of bradykinesia. Brain. 1980 Jun;**103**(2): 301-314

[59] Flowers KA. Visual "closed-loop" and "open-loop" characteristics of voluntary movement in patients with parkinsonism and intention tremor. Brain. 1976 Jun;**99**(2):269-310

[60] Marsden CD, Parkes JD, Quinn N. Fluctuations of disability in Parkinson's disease: Clinical Aspects. In: Marsden CD, Fahn S, editors. Movement Disorders. London, UK: Butterworth Scientific; 1982. pp. 96-112

[61] Giladi N. Medical treatment of freezing of gait. Movement Disorders. 2008;**23**:S482-S488

[62] Stacy M, Bowron A, Guttman M, Hauser R, Hughes K, Larsen JP, LeWitt P, Oertel W, Quinn N, Sethi K, Stocchi F. Identification of motor and nonmotor wearing-off in Parkinson's disease: Comparison of a patient questionnaire versus a clinician assessment. Movement Disorders. 2005 Jun;**20**(6):726-723

[63] Stacy M, Hauser R. Development of a patient questionnaire to facilitate recognition of motor and non-motor wearing-off in Parkinson's disease. Journal of Neural Transmission. 2007 Feb;**114**(2):211-217

[64] Zijlstra W, Hof AL. Assessment of spatio-temporal gait parameters from trunk accelerations during human walking. Gait & Posture. 2003 Oct;**18**(2):1-10

Chapter 3

Wearable Technology as a Tool to Motivate Health Behaviour: A Case Study

Venere Ferraro, Mila Stepanovic and Silvia Ferraris

Additional information is available at the end of the chapter

http://dx.doi.org/10.5772/intechopen.77002

Abstract

According to the Scientific Committee on Occupational Exposure Limits, work-related exposures are estimated to account for about 15% of all adult respiratory diseases. Today, the use of personal protective equipment (PPE) is the only way for workers to prevent disease. Nevertheless, its use is highly sparse. Currently, products and systems embedded with wearable technologies are able to protect, motivate and educate users. The authors then suggested the development of a novel wearable system following the beliefs that wearable technology can be persuasive and elicit a conscious behaviour towards the use of the PPEs by consequently improving their health condition. The authors here describe the result of a Transnational Research Project named "P_O_D Plurisensorial Device to prevent Occupational Disease." The chapter describes the findings achieved so far, the research phase and the new wearable system conceived as a possible example of how to use wearable technology as a useful tool to influence behavioural change.

Keywords: wearable technology, persuasive technologies, behaviour change, wearability, health, human-centred approach

1. Introduction

The World Health Organization and Europe Mortality database in 2011 in Europe, estimated that 7200 cases of respiratory diseases are related to occupational exposures to Volatile Organic Compound (VOC) and dust. The Annual Inail (Istituto Nazionale Assicurazione Infortuni sul Lavoro) Report from 2015 confirms the European trend, putting in Italy the respiratory diseases at the third place (13.5%) between the occupational ones, also stressing the severity of their consequences. Volatile organic compounds (VOC) are—within working environment—one of the highest causes of asthma, lung cancer, chronic obstructive pulmonary disease

(COPD) and respiratory tract infections. Moreover, the occupational disease percentage increment, in particular respiratory related diseases, is typically present in sectors such as agriculture, transportation and manufacturing which is considered also as the most dangerous with 5.172 cases of respiratory diseases in 2014 (INAIL). This text is funded on results gained from Transnational Research Project (SAF€RA Joint Call 2014) called "POD-Plurisensorial Device to Prevent Occupational Disease" which was developed by a consortium made up of Department of Design of Politecnico di Milano, Department of Design Engineering from Delft University of Technology and the Italian company Comftech. This project took place from July 2015 to August 2017. The data about the respiratory diseases shows the importance of wearing personal protective equipment (PPEs) preventing the users from the inhalation of dangerous particles. Starting from those data, the research project targets the development of a protective wearable system concept that is able to prevent respiratory diseases inside the environment of a coating plant.

The best method to reduce work-related diseases and improve general health of workers is prevention (International Labour office). Prevention can be done on two levels, which are (1) environmental monitoring and (2) the use of Personal Protective Equipment (PPE) which is also mandatory but often neglected by workers because they do not see the benefits of wearing it. A way to encourage the use of the PPEs could be to make workers aware of work-related risks they undertake and their health status by exploiting persuasive technologies in the shape of wearable devices.

Nowadays, the emerging smart technologies are able to support humans and can be persuasive and change humans' habits and behaviour without coercion. Intelligent technological products are able to change users' behaviour and educate, rather than only inform or do something for humans [1]. Technology might resolve the problem of subjectivity and precision by collecting and providing objective data that are based on real values of monitored subjects, objects and environments. In this perspective, technology becomes a sort of a prosthetics for human beings; it empowers human beings. Persuasive technologies are defined as "computer-based tools designed for the purpose of changing people's attitudes and behaviours" [2].

The authors investigated in particular the use of smart technology that belongs to the class of wearable technology. Wearable sensors and systems are defined as wearable sensors/actuators and sensor-based communicative systems that can monitor and/or stimulate, treat and replace biophysical human functions. Wearable technology's purpose is to facilitate everyday life and also protect and inform users in order to avoid human errors, as those related to subjectivity of senses and perception [3–5].

Development of this research project consisted also in investigating technological solutions from other fields of study that have a similar purpose, as those in particular: medical (i.e. wearable monitoring systems for physiological parameters) and military solutions (i.e. the electronic nose developed by NASA). The main aim of the project was to create a meaningful interaction with the device able to motivate users to wear PPE by enhancing their consciousness related to the risks in the working place. Here the technology is supposed to make risks perceivable to users by obtaining valuable interaction [6]. The entire research was based on human-centred approach that relates wishes and needs of users, considering engagement of them in design process. Engagement of the user in design process was executed in repetitive cycles divided in five design phases: empathy, define, ideate, prototype and test [7, 8]. In this chapter, the authors describe the entire design process that led to the generation of a new novel wearable system by involving and emphasising with the user in every step to define his/her expectations. This is in order to translate those expectations in a way where wearable technology is shaped purposefully for the final user by becoming a powerful tool to change their habits in the use of the PPEs.

2. Wearable technology to persuade

Nowadays technology is taking over in different disciplines and with a different purpose: education, medicine and in general wellbeing, working environment, smart home, and so on. (Smartphone, smart watches, smart meter, run trackers, etc.). New Technologies have become deeply embedded in our ordinary everyday experiences by improving the way products and systems help, inform, engage and/or entertain us, coding new languages of communication and interaction. Different technologies change the way the people behave (interact) and also define the behaviour of artefacts, environment and system [9]. Technological products are able to motivate users and change their attitudes and behaviour but also to act beyond human capabilities. Donald Schön [10] defined "extension of human capabilities" as a main goal of technology.

On the other hand, as Fogg stressed, technology can be persuasive and change user behaviours. "Persuasion is a non-coercive attempt to change attitudes or behaviours" [11]. Interactive technologies have been—and will continue to be—created to influence (persuade or motivate) people in different fields of application. Interactive products able to persuade and motivate users are evolving in parallel with a technological development and level of adeptness of users to different forms of interaction with advanced products, systems and environments. Recently, persuasive technologies are developing also with a purpose of improving health and wellbeing, among those commercial based on computers and Web applications as a principal tool of interaction (i.e. selling products or services).

Advanced technologies and solutions and information communication technology (ICT) have potential to be a tool for promoting positive health and lifestyle habits [12–16], giving social support [17, 18] and helping with lifestyle change [19, 20]. Furthermore, research on persuasive technology [21] and affective computing [22] is providing technological (partial) solutions for the development of systems and devices able to cope with health issues and awareness. There are many factors to consider when designing products or systems for persuasion. Some people are able to act and be motivated by themselves, while others need to be involved and motivated by others. Motivation is a key element in any system designed to assist users in changing their health behaviour [23].

These kinds of technologies are supposed to persuade user in a logical and reasonable way. If designed well, digital persuasion has way more advantages with respect to the traditional one based on human-human persuasion. Fogg [2] underlines a few advantages of digital persuasion compared to traditional ones and those are related to the fact that the "computers" are much more interesting than people in the long-term persuasion process, they offer more anonymity, and they arrive at the contexts where human beings might not arrive.

Among emerging—persuasive—technologies, the wearable ones have the potential to build motivation by increasing health awareness through timely feedback [24]. A multidisciplinary approach in developing wearable technology based on biology, physiology, nanotechnology, material science and other numerous disciplines is influencing rapid increase both in product and in research development. These devices are often based on sensing technology, communication and persuasion [25].

So far, in the field of health, the wearable technologies have predominately focused on diet and physical activity, such as weight loss, motivating physical activity, maintaining exercise routines and an overall better understanding of an individual's health. In the commercial sector, one of the earlier more successful devices includes Nike+iPod (nikeplus.com), which employs a sensor in the runner's shoe and an iPod to keep track of the current pace and distance of a workout. In addition to tracking personal workout statistics, the runner can connect to an online community through the Nike+ website where he/she can track goals, compare running times with others and challenge friends. While Nike+ is designed for the specific activity of running, FitBit (fitbit.com), BodyMedia Fit (body-media.com) and Jawbone UP (jawbone.com) track everyday activities like walking (steps taken), climbing stairs (number of floors) and sleeping (hours and quality of sleep). Much like Nike+, all these wearable devices sync with an e-health application through a wireless base station. Additionally, they provide users with the ability to log individual foods and workouts from a smart phone or computer to track diet and exercise. BodyMedia Fit is geared towards weight loss and is worn as an armband, while FitBit and UP focus on smaller lifestyle changes and are clipped to the waist and worn on the wrist, respectively. Pebble time instead (www.pebble.com/health) automatically tracks when you go to bed, displaying sleep, deep-sleep, and the times when you fall asleep and wake up.

Beside previous examples, wearable systems can be designed with a purpose to provide more control to the users and focus on more health consciousness increasing applications rather than just monitoring because presumably the prevention and health sector is one of the strongest areas of innovation [26].

There are activities and environments for which are required external incentives and motivations. As regards, the objective of our research project was in designing a protective and preventive wearable system that consists basically in monitoring of the user's physiological parameters (1), environmental monitoring and hazard detecting (2), real-time feedback providing (3) and allowing data transmission.

3. Method

Since when designing wearable technologies it is very important to focus on user's needs, development of research project was funded on human-centred approach, aimed at meeting not only technological requirements but also and especially user's needs and meaningful experience.

Value of wearables is not in having a device that is carried on or held but in its possibility to be worn and integrated in everyday life and activities. A wearable device is not supposed to create the barrier between the wearer and outer world; thus, it needs to have a certain aesthetical quality and be approached as a part of clothing in order to achieve requirements of wearing it everywhere, by everyone and anytime [27]. Despite the fact that commercial wearable devices have recently become more common, up till now there are only few products from this category that satisfy all requirements related to users' needs.

User doesn't really understand the advantages of wearing such devices because often these devices are perceived as uncomfortable and alien to users. This occurs especially when the requirements referring to the formal and ergonomic characteristics are not well accomplished well as adaptability and adhesion to the body and dimensions of the housing [28]. There are different theoretical definitions about comfort, but all of them embrace the concept of creating harmony between the individual and environment and avoiding unpleasantness on all levels of existence. Any good design should consider both a user's need—as in this case is the comfort and body anatomic—and also the aspiration—which would be the designer's know-how. In this way it is possible to design devices that meet both perceptual needs and needs related to wearability of the user [29].

The word "wearability" literally means ability to wear and concerns the physical shape of wearables and their active relationship with the human form. The target is to define the correct/pleasant interaction between the human body and the wearable object by trying to figure out a flexible shape without interfering with human daily activity [30].

Applying human-centred approach in designing means digging in worker's behaviour, perception, working environment and generally social aspects of the work. This does not exclude identifying of product's technical requirements and issues related to wearable technologies. Merging these two in designing process makes possible the achievement of an appropriate design solution both from the point of technological adeptness and general comfort related to wearability [31].

In the development of the new wearable system, authors had to take into account the product language and the correct pleasant interaction between the product and the user but also to design for motivation. Indeed, authors found important emphasising the factor of influencing more responsible users' behaviour towards their health conditions through the method of personal and environmental monitoring. Introducing the technology and technological solutions in this context is a way to provide objective data to the user and raise awareness and motivation.

When designing devices for persuasion, there is also a high demand on the way the feedback is transmitted to the user. For example, the reason why health awareness is accomplished with this kind of devices or applications is because it is in the real time—the gap between action and information (feedback) is denied [24].

For some applications, there is the necessity to create a real-time feedback. If we need a device to persuade on longer periods, feedback out of the context is a valid consideration. One of the often-used methods for creating persuasive technologies is personalisation. Personalised products are more likely to be accepted by the users. Designing personalised product does not have to be realistically addressed to a single person, but that it gives a feeling of being personal [2].

In the following paragraphs, the authors show the design process applied in this research project and development that consists in few phases: (1) the Empathise phase based on design thinking approach and user centred; (2) the Define phase where the project brief was imposed; (3) the Ideation phase in which concepts were developed, followed by the concept evaluation; (4) the Developing phase that is the development of the chosen concept and finally (5) the testing phase to figure out if the new wearable system satisfies user demands and project objective.

3.1. Participants

For the user analysis, the choice of the proper participants was relevant. We decided to choose workers of coating plants. This choice is strategic for the project execution because workers of this environment are highly exposed to the inhalation of several damaging agents (volatile organic components, organic dust, chemical contaminants, paints, varnishes) that cause respiratory disease. Our research activity was supported by Anver (Associazione Italiana Verniciatura) in finding three SMEs available for participating in the user analysis and observation and helping us understand how the process of painting is being executed, who and how is performing this activity and what are the general conditions, rules and norms in coating plants. Three companies with less than twenty employees specialised in finishing and lacquering of metals, polymers, wood and steel were chosen.

3.2. Identifying the users and general problematics

In order to understand the users and what their activity actually consists of we proceeded with organising the user sessions in all three companies. Sessions consisted of two phases: (1) observation; (2) semi-structured interviews. Observation was focused on identifying characteristics of the working activity performed by the worker, his/her behaviour and usage of PPE. Semi-structured interviews were done both with workers to understand the level of perception about the risk and what they find as motivating or demotivating about wearing the PPE, and employers with a purpose to understand how the question about providing and wearing the PPE in this kind of environment is regulated, and in both cases we wanted to understand whether they are positive about introducing advanced technology based on monitoring in their ambit. In this session, all 20 workers were engaged in the period between December 2015 and January 2016. The user session was essential for this research because it was necessary to get to know the users well, their behaviour, if their activity is supposed to take place in coating cabin and with an aspiration system or not and typology of PPE that they are provided. Findings about the PPE show that the workers are generally provided with two different types of masks: (1) Disposable Protective Mask (with a small valve that helps the respiration) and (2) Reusable Half Mask with carbon filters (solid or liquid particles gas and vapour); the selection of these is certainly dependent on technical and chemical parameters such as: Typology of coating, Typology of pollution, Density of Coating Powder. The mask has its expiry date and it is dependent on its typology. Disposable masks last one day and have to be immediately thrown away, while the reusable ones have temporary bodies and filters that are interchangeable and last no more than 1 month. The filter expiring frequency is not dictated only by the hours of wearing but also whether it was protected in the box or exposed long time in the environment without being used. Reusable mask should be placed in a protected place and closed when not used.

From 1 h observation (**Figure 1**), it was observed that the workers are wearing PPE only occasionally when there is a presence of overspray that is mostly present in complex geometry painting. These findings brought us to the conclusion that the decision of wearing the PPE and the frequency of wearing it is strongly dependent on subjective perception (principally olfactory or visual).

Even though not all companies have the same equipment and typology of the activity, helpful conclusions were brought from both phases of user analysis. Observation phase results were reinforced with those gathered from interviews which confirmed all our doubts. The Interviews were organised under a few topics: Working activity and protective equipment; Mask's aesthetic and comfort; Safety perception; Personal devices and executed with ten workers (four from the first, four from the second and two from the third company).

All interviewees stated that the decision about wearing the mask is autonomous and personal and that the mask is stored in a drawer when not used. Since the decision of wearing the mask is personal, it is based on the subjective feeling of the worker, or rather on olfaction perception. This causes sparse wearing of the mask that considers time that does not exceed 1.5 h per day. The level of odour presence and overspray is dependent both on the geometry of painted piece and on efficacy of aspiration system. Workers do not believe that the PPE is not useful; they also know that they are suggested to wear it, but they simply do not perceive well the situations for which and for how long it is necessary to wear it.

Beside low comprehension about PPE usage due different factors, workers consider that the mask is uncomfortable because it is too rigid, not enough breathable space and laces for regulations are rather bothering due to being too tight. Workers suggested that the application of new materials and particular attention in designing details may increase wearability and comfort of the mask. Moreover, we wanted to investigate also the safety perception of workers, where only 8 of 20 claimed to be aware of it and understood that the risk comes from the hazards. Additional information that we obtained confirm that according to workers protection level of the mask is good and that they are positive about having information both about personal health parameters and environmental ones that may consider implementation of sensors and advanced technology. With regard to this, workers were delighted about the idea of integration of sensor in

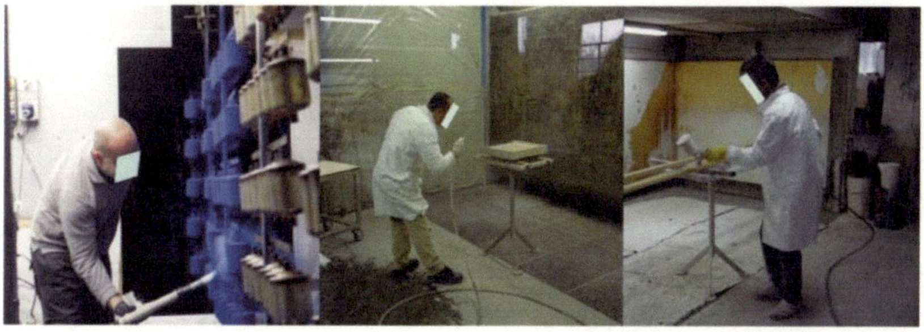

Figure 1. Workers in a cabin.

the mask, wearable device equipped with a sensor able to detect breathing and heart rate parameters and provide information through a Mobile App. Moreover, they would like to know when it is necessary to wear the mask (hazardous condition). They would like to know about daily data (environment and personal health) but also seasonal ones [32–37].

3.3. Delineating project requirements

Results gathered from the user analysis showed a few facts that are very important for project development: (1) Workers prefer not to wear the mask because the environment is not perceived as dangerous, except in some cases where the smell of paint is too evident. (2) Another reason that is influencing sparse wearing of mask is that workers state that generally the mask is uncomfortable. (3) Beside being pessimistic towards existing PPE, workers claimed that they would like to know more about their health condition, activity and working environment. There are some characteristics of the working activity that are to be respected when designing for these users, such as moving easily, having a good and vide visibility, sweating less and maintaining concentration. Thus, the most important factor for the workers was *comfort*.

Related to previously described points, design requirements were defined in the following way: (1) Form has to respond to the anatomical characteristics of the users, considering dimensions that are supposed to remain relatively small; (2) Apply materials that are breathable, anti-allergic and light; (3) Consider to provide filtered air (mask) and (4) Design the device, portable preferably, able to monitor, sense and eventually alert the user.

Making hypothesis about the potential appearance of the monitoring system, we must consider the possibility of designing a small portable device or apparel, or even integrate both. It seems like the only possibility to protect respiration is in the form of mask but integrating sensor inside it and its modus operandi may have different application possibilities. Information provided to the worker should be simple and comprehensive in order to gain motivation and reduce frustration factor. The principal tasks and requirements for designing wearable system were imposed with the intention to raise motivation about wearing PPE in workers by exploiting persuasive technologies, which also considered in improving comfort and providing real-time information about those that can be defined work-related risks, as in this case air quality and personal health condition.

3.4. Ideation phase and focus group

The objectives enhancing the user's consciousness and providing better comfort considered designing the wearable system based on requirements able to satisfy these needs. Related to this were suggested a few strategies as those that consider creating the natural interaction language between the user and device funded on comprehension and engagement able to influence motivation and perception related to personal health. For this purpose, three concepts were generated and each of these was approached by following imposed requirements and design brief regarding materials and technology (soft technology, smart textiles, electronics, etc.). These concepts were studied trough Mock-ups and as a system rather than separate parts (see **Figure 2**).

Among all the elements of the system, the Protective Mask was the most diversified one not only on a technological level but also on a formal level; the general idea was to introduce padded fabric instead of rubber material in order to increase comfort and provide more transpiration between skin and mask and perhaps introduce a new mode of use. Beside these design requirements, the mask was supposed to contain temperature and humidity sensor for human-presence detection. Based on these three concepts for the mask that were generated, the first concept represents familiar architecture of the protective mask with a rigid central part and soft structure around the face. Here the proposed improvement is in the material that is in contact with a face that is designed to be in spacer textile. The mode of use is as in traditional masks with standard regulation laces. Beside the rigid part in front of the nose is provided the space for sensor mounting. Second concept is principally focused on suggesting completely new mode of use that is using a lateral Velcro® closing of the mask body and regulation of laces with a small rotating wheel which makes mask regulation more immediate without a need of taking it off. Here the biggest part of the mask is designed in spacer fabric, leaving rigid part for sensor and filter mounting. The spacer is taught to be coupled with a polymeric membrane in order to isolate from outer hazard agents. The third concept puts major focus on the comfort and here we are proposing a mask completely made in thermoformed laminated spacer textile. The idea is to exploit the process of fabric thermoforming, its possibilities to create the space for sensor and filter cartridges mounting. Besides making the structure lighter and providing more comfort, the filter cartridge mounting facilitation by creating the slide-in system was also thought of here.

For Electronic Nose Device, three solutions based on different visual interactions—graphical form of the information—were generated. The information about air quality is communicated in the following way: (1) *Concept 2:* very simple with two phases of communication that consider only turning on and alarm; (2) *Concept 1:* more complex solution based on three phases that are turning on, warning and alarm; (3) and the last one—*Concept 3*—where there are also three phases as turning, warning and alarm but shaped in the form of a face expressing the emotions from happiness to sadness. All devices combine vibration and LED response. Different modes of use were studied for each device from the point of general wearability (harmony with the human body) that considers putting on and taking off the device and visibility of the feedback due to its position on the clothes or body (see **Figure 2**). Hypothesis about the mode of use of device was based also on the sensitiveness of the body part because beside the visibility of the feedback it is also important that the user feels vibration. Concepts propose the device be positioned in the zone of abdomen in particular arms and chest or making a more flexible form that can adjust to

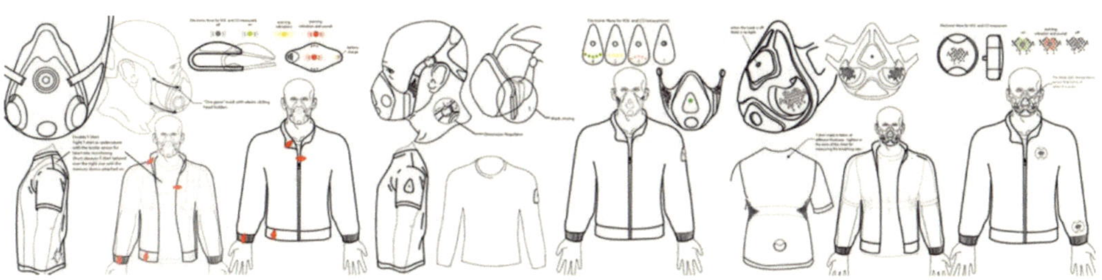

Figure 2. Three concepts of the system.

any part of the clothes. The first concept (see **Figure 2** on left) proposes a flexible clip that can be put anywhere from the T-shirt to the pockets of pants, that can fit to any kind of working uniform. The feedback for this device is in the form of LED that is blinking as alarm passes through four LED colours. Second and third concept (see **Figure 2** on right) are thought to be positioned and fixed on the T-shirt by a push button in order to take it off when the uniform has to be washed. These two are not versatile; they stay always in the same position. Second concept is showing the interface based on mood expression as happy, not so happy and sad as a way to immediate the information related to the air quality that is equal to good, becoming hazardous and hazardous. The last concept is not using specific interface but just colour and vibration. In this case, the device adopts to the protective mask in sense of formal language. Even though the last concept is the one where the system language is empathised, other concepts are also adopting more or less formally to the protective masks next to which they are represented.

Technological principle is the same for each proposed device which consider Volatile Organic Compound (VOC) sensor, vibration motor and RGB LED. All of them have four steps of feedback where the first one indicates ignition signal and then passes to air quality detection that shows three levels: good air quality (green), air starting to be polluted (yellow) and hazardous air (red). Also, the vibration is set on three levels that correspond to LED signal for which when it is yellow the vibration is lower while when red it gets strong. It is possible to stop the alarm by pressing the delay button on device.

Initially, a T-Shirt with textile sensors on the inner part was thought of for the breathing and heart rate measuring. Two solutions were proposed where in the first one T-Shirt was tightened in the zone of chest in order to adhere better and in the second one was proposed undershirt with textile sensors that were supposed to be worn under the clothes or uniform.

Information gathered from the protective mask and electronic nose device are stored in the last part of the system that is the Mobile Application for which are showed three different concepts. Two of these are designed for workers and one for employer (see **Figure 3**). The idea of proposing few solutions, as in the case of Electronic Nose Device, is in the possibility to understand the level of technological and information engagement of users, starting from the basic communication to more complex one. Here were included options of personalisation of the Mobile App profile and various settings related to the working environment for the more complex interaction (*Concept 1*), while for the simple one, only parameters that are results of monitoring are represented (*Concept 2*).

The generated concepts were also realised in form of mock-ups for the user session. Those mock-ups were presented since the early stage of the design development both to engage the user in design process, understand the overall dimension of the product and evaluate its wearability. Indeed, all of these were tested and evaluated by the users. We wanted to investigate how created concepts satisfy needs of users related to previously mentioned parameters. Objective of the session was defining the solution that will be a starting point for final design development. Workers were invited to evaluate each part of the system, for each concept, on a scale from one to five for different objects' qualities, funded on parameters such as aesthetics, function, comfort, mode of use, comprehension of interaction. User session took a place in the same companies where our analysis started, and it was organised with a group of workers and employers that were wearing and trying all mock-ups after which they were given a

http://dx.doi.org/10.5772/intechopen.77002

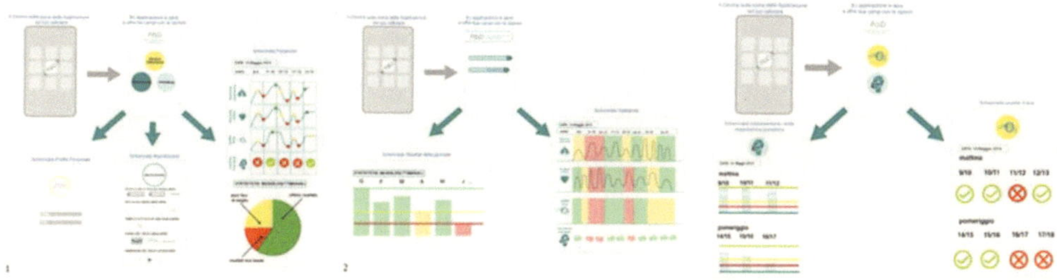

Figure 3. Two concepts of the application for the workers (left); the application for the employer (on the right).

questionnaire to evaluate their experience with each part of the system. During the session, participants' voices were recorded and photos were taken (see **Figure 4**).

During the session workers were able to give general comments and discussed with each other. The difficult part of the session was related to the evaluation of the Mobile App that was provided in form of screen's graphics and it was explained to them step-by-step. Nevertheless, they were also able to validate the App in the given questionnaire. After the session concluded, we gathered results and organised it in order to detect more easily and precisely the main strengths and constrains of proposed concepts. As it was accented previously, the main objective was defining the final concept and the starting point for final design definition. Since concepts proposed for protective mask also integrated application of textile, it was particularly important to have a feedback from workers on whether they liked it or not. **Figure 5** shows the preferences that workers expressed for all concepts.

The Mobile Application that workers preferred was *Concept 2* because they found it the most comprehensive and simple to use. For the mask they indicated in each concept something that was adequate but the most liked one was *Concept 1*. One of the most important parameters for the mask designing was the aspect of comfort and it was evaluated very well for *Concept 1* and *Concept 2*, while *Concept 3* was generally evaluated with lower points. General image of concept evaluation is given in the charts below. The figure does not show results of the smart shirt because it was not appreciated as a solution, so we decided to eliminate the undershirt and leave just a band because it was something that was already familiar to users and considered as more comfortable.

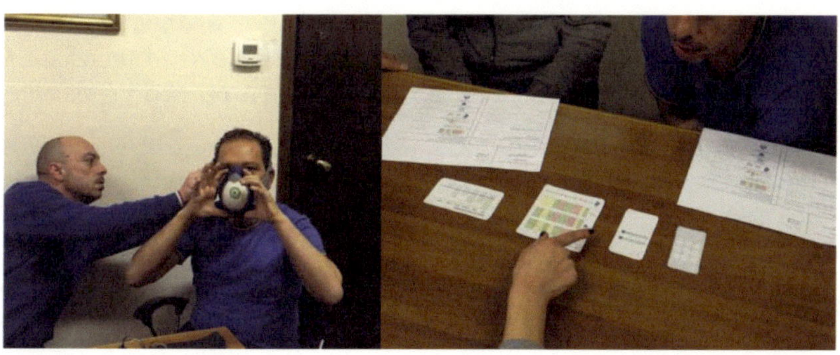

Figure 4. Some pictures from the Focus Group.

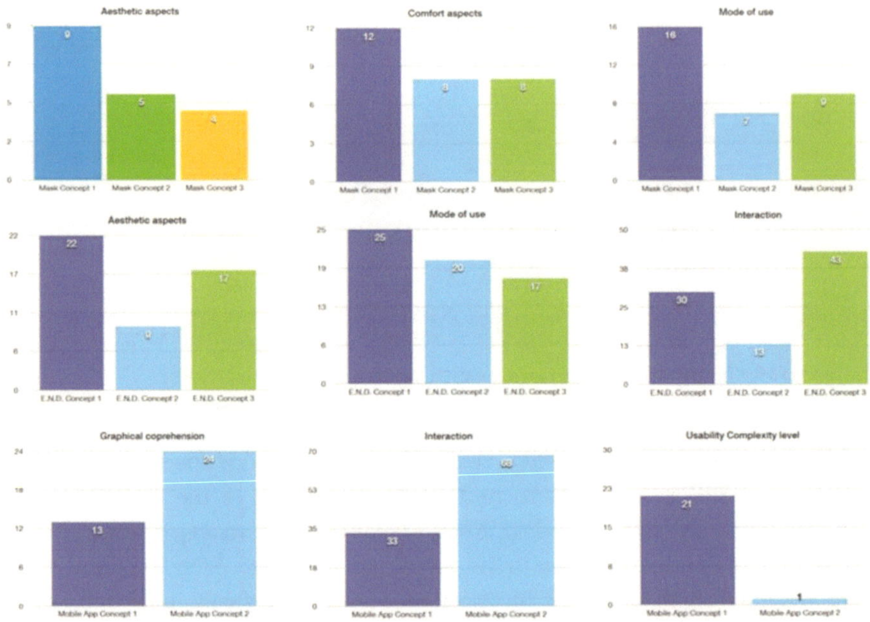

Figure 5. System evaluation.

Concept 1 of Electronic Nose Device was appreciated the most for its mode of use, while *Concept 3* for the mode of interaction by being the more comprehensive one.

About the Mobile App, the most liked one was the *Concept 2*. Mobile App for employers showed approval as being retained as very useful. All workers were positive on wearing the more comfortable mask regularly and being informed about the air quality. These factors may help not only in personal protection but also in controlling the ventilation and aspiration system, considering the employers' plants. After studying new forms, functions and communication, we came up with reviewing the previous concept and created the mask with new formal features, with an outer shell in spacer textile and filters made in thermoformed non-woven textile with a purpose to make the mask lighter. Electronic Nose Device gained a simpler form with a flexible silicon clip. Since the Mobile App was very comprehensible to the users, we maintained the same concept and added the App for employer based on showing the results about the air quality in time and percentage of mask wearing during the day. Since the data about the mask wearing are strictly personal, employers cannot see who is not wearing it but have a general image about the mask wearing expressed in percentage scale.

Once we defined these details and developed new prototypes (**Figure 6**), we organised another focus group session with users after which they approved the final concept and showed rather positive about a possibility to have this kind of devices.

3.5. Developing phase

Further, we proceeded with engineering parts and the prototype of the system (**Figure 7**) which brought us to reviewing again some parts of the project. In particular, we intervened on the Electronic Nose Device design. For the reason of comfort and versatility that we want to obtain, the form of device changed and the form of the graphical feedback as well. We simplified even more the product by creating the abstract graphical representation that excluded the facial expression as a communication and maintained colours because both culturally and perceptually it communicates desired information. The new system consists of a Smart Personal Protective Reusable Mask, a wearable alert device called Electronic Nose Device, a Chest Band t and a Mobile Application.

In order to encourage workers to wear the Protective Mask, its re-design followed the user requirements about aesthetic and wearability. Hereafter, we can see a comparison between the traditional mask and our new design (**Figure 8**).

In the traditional mask, the parts in direct contact with the user's face are made in rubber that causes the mask to be uncomfortable and not worn regularly by the workers. In our new product, the rubber is substituted by a thermoformed spacer textile padded with soft foam. Spacer textile has a structure that is forming exactly to the face and its property to be transpired has the value of preventing from sweating even during the summer days. The Spacer textile chosen for this application has smooth external surface that is very pleasant in contact with skin. Since the spacer textile has a smooth surface, there was necessity to create more friction between the skin and the material surface. This was resolved by printing the silicon pattern in the zone of nose, chin and cheekbone. The layer of the silicon pattern is so thin that it is almost imperceptible, but it is giving enough adherence and friction. Workers were indeed very positive about the idea to replace traditional materials with textile.

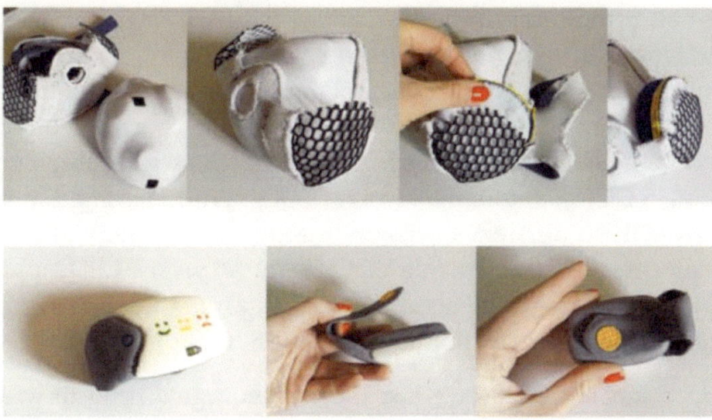

Figure 6. System refinement.

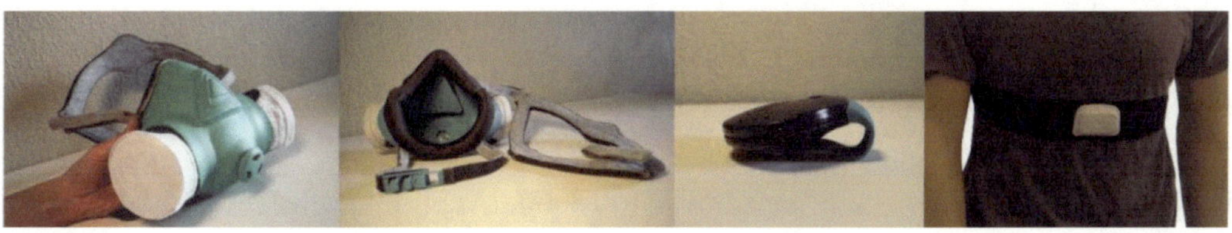

Figure 7. New wearable system prototype.

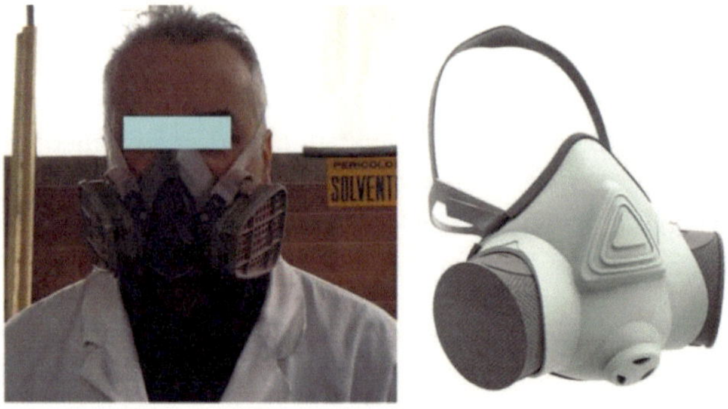

Figure 8. Comparison between traditional mask and new design.

In our re-design, the thermoformed fabric has been chosen for two specific parts (both grey in the image): the part that adheres between the mask's body and the user's face and the laces staying around the user's head. In our intention, in addition to improving the wearability of the mask, the thermoformed fabric will also reduce the risk of skin irritation.

The laces around the head are traditionally manufactured in plastic with strong edges while we decided to use the thermoformed fabric because it is softer and smoother even in contact with the head. Moreover, we used non-woven thermoformed fabric for the filters that are then welded to the mask's base. This choice was aimed at decreasing the mask's weight reducing the use of plastic pieces. The mask is equipped with a temperature and humidity sensor aimed at monitoring the user 's breath (humidity and respiration frequency) but also to check if the mask is worn or not: if the mask comes in contact with the face, the sensor automatically detects the human temperature, or rather presence, and this information goes to Mobile App. User has a real-time feedback about the quality of the air in coating plant trough the Electronic Nose Device. This function is executed with a sensor that is giving objective data about pollution, which are not perceivable by the human nose. The sensor used for Electronic Nose Device (USM-MEMS-VOC) monitors the level of VOCs. While sensing the environment, the Electronic Nose gives three different feedback to the user: (1) green LED on: the sensor is operative and the air contains an acceptable VOCs' level; (2) yellow LED on and low vibration: the level of VOC is starting to exceed acceptable value; (3) red LED on and strong vibration: VOCs concentration in the air is risky for the user health (**Figure 9**).

The real-time feedback alerts the user when there is an urgent need to wear the Protective Mask. This might be the first step in increasing the benefits of wearing the mask regularly. Indeed, in the third-use case, if the user properly feels the vibration, he/she should press the button on the Electronic Nose and wear the mask. Therefore, the nose stops vibrating but in the meanwhile the air quality parameters are constantly monitored and transmitted to the Mobile Application via Bluetooth. Finally, the Chest Band (provided by the Company Comftech) monitors through the working day the user's breathing frequency. It is a textile band sensor that is worn under the T-Shirt (**Figure 10**).

Breathing frequency values are transmitted to Mobile Application to let the user be aware about his/her health every time (when he wears or not the mask). The Mobile Application is a storage of the system; it collects and organises data from the device, mask and chest band. The

Mobile Application is based on simple graphical representation of three parameters that are comparable in time: breathing rate, air quality (related to the level of VOC) and mask-wearing frequency. Application provides the worker with a complete image of the condition at the working place and frequency of his/her mask wearing and possibility to compare these data in sense of understanding whether the mask was worn when the air was hazard and how physiological parameters were respect to this event. Mobile application was designed to organise data on daily basis and provide weekly and monthly statistics as a sort of archive. Statistical data are intended to increase the visible benefits in wearing the Protective Mask: the worker can see the health progress made thanks to the wearing of the mask. Indeed, he/she will realise (see following section) how his/her breathing frequency improves while wearing the mask. This way, we created a wearable system with several useful features: (1) real-time sensing of the environment; (2) monitor the user respiration giving feedback useful to increase awareness about personal health; (3) show statistical data of the Mask-wearing frequency in relation to the presence of dangerous particles. Nowadays the use of PPEs in dangerous working environments is the only way for workers to prevent serious disease. Existing

Figure 9. Electronic nose device feedback.

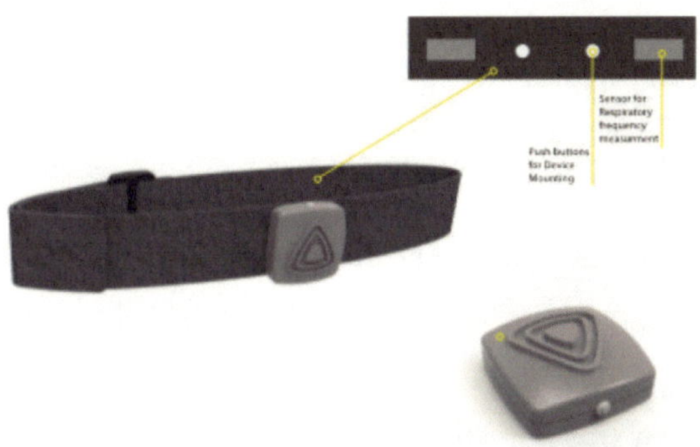

Figure 10. Chest band and its configuration.

personal equipment are "static" products. We instead developed an interactive system made up of "dynamic" products empowered with sensors and technology.

3.6. Testing phase

The testing of the prototype was framed into two phases: Phase 1: The observation of the Practical Use of the System in two steps: (1) the workers were asked to wear the chest band and then to use the electronic nose for 30 min, this was useful to figure out if they understood how the electronic nose works; (2) They wore both the electronic nose and the mask; Phase 2: An unstructured Interview aimed at understanding: (1) Comfort and wearability of the overall system (Mask and Nose); (2) System Feedback Understanding; (3) Application Understanding; and (4) P_O_D_ Efficacy. The user session was organised as a general observation coupled with sensor data collected in the working space and was carried out in May of 2017 on five workers. For the test, we choose the first one of the three companies analysed during the first user session because we met a very open-minded owner willing to innovate, experiment and to use the results of the project afterwards. Moreover, the workers of this company were the most collaborative and proactive ones.

Firstly, workers were given the prototypes and they were supposed to work with it on. This user session had two objectives. The first one was testing the products with users and the second was to understand the accuracy of the system in sense of monitoring data, both vital and environmental. In the image below, we can see the users in the painting cabin wearing the prototype (**Figure 11**).

At the same time, while workers were performing their activity we were controlling the parameters about the breathing rate and air quality through Apps on the mobile phone. Electronic part of the product showed as accurate, the quality of the air was changing respect to the presence of the overspray and distance from the object that is painted. It was giving a feedback to the user in real-time and the user found it quite comprehensive. The same is for breathing data; they were changing with respect to the level of presence of overspray and whether the mask was on or off. In the image below are shown the results obtained from air quality measurement and heart and breathing parameters (**Figure 12**).

After the observation in Coating Plant, users were engaged through an unstructured interview where they explained what they experienced while working with a P.O.D. system and if we succeeded in improving the wearability of the system. Workers were rather positive about the new mask; they appreciated the application of spacer textile instead of rubber and said that they were willing to wear it. They also appreciated Electronic Nose Device very much; the workers appreciated the possibility to wear it a versatile way: on the pants' pocket, on the shirt's collar, on the shirt's sleeves; the feedback and its general functioning was clear to them. They did not complain about the comfort of chest band, they found it unobtrusive and easy to use. In the end, they were given the Mobile App simulation to try to use it. Workers found it simple to use, the complexity of the app was proportional to their level of operating with this kind of medium and the graphical representation was immediate for them. Moreover, the employer showed interest in the system, but in particular he found very useful Electronic Nose Device and his Mobile App. Finally, observing the system as a whole, we can conclude that the improvement of the mask

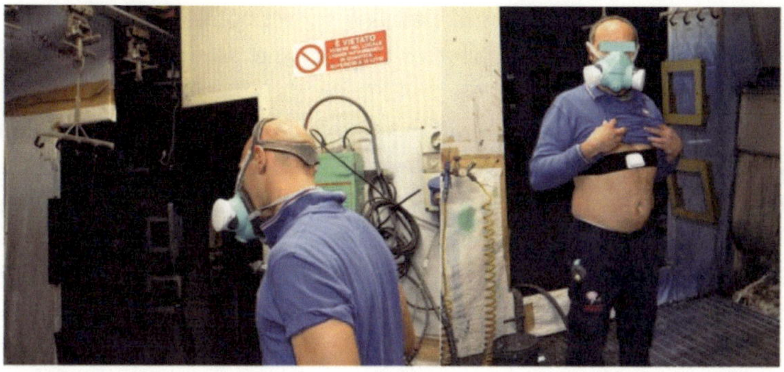

Figure 11. Users engaged in the final testing of the product.

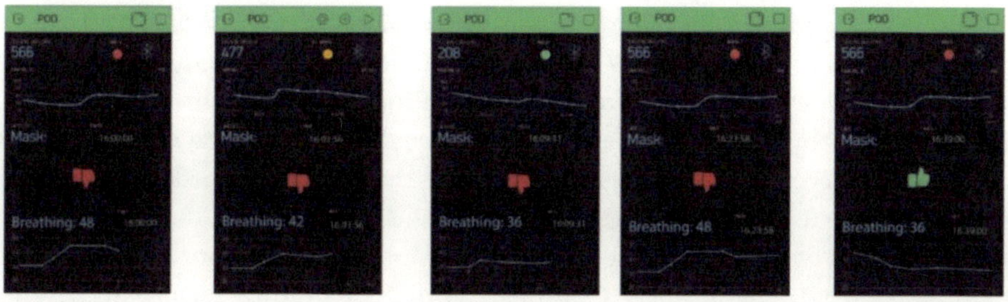

Figure 12. Simulation measurement of air quality and vital parameters.

respect to the user's needs defined in the beginning of the project, raised the benefits related to the wearing of the mask and by this the willingness of workers. Giving a real-time feedback showed that when there is the visible input about the risk, they remember to wear the mask and their belief in the air quality changes. Having the data in Mobile App certainly impact motivation of workers by showing the progress of wearing the mask in time.

4. Conclusions and further development

A high number of respiratory diseases related to working environment are mostly present in sectors such as agriculture, manufacturing and transportation. Data coming from INAIL show that the manufacturing sector is the most dangerous among these where just in 2014 were detected 5172 cases. The best method of prevention and general increment of health in this case is wearing of PPEs. Results from our user analysis, obtained from three coating plants, approve that the workers are not willing to wear the mask. What we suggested as a solution to this problem was to encourage the workers to wear PPE by raising work-related risk awareness which they often undertake. This was thought to be executed by exploitation of persuasive technologies. Persuasive technologies may be defined as a tool able to influence or indeed change people's attitudes and behaviours [2]. Fogg [21] believes that successful products, the ones user will continue to use after the novelty, will be those that incorporate "hot triggers" to help people change their behaviours for the better. Behavioural Model developed by Fogg, (FBM), is funded on the thought that any target behaviour occurs only if the person has sufficient motivation, ability and effective trigger; and these are supposed to occur at the same time. This model suggests that the motivation is a trigger of an action or behaviour. Motivation as a characteristic of human action is stimulating to act and demonstrate one kind of behaviour and not the other. It is not always oriented towards positive but also towards negative or not acting decision. Motivation is certainly the reason of our actions and it shapes behaviour with respect to what is desired or how much willingness we have for acting. We use our motives as inner factors that are directing towards action [23]. In the field of working environment, we faced the behaviour problem related to the responsibility and the willingness of wearing the protective equipment. We had few questions to answer: Could an individual approach like self-motivation generate consciousness about risks the workers are at? How the sense of hazards and injuries could let them understand health risks and change behaviour? To answer these questions, we proposed a Wearable System able to address the main problem of "motivating" by improving the comfort and wearability of the system and raising self-awareness about health issues. Besides the theoretical literature, the analysis on the direct user brought to the development of a system composed of four devices: a smart mask, two devices that are monitoring physiological parameters and one dedicated to the environmental monitoring of working place and real time alerting about the level of hazards in the surrounding (i.e. visual and tactile feedback when the environment is hazardous). The system suggested in this research consists of three wearable devices and Mobile Application. In such a system, the motivation starts from a new mask (equipped with temperature sensor) redesigned with attention to the wearability and comfort, evaluated as high by the workers, while the Electronic Nose Device maintains the motivation of the user through information, or alert, about air quality (gas, CO_2), and even though the user generally does not want to wear the mask, he/she will be more encouraged to do it by changing the perception about the risk and by this also the behaviour. Beside environmental monitoring and alerting, also physiological parameters have a purpose to inform the user about the health condition when exposed and not to hazards, and here Chest Band has an important role. Data gathered from all devices are organised in mobile app and are communicated through simple graphical information: if the mask is worn or not; the level of air quality; and the quality of breathing. Seeing how these parameters are related will impact one's self-awareness and increase consciousness. This kind of system operates during the time, by giving knowledge and—as a result of it—conscious behaviour, that on long distances improves health condition. The results of the performed user test are promising: they gave us a first confirmation that the designed system could be an acceptable and effective proposal. Nevertheless, more research is required to demonstrate both that the behaviour change by workers will last longer than the duration of an observation session and that the improved mask's wearability will be enough also after several hours and days of usage. We are currently working on the research project follow-up in collaboration with a SME based in the northern area of Milan that develops and sells respiratory protection products; the collaboration is aimed at improving the system. The company is interested in testing the entire system on a wider population of users not only in Italy but also in Spain and Netherlands where it has other companies' branches. The company also wants to test it in different working environments such as body shops, agriculture, and restoration and is very committed to translating the prototyped system into a commercial product.

http://dx.doi.org/10.5772/intechopen.77002

Acknowledgements

We acknowledge funding provided by INAIL in the context of the SAF€RA under the ERA-NET actions of the 7th Framework Programme for European Research and Technological Development.

Conflict of interest

In accordance with our ethical obligation as researchers, we declare that we have no financial, commercial, legal or professional relationship with other organisations, or with the people working with them, that could influence our research.

Author details

Venere Ferraro*, Mila Stepanovic and Silvia Ferraris

*Address all correspondence to: venere.ferraro@polimi.it

Department of Design, Politecnico di Milano, Milan, Italy

References

[1] Fogg BJ. Creating. Persuasive Technologies: An Eight-Step Design Process. Persuasive '09, April 26–29

[2] Fogg BJ. Persuasive Technology: Using Computers to Change What We Think and Do. San Francisco, CA: Morgan Kaufmann; 2003

[3] Gill SP. Cognition, Communication and Interaction: Transdisciplinary Perspectives on Interactive Technology. 1st ed. London: Springer-Verlag; 2007

[4] Hendrik NJ, Schifferstein, Özcan E, Rozendaal MC. Towards the maturation of design: From smart to wise products. Design and semantics of form and movement in DeSForM 2015. Aesthetics of interaction: Dynamic, multisensory, wise. 2015

[5] Motti V, Caine K. Human factors considerations in the design of wearable devices. HFES 2014 Annual Meeting

[6] Brown T. Design thinking. Harvard Business Review. 2008;**86**(6):84

[7] Leavy B. Design thinking—A new mental model of value innovation. Strategy & Leadership. 2010;**38**(3):5-14

[8] Cross N. Design thinking. Berg; 2011

[9] Forlizzi J, Battarbee K. Understanding experience in interactive systems. In: Proceedings of the 2004 Conference on Designing Interactive Systems (DIS 04): Processes, Practices, Methods, and Techniques. New York: ACM; 2005. pp. 261-268

[10] Schön D. Technology and Change—The New Heraclitus. Delacorte Press; 1967

[11] Fogg B, Cuellar G, Danielson D. Motivating, influencing, and persuading users: An introduction to captology. In: Jacko J, Sears JA, editors. The Human-Computer Interaction Handbook: Fundamentals, Evolving Technologies and Emerging Applications. New York: Taylor & Francis; 2009. pp. 133-146

[12] Kidd CD, Breazeal C. Designing a sociable robot system for weight maintenance. In: IEEE Consumer Communication and Networking Conference; IEEE, Piscataway; 2006

[13] Ruttkay ZM, Zwiers J, Van Welbergen H, Reidsma D. Towards a Reactive Virtual Trainer. Berlin: Springer-Verlag; 2006. pp. 292-303

[14] Bickmore T, Gruber A, Picard R. Establishing the computer–patient working alliance in automated health behavior change interventions. Patient Education and Counseling. 2004;**59**:21-30

[15] Goetz J, Kiesler S, Powers A. Matching robot appearance and behavior to tasks to improve human–robot cooperation. In: IEEE Ro-Man 2003; IEEE, Piscataway. pp. 55-60

[16] Gockley R, Mataric MJ. Encouraging physical therapy compliance with a hands-off mobile robot. In: ACM SIGCHI/ SIGART Human–Robot Interaction 2006; New York: ACM. pp. 150-155

[17] Kidd CD, Taggart W, Turkle S. A sociable robot to encourage social interaction among the elderly. In: ICRA 2006. IEEE, Piscataway; pp. 3972-3976

[18] Kriglstein S, Wallner G. HOMIE: an artificial companion for elderly people. In: Conference on Human Factors in Computing Systems 2005; ACM, New York. pp. 2094-2098

[19] Bigelow JH, Cretin S, Solomon M, Wu SY, Cherry JC, Cobb H, et al. Patient Compliance with and Attitudes Towards Health Buddy. Santa Monica: RAND Corporation; 2000

[20] Looije R, Cnossen F, Neerincx MA. Incorporating guidelines for health assistance into a socially intelligent robot. In: Ro-man, 2006. Piscataway: IEEE. pp. 515-520

[21] Picard RW. Affective Computing. Cambridge, MA: MIT Press; 1997

[22] Fogg BJ. Computers as Persuasive Tools in Persuasive Technology. Standford: Morgan Kaufmann Publishers; 2003. pp. 31-59

[23] Ananthanarayan S, Siek K. Persuasive wearable technology design for health and wellness. In: 2012 6th International Conference on Pervasive Computing Technologies for Healthcare (PervasiveHealth). pp. 236-240

[24] Ferraro V. The Designer Approach to Wearable Technologies—A Practice-based Approach. Milan: Maggioli Editore; 2012

[25] IJsselsteijn W et al. Persuasive technology for human well-being: Setting the scene. In: Persuasive Technology. Berlin, Heidelberg: Springer; 2006. pp. 1-5

[26] Marculescu D, et al. Electronic textiles: A platform for pervasive computing. Proceedings of the IEEE. 2003. pp. 1995-2018

[27] Fiorani E. Abitare il corpo: la moda. Lupetti; 2003

[28] McCann J, Bryson D, editors. Smart Clothes and Wearable Technology. Oxford, Cambridge, New Dehli: Woodhead Publishing in Textiles; Feb 2009

[29] Gemperle F, Kasabach C, Stivoric J, Bauer M, Martin RR. Design for Wearability. Digest of Papers. Second International Symposium on Wearable Computers; 1998

[30] Ferraro V, Ugur S. Designing wearable technologies through a user centred approach, DPPI 11, Designing pleasurable products and interface. Milan: Politecnico di Milano; June 2011. pp. 22-25

[31] Argyle M. The Psychology of Interpersonal Behaviour, a Pelican Original. Bristol, SOM, United Kingdom: Penguin Books; 1976

[32] Torning K, Oinas-Kukkonen H. Persuasive system design: state of the art and future directions. In: Proc. 4th International Conference on Persuasive Technology; New York, NY, USA; ACM. 2009

[33] Purpura S, Schwanda V, Williams K, Stubler W, Sengers P. Fit4life: The design of a persuasive technology promoting healthy behavior and ideal weight. In: Proc CHI '11. ACM; 2011. pp. 423-432

[34] Shiraishi M, Washio Y, Takayama C, Lehdonvirta V, Kimura H, Nakajima T. Using individual, social and economic persuasion techniques to reduce CO_2 emissions in a family setting. In: Proc Persuasive '09; ACM, Article 13; 2009. 8 pp

[35] Tromp N, Hekkert P, Verbeek P-P. Design for socially responsible behavior: A classification of influence based on intended user experience. Design Issues. 2011;**27**(3):3-19

[36] Oinas-Kukkonen H. Behavior change support systems: A research model and agenda. Lecture Notes in Computer Science, Vol. 6137. In: The Fifth International Conference on Persuasive Technology; June 7–9, 2010; Copenhagen, Denmark. Berlin/Heidelberg: Springer. 2010

[37] Consolvo S, McDonald DW, Landay JA. Theory-driven design strategies for technologies that support behavior change in everyday life. In: Proceedings of the 27th International Conference on Human Factors in Computing Systems; CHI '09; April 4-9; Boston, MA, USA. New York, NY, USA: ACM; 2009

Chapter 4

Wearable Neuromodulators

Arsam N. Shiraz, Brian Leaker and
Andreas Demosthenous

Additional information is available at the end of the chapter

http://dx.doi.org/10.5772/intechopen.76673

Abstract

In neuromodulation, by delivering a form of stimulus to neural tissue the response of an associated neural circuit may be changed, enhanced or inhibited (i.e., *modulated*) as desired. This powerful technique may be used to treat various medical conditions as outlined in this chapter. After a brief introduction to the human nervous system, key example applications of electrical neuromodulation such as cardiac pacemakers, devices for pain relief, deep brain stimulation, cochlear implant and visual prosthesis and their respective methods of deployment are discussed. Furthermore, key features of wearable neuromodulators with reference to some existing devices are briefly reviewed. This chapter is concluded by a case study on design and development of a wearable device with non-invasive electrodes for treating lower urinary tract dysfunctions after spinal cord injury.

Keywords: lower urinary tract, electrical stimulation, nervous system, neuromodulation, spinal cord injury, TENS, transcutaneous stimulation

1. Introduction

In neuromodulation by applying a form of stimulus to neural tissue, the response of the underlying neural circuits is modulated and a response is evoked. In this chapter, we solely focus on electrical stimulation of neural tissue. Neuromodulation is used to suppress pain, assist locomotion, treat various conditions and rehabilitate patients. Current neuromodulators are primarily used for medical purposes. Neuromodulators may be classified as invasive, where a form of surgical operation is required to implement the technique, or non-invasive, where surface electrodes are used. Due to the risks associated with surgery and high cost, non-invasive solutions are more desirable. Both classes of neuromodulators may be controlled by the user or be completely autonomous and this depends on the application. Most of them are portable and may be on the person while they go about their daily life. Invasive neuromodulators may not be removed when desired. However, if they are composed of external and implanted units, the external part may be detached from the patient if needed. While connectivity can be readily incorporated in such devices, this is not at this stage a fundamental part of such solutions.

After an introduction to human nervous system, fundamentals of neuromodulation are described in this chapter. Key examples of neuromodulation devices are briefly discussed, followed by a note on wearable neuromodulators. This chapter is concluded by a discussion on the design and development of a wearable neuromodulator to treat lower urinary tract (LUT) dysfunctions after spinal cord injury (SCI).

2. Human nervous system

This section is a brief and general introduction to human nervous system, aiming to assist readers in understanding what follows in the chapter.

2.1. Central and peripheral systems

The human nervous system may be divided into central and peripheral systems. The central nervous system (CNS) comprises the brain and the spinal cord and the peripheral nervous system (PNS) mainly includes the nerves that connect the periphery of the body to the CNS. The brain comprises the cerebral hemispheres, diencephalon, cerebellum, and the brainstem. The spinal cord connects lower parts of the body to the brain and can be divided into four sections, each of which has various segments from top to bottom: cervical (C1–C8), thoracic (T1–T12), lumbar (L1–L5), and sacral (S1–S5) [1].

The PNS comprises the spinal nerves, originating from the spinal cord, and most of the cranial nerves, originating from the brain, above the spinal cord. The peripheral nerves originating from a specific segment of the spinal cord (C-S) generally mediate a specific functionality. The PNS may be subdivided into the autonomic nervous system and somatic nervous system. Generally, the former oversees the involuntary control and the latter provides the voluntary control. The autonomic nervous system may be further divided into sympathetic and parasympathetic nervous systems. The sympathetic nervous system may be perceived to oversee the fight-or-flight response while the parasympathetic nervous system is responsible for the rest-and-digest functions.

2.2. Key elements

There are two main classes of cells in the nervous system: nerve cells (neurons) and glial cells (glia). Generally, the former are the ones involved in neural signaling and the latter function as support cells. Typically, a neuron may consist of four distinct regions: a soma, dendrites, an axon, and presynaptic terminals [2]. The main function of neurons is to convey the neural signal (i.e., action potential (AP)).

Soma houses cell genes and proteins are synthesized in it. A soma may give rise to two types of projections, also referred to as processes. There may be one long axon, also referred to as the nerve fiber, which acts as the main conducting unit of the neural signal (i.e., AP) to other neurons. Other projections, which are generally shorter and branch-like, are dendrites. Dendrites may be thought of as the main neural signal receiving ports in a neuron. Near the end, an axon divides into these branches which communicate with other neurons through the presynaptic terminals. The point of communication between a neuron and other neurons is called a synapse. The cell transmitting the signal is called a presynaptic cell and the one receiving it is called the postsynaptic cell. Presynaptic terminals land on the soma and/or the dendrites of a subsequent cell. Depending on the functions of the cells their morphology may vary [1, 2].

Glial cells surround neurons and have various functionalities including acting as the general and structural support for neurons. Schwann cells and oligodendrocytes insulate axons in the PNS and CNS, respectively [3]. Schwann cells form the myelin sheath which essentially acts as insulation around a nerve fiber, referred to as myelination. Fibers may be myelinated or unmyelinated and, consequently, they show different characteristics. In a myelinated fiber, large segments along its length are wrapped by layers of myelin while only small segments between the wrapped segments are left exposed at regular intervals. The exposed segments are referred to as the nodes of Ranvier. There are other types of glial cells that are thought to bring nutrients to cells, provide the blood-brain barrier and help maintain the right ion concentration. They also absorb neurotransmitters and some are only recruited during infection, injury and seizure [1].

In PNS, generally, there are two roots originating from the spinal cord. The dorsal root conveys the sensory information while the ventral root conveys the motor signaling. The dorsal root then enters the dorsal root ganglion, a local accumulation of cell bodies and supporting cells. Peripheral axons may be bundled into what is called nerves. The nerves from the dorsal and ventral roots merge to form the spinal nerve. In CNS, cell bodies may be arranged in two ways. Nuclei are the local accumulations of neurons with similar connections and functionalities, such as the collections found in the cerebrum, brainstem, and spinal cord. The second arrangements are cortices which are sheet-like arrays of cells such as those found in the cerebral hemispheres and cerebellum. The axons in the CNS are gathered into tracts, analogous to nerves in the PNS. Tracts that cross the mid-line of the brain and spinal cord are referred to as commissures. The term gray matter identifies any accumulation of cell bodies and neuropils (synaptically dense regions of dendrites, synaptic terminals, and glial cell processes) in the brain and spinal cord (e.g., nuclei or cortices) while white matter is essentially tracts and commissures.

2.3. Afferent and efferent neurons

In terms of functionality or the direction of conveying the neural signal, neurons are classified into three major groups: afferents, efferents and interneurons. Afferents carry information from the periphery of the body to the CNS. These neurons are very often interchangeably referred to as sensory neurons although not all afferent communications may lead to a form of sensation [1]. Efferents or the motor neurons carry the neural signal and commands from the CNS to muscles and glands. Interneurons, which are neither sensory nor motor, constitute the largest group of neurons [1]. Interneurons convey APs over long or short distances from a neuron to the other.

2.4. Neural circuits, receptors and transmitters

Neurons are organized to form neural circuits and reflexes which may, for instance, process the sensory information and elicit motor responses. A typical circuit on the afferent end may be terminated by sensory receptors. These are nerve endings which transduce specific stimulus energies into the corresponding receptor potentials. Receptors are specialized in terms of the energy they transduce which may be electromagnetic, mechanical, thermal or chemical [4]. A complex sensation such as pain may recruit different receptors. The receptor potential then travels along the attached afferent fiber with a specific AP firing pattern. The attached afferents may covey a specific mode of sensation as well. The cell bodies of these afferents may be in the corresponding ganglia. The central axon of the mentioned cell might enter through the dorsal horn of the gray matter of the spinal cord and synapse with central neurons. Namely, it may synapse with interneurons or motoneurons in the spinal cord.

The basic forms of the synaptic transmission between neurons are electrical and chemical while most of the transmissions are chemical [4]. While at an electrical synapse a low resistance electrical path exists between the two associated cells, at chemical synapses neurons are separated completely by a small space called the synaptic cleft. The entities that essentially pass the neural signal from a presynaptic cell to a postsynaptic cell are called neurotransmitters. The arrival of an AP at the presynaptic side of a synapse leads to the release of chemical neurotransmitters which then diffuse across the cleft and interact with the receptors of the postsynaptic side. This interaction leads to either excitation or inhibition. The former leads to the suppression of the AP generation while the latter leads to the generation of APs. The type of the resulting action, primarily, depends on the type of the transmitter and receptor involved. Also, the synaptic input may add up from various presynaptic cells.

The sensory information may be conveyed through the white matter of the spinal cord to different regions in the brain for further processing. In which case, the response of the brain may be conveyed back to the target cells through different regions of the white matter of the spinal cord. A spinal reflex on its own may result in a motor activity without any modulatory input from the brain [1]. Through the ventral horn of the gray matter, efferent fibers leave and land on target muscles or organs. At a neuromuscular junction, neurotransmitters are released and the corresponding contraction or relaxation effects are mediated.

2.5. Action potential

Having presented an overview of different elements and aspects of the human nervous system, we now discuss the AP as the most fundamental phenomenon in the nervous system further. An AP is a rapid (~1 ms) variation (~100 mV) of the transmembrane potential [1, 3]. APs are naturally triggered in the initial segment of the attachment of an axon to its corresponding soma and travel down the axon at velocities ranging from about 1-120 m/s depending on the fiber type [4]. Different activities are mediated via this signal throughout the nervous system by the neural circuits mentioned before. The actual command or information is not perceived by the variation of the signal (i.e., AP) but different effects are mediated through the paths APs travel and their firing pattern.

Generally, there are three main temporal phases in an AP. The resting potential, the transmembrane voltage from inside to outside in the absence of any stimuli, is a negative value. Upon an increase in the transmembrane potential beyond a threshold, it rapidly increases to a specific positive value. This phase is called the depolarization period. The AP generation is an all-or-none process, meaning that any rise smaller than the said threshold does not yield such a response. The depolarization phase is followed by the subsequent reduction in the transmembrane potential which is called the repolarization phase. The transmembrane potential reduces even below its initial resting potential for a period which is referred to as the hyperpolarization phase. There exists the absolute refractory period in which no action potential may be triggered in the segment of the membrane under question. This is followed by the relative refractory period during which period a stronger than normal stimulus may produce a new AP [5].

The generation and propagation of APs are mediated by voltage-gated ion channels in the membrane and the subsequent influx and outflow of ions. These channels are ion specific, meaning that when they are open they are only permeable to a specific ion. Also, the state that these channels are at depends on the transmembrane potential. Supposing that two media are separated by an ion specific channel and the density of that ion is different in the two media, ion flows against the concentration gradient. It is noted that ions do not exist in isolation, meaning that a positive ion is paired with a negative one. Thus, as ions that the barrier is permeable to diffuse, assuming a positive ion is diffusing, an electric field is built up in the direction of the concentration gradient. The opposing forces of the concentration-dependent diffusion and electric field reach an equilibrium, at which stage there is no net ion flow. By equating these opposing forces, the transmembrane electric potential at which the equilibrium is reached can be calculated. This is referred to as Nernst voltage. Nernst voltage is measured from intracellular to the extracellular space. Thus, a higher concentration of a positive ion outside the cell results in a positive Nernst voltage for that ion. Where several ion species are present, the transmembrane potential can then be calculated using the Goldman-Hodgkin-Katz formulation. Building on the Nernst equation, if the membrane becomes more permeable to a specific species, the equilibrium potential tends to the Nernst potential of that species. The resting potential is not a case of equilibrium but that of the steady state.

The permeability of channels is governed by the voltage-gated channels mentioned before. The pioneering work of Hodgkin and Huxley (HH) [6] laid the foundation for the existing systematic understanding of the channel mechanisms. By setting the transmembrane voltage to a predetermined value and measure the transmembrane current for the said voltage in the giant axon of loligo, HH formed a set of curves of the conductivity of the membrane to potassium and sodium versus time for various depolarization voltages. Apart from the overall time trace, they noticed that as the clamp voltage increases, the rate at which the conductance varies increases too. They decided to use first order kinetics (multiple processes if necessary) to fit the experimental data and consider activation and inactivation variables to explain the cessation of conductivity following its initial rise. Although HH developed a merely empirical formulation for the observations but the overall macroscopic system of equations they defined is compatible with the microscopic understanding which exists today [1]. The channels have gates that depending on the voltage across them the rate at which they open or close varies. For a channel to conduct, all its gates should be open. The concentration of the sodium ions in the extracellular space is much more than those in the intracellular space while that of potassium ions is the other way around. Therefore, the opening of the sodium channels results in a rise of the transmembrane potential while the opening of the potassium channels results in its reduction.

Assuming only sodium and potassium ions are involved, a full cycle of the AP generation is described as follows. The inputs in an axon hillock raise the transmembrane potential beyond a threshold at which point the permeability of the membrane to sodium and potassium ions increases. The conductivity of sodium channels increases faster than those of potassium ions. Thus, the influx of sodium ions brings the transmembrane potential closer to the equilibrium potential of sodium ions. At the higher transmembrane potentials, the conductivity of the membrane to potassium ions increases while sodium channels start to inactivate, reducing the membrane permeability to sodium ions. This results in repolarization of the membrane. The inactivation of sodium channels and the prolonged permeability of the membrane to potassium ions result in the lowering of the membrane potential below the resting potential and closer to the equilibrium potential of potassium ions. Several more channel types may be involved in different species but the overall process is generally similar.

When an AP is generated, it automatically induces the generation of other APs in its vicinity [1, 4]. Considering unmyelinated fibers, the extracellular and intracellular current flows along the fiber induce an increase in the transmembrane potential near the depolarized segment. When the AP is induced in a segment of the fiber, the preceding segment is in the hyperpolarization phase, thus, the AP travels in a specific direction. In the case of a myelinated fiber, as the current loops are relatively weak where the myelin insulation is present, the loops are at their highest intensity from a node to the other, resulting in the AP essentially hoping from a node to the other, rendering the transmission much faster. Apart from myelination, fiber diameter also affects the conduction velocity. Larger fiber diameters generally have higher conduction velocities. Depending on the size and conduction velocity of fibers, Erlanger's (Aα, Aβ, Aδ, B and C) and Lloyd's (I, II, III and IV) systems have been proposed for a classification of fibers [1, 4].

3. Fundamentals of neuromodulation

Neuromodulation has two components. The first component involves the delivery of the stimulus current to produce APs in a target neural tissue and the second one is the way this leads to a response. If a nerve is pierced via an electrode and the transmembrane voltage is forced more positive, then APs may be generated. Often, fibers are stimulated by merely perturbing

http://dx.doi.org/10.5772/intechopen.76673

the extracellular potential. Assuming a cathode–anode pair is placed along a fiber at a distance away from it, the extracellular space near the cathode may become more negative and this forces transmembrane voltage there more positive. This may lead to the generation of a propagating AP in that position. This yields a very simplistic view of electrical stimulation. The actual process is far more complex and requires an understanding of HH type models of the fiber.

In an electrical conductor, charge carriers are primarily electrons while in the tissue they are ions. Thus, electrodes may be thought of as the interface between these two media. It is, therefore, understandable that the processes involved are electrochemical and reduction and oxidation processes may happen at the electrodes which are not infinitely fast and unimpeded. Furthermore, due to the adsorption of charges and redux processes, the interface impedes the flow of current. The contact impedance due to these plays an important role in electrode material selection and design. Enough charge should be pumped into the medium to sufficiently perturb the extracellular potential. When applied as a pulse, the strength and duration of the pulse dictate how much charge is delivered. This leads to the formation of the strength-duration curves that identify a balance between the amplitude and the duration of stimulus pulses. The electrode design, its relative position with respect to the nerve and characteristics of the surrounding tissue, among other features, play a significant role in defining the stimulus pulse design. Moreover, there are safety considerations when charge is injected into the tissue. To understand the dynamics and safety considerations when injection stimulus current through electrodes the following description should be considered.

As an electrode is put into contact with an electrolyte (tissue) a potential develops which is primarily governed by the Nernst formulation and is referred to as the equilibrium potential of the electrode. As the current is injected through the electrode, the potential of the electrode rises above the equilibrium, at which time the potential dependent reversible and irreversible Faradaic processes may occur. Such reactions are dependent on the fastness of the kinetics of the said reactions and the speed of the mass transport [7]. A kinetically fast process with respect to the mass transport requires a small over-potential, defined as the variation of the electrode potential from its equilibrium potential, to lead to a significant amount of current flowing. As the reactants, consequently, do not move away from the surface, applying current in the reversed direction may force the reactions to be reversed (i.e., biphasic charge balanced stimulation). Kinetically slower reactions are limited by the mass transport. Thus, the reactants move away from the surface as the reactions occur at higher over-potentials [7]. These reactions are referred to as the irreversible Faradaic processes while those involving kinetically fast processes are reversible. One of these irreversible processes is the electrolysis of water. This may happen above and below a certain positive and negative potential, respectively, for a given electrode material. The range of the potential below these limits (i.e., safe limits) is referred to as the water window. The electrolysis of water results in pH changes and gas formation which may lead to tissue and electrode damage. It is noted that electrode material forms a very important feature of the design. The safe limits of the charge injection and the general electrical properties of the electrode should be considered. Under most neural stimulation conditions, the Faradaic processes dominate [8], noting the existence of non-Faradaic currents at the electrode interface as well. Thus, it is important to ensure a given material can safely deliver the required charge. Electrodes may be implanted or surface electrodes may be used. The former is invasive but the delivery of charge is direct while the latter may be non-invasive but the current have a large spread. Unfortunately, it has been shown that risk versus expected benefit has a linear relationship, the higher the risk the more substantial reported benefits, for different methods of stimulation [9]. This may be attributed to efficacy and selectivity. While non-invasive electrodes are simply conductive patches attached to the surface of the body, minimally invasive electrodes may be designed as needles piercing the skin to deliver the stimulus current. General classes of invasive electrodes for PNS in increasing order of invasiveness include extraneural, interfascicular, intrafascicular and those relying on the regeneration of the neural tissue in them. Examples of these electrodes are cuff electrodes, flat interface nerve electrodes (FINESs), longitudinal intrafascicular electrodes (LIFEs), transverse intrafascicular electrodes (TIMEs) and various multi-channel electrodes [10]. For the CNS, the general classification of electrodes is superficial/distal and those for deeper structures.

The second component of neuromodulation is the way the stimulus current leads to a desired response from the associated neural circuits. This may be a simple motor response by stimulating the very same motor fibers or may involve much more complex circuits involving PNS and CNS. Thus, in designing the stimulus signal one should also consider the appropriate features that lead to the desired response. This may include, but is not limited to, stimulus waveform and frequency.

4. Example applications of neuromodulation

A very well-known example of neuromodulators is the cardiac pacemaker. Electrodes are inserted transvenously and placed at the targeted chamber of the heart. The pulse generator is placed in a subcutaneous or submuscular position. Electrical stimulus may be applied in a unipolar or bipolar fashion and the process can be synchronized with hearts pace [11].

One of the applications of neuromodulation is for chronic pain management. Transcutaneous electrical neural stimulation (TENS) owing to its simple delivery method has been investigated for many years. Various waveforms and frequencies have been used. The placement of electrodes may be directly over the painful area, its peripheral neural supply, the dermatome of the associated spinal roots and known trigger points. It is thought that the resulting analgesia is due to the activation of central inhibitory mechanisms. For instance, stimulation of Aβ afferent fibers are thought to lead to inhibition of nociceptive C and Aδ afferent fibers in some cases. Examples of devices developed based on TENS concepts include NEMOS (Cerbomed GmbH, Erlanger, Germany), a portable device that targets the auricular branch of the vagus nerve via electrodes placed on the concha of the outer ear, Cefaly (CEFALY-Technology, Seraing, Belgium), a wearable device for supraorbital stimulation by a bipolar electrode patch on the forehead and gammaCore (electroCore LLC, NJ, USA), a handheld noninvasive vagal nerve stimulator, all for the treatment and management of migraine. Spinal cord stimulation using implanted electrodes [12], occipital nerve stimulation using implantable electrodes [13], transcranial magnetic stimulation [14] and transcranial direct current stimulation [15] have also been used to treat various neuropathic pain patients.

http://dx.doi.org/10.5772/intechopen.76673

Deep brain stimulation (DBS) is one of the key applications of neuromodulation that has found applications in treating Parkinson's disease [16], essential tremor [17], dystonia [18], Tourette syndrome [19], chronic pain [20], epilepsy [21], Huntington's disease [22] and even mental disorders [23] with different sites of stimulation. A linear electrode array is implanted deep in the brain. In many of the above cases the exact mechanisms that lead to desirable outcomes using DBS are not known due to the complexity of neural circuits in the brain. However, its efficacy has promoted its use and further research.

A very successful example of neuromodulators is the cochlear implant [24] in which the auditory nerve is stimulated via electrodes placed on a flexible platform that is placed in a layer of the cochlea (basilar membrane). Sound is recorded externally and upon processing the stimulation paradigm is inductively transmitted to the electrodes. Sound as a mechanical wave hits the tympanic membrane (i.e., eardrum). Ossicles, three bones connected to the tympanic membrane, convey this motion to the cochlea which is filled by fluid. At a layer of the cochlea there exists the basilar membrane which is stiff and narrow near the ossicles and wide and flexible on the other end. Due to this structure, the thinner segments *resonate* with higher frequencies. The hair cells on the basilar membrane then translate this to a neural signal which is conveyed through the auditory nerve. The basilar membrane may be appropriately stimulated to induce hearing perception.

Visual prosthesis is another example of using neuromodulation to restore function. One of the most common forms of blindness in the developed world is neural blindness. This is generally due to the loss of photoreceptors. In many cases part of the neural circuits between the receptors and the brain survive and this may be harvested to restore vision. Light passes through the cornea, pupil and lens and is shone on the retina. Retina is covered by photoreceptors (rod cells, more populous and specialized for dim light and cone cells, for bright light and colors) that transduce light to a neural signal which goes through the visual pathway to be perceived in the visual cortex of the brain. Electrical stimulation may be delivered in various places [25]. The stimulation may be delivered via epiretinal [26], subretinal [27] or suprachoroidal [28] electrode arrays. Stimulation may be delivered by deriving the appropriate pattern via an external sensor like camera and transmitted via an inductive link to the array or the photo sensors may be integrated with the electrodes. Stimulus current may also be delivered to optic nerve [29], lateral geniculate nucleus [30] or even the visual cortex itself [31].

Electrical stimulation may be applied to restore or rehabilitate motor function. In this case it is referred to as functional electrical stimulation (FES). While not in all cases of FES a neural reflex may be modulated, due to its significance we briefly discuss it here. FES may be delivered via implanted electrodes [32], percutaneously [33] or transcutaneous [34] in various places. FES have been used in a variety of applications such as improving activity after stroke [35], chronic heart failure [36] or control and rehabilitation of extremities following spinal cord injury [37] or other forms of neural disease.

Neuromodulation may be used to treat various lower urinary tract dysfunctions that arise upon spinal cord injury. A more elaborate description of such interventions is presented as a case study in designing a wearable neuromodulator later in this chapter.

5. Common features of wearable neuromodulators

The introduction to key underpinning concepts surrounding neuromodulation and the key applications presented in this chapter reveal the great potential of the method in improving the quality of life for patients in a variety of patient groups. However, the concept of wearable neuromodulators in terms of their defining features should be further discussed.

In most devices currently advertised as wearable neuromodulators such as Senza system (Nevro Corp., CA, USA) for spinal cord stimulation for pain relief, a range of products from Bioness (Bioness Inc., CA, USA) for improved mobility, Stimwave (Stimwave Technologies Inc., FL, USA) for pain relief, Bionode (Bionode LLC, IN, USA) for glaucoma, Reliefband (Reliefband Technologies LLC, PA, USA) for nausea treatment, various products from Bioelectronics (BioElectronics Corporation, MD, USA) for pain management, Quell (NeuroMetrix Inc., MA, USA) for pain relief and Cefaly (CEFALY-Technology, Seraing, Belgium) for migraine, portability, in a sense that they do not impede the normal life of the user, appears to be the primary common factor. The other common feature of these technologies is that they have a controllable interface with the user, even if this is simply for turning the device on or off. While most of the wearable neuromodulators are non-invasive but this is not necessarily a common feature among all the neuromodulators referred to as wearable. For devices whose electrodes are implanted higher efficacies are expected but various challenges including the risk of surgery, possible electrode migration and power delivery to internal unit should be tackled.

6. A case study

This section presents a brief case study on designing a wearable neuromodulator for treating LUT dysfunctions after SCI.

6.1. Functions of the LUT

The main functions of the LUT are to store and periodically void urine. These functions are performed by the main functional units of the LUT. These functional units include the urinary bladder and an outlet consisting of the bladder neck and urethra. The urinary bladder is engulfed in a smooth muscle called the detrusor while the urethra, the projecting tube from the bladder, is composed of smooth and striated muscles. During the storage phase, the urinary bladder relaxes while its outlet contracts and during the voiding phase the bladder contracts while the outlet relaxes. These synergic and reciprocal activities of the functional units are controlled by complex neural circuits [38].

6.2. Neural control of the LUT

Current understanding of neural control of the LUT is based on studies on various species. This section aims to draw a simple picture of the way the circuits are thought to be formed. The associated neural circuits are primarily controlled in the brain, spinal cord, and peripheral ganglia. The intricacy of the neural circuits is partly due to LUT functions being under the somatic as well as autonomic control and partly due to the required switching action between the reciprocal activities. Four types of neurons are thought to be involved in the neural control of the LUT functions: the primary afferent neurons, spinal efferent neurons, spinal interneurons and neurons in the brain. These neurons activate or modulate the reflexes [38].

The autonomic nervous system controls the LUT through the sympathetic nervous system, primarily via the hypogastric nerve, and parasympathetic nervous system, primarily via the pelvic nerve. The somatic control of the LUT is mediated by the pudendal nerve. It is generally thought that the activation of the sympathetic nervous system is responsible for maintaining continence while the activation of the parasympathetic system facilitates voiding [39]. Various excitatory and inhibitory neurotransmitters and the associated receptors are involved in the healthy function of the LUT.

Intrinsic properties of the detrusor, leading to low and constant intravesical pressure during the storage and filling, inhibition of the parasympathetic outflow to detrusor, activation of sympathetic efferents to the outlet and pudendal efferent activities are some of the features that might contribute to continence based on studies on various species. Spinal and supraspinal reflexes activated by afferent activities are thought to mediate these. Upon voluntary decision for voiding in healthy adults, the sympathetic and somatic outflows are inhibited and the parasympathetic efferents are activated to facilitate voiding by an initial relaxation of the outlet and the subsequent contraction of the reservoir [38, 39].

6.3. LUT after SCI

The spinal cord is the major pathway through which efferent and afferent command and information travel between the brain and the body. The spinal cord contains longitudinally oriented spinal tracts (white matter) surrounding its central areas (gray matter) where most of the spinal nerve cell bodies are located. The gray matter is organized into segments that comprise sensory and motor neurons. As briefly discussed earlier in the chapter, every segment of the spinal cord is generally in charge of controlling a specific part of the body and functionality. Thus, damage to a specific segment of the spinal cord may result in the loss of control in the corresponding part of the body.

SCI may be divided into two major groups. Tetraplegia is the impairment or loss of motor and/or sensory functions in the cervical segments of the spinal cord due to the damage of the nervous system within the spinal canal. Paraplegia is the impairment or loss of motor and/or sensory functions in the thoracic, lumbar or sacral segments of the spinal cord. Neurological level of injury refers to the most caudal segment of the spinal cord with normal sensory and anti-gravity motor function on both sides of the body provided that there is an intact sensory and motor function rostral to that position [40]. SCI may also be divided into incomplete and complete. In an incomplete injury there is a preservation of some sensory and/or motor function below the neurological level that includes the lowest sacral segments. In contract, a complete injury refers to the situation when there is an absence of any sensory and motor function below lesion [40].

Given the intricate neural circuits involved and the switching-like somatic control of the LUT functions, it comes as no surprise that after SCI both functions of the LUT may be significantly disrupted. Depending on the location of the lesion and its completeness the degree of LUT dysfunctions and their nature may vary. After SCI above a lumbo-sacral level, understandably, the input from higher orders in the brain and, consequently, the voluntary control of the LUT may be disrupted. After a period, new sacral spinal reflexes develop. It is thought that these reflexes are mediated by C fibers which are generally thought to be mechanically silent in healthy individuals but become sensitive to low volume bladder activities after injury. Thus, contractions occur during the filling phase of the bladder [41]. Major conditions appearing after SCI include the neurogenic detrusor over activity (NDO) and detrusor sphincter dyssynergia (DSD). The NDO involves the existence of bladder contractions at low volumes and the DSD is the condition in which the reciprocal activities of the bladder and the outlet are not maintained, hence bladder contractions are mirrored with contractions in the urethra [41]. If untreated, these conditions may have severe consequences for the patient.

6.4. Existing solutions

Solutions to the NDO and DSD may be divided as pharmaceutical, mechanical and surgical. Mechanical devices may be subdivided into passive and active devices. The latter are primarily the kind of devices used for the neuro-muscular stimulation. In the case of the NDO, typically, the first devised treatment involves anticholinergic drugs [42]. High doses of prescribed anticholinergic drugs may lead to troublesome side effects including dry mouth, blurred vision, constipation and cognitive impairment [41]. If such drugs do not yield the desired response, more intrusive interventions, including surgical operations, may be devised. Bladder augmentation, an operation in which a section of the intestine is used to increase the bladder volume, can be used to partially suppress the NDO [43]. For the DSD, catheterization, pharmacologic agents, injection of botulinum A toxin, urethral stents, artificial sphincters, sphincterotomy, which is a complex procedure in which urethra is cut into several times to incapacitate it, are some of the key options [42, 43]. Intermittent catheterization, developed in the twentieth century, is the predominant solution used in tandem with drugs for NDO and voiding. The major issue with catheterization is that it is the foremost cause of re-hospitalization due to infection in the people with SCI [42, 43].

The application of electrical stimulus current to treat LUT dysfunctions has been extensively explored at various sites in the body. The stimulus current may target the excitable tissue of the bladder directly in the form of an intravesical stimulation or bladder wall stimulation to facilitate voiding. Such solutions may be used in patients with damaged lower motor neurons in which cases the neural stimulation is not possible [43]. As an example for stimulating other sites, Finetech-Brindley system (Finetech Medical Ltd., Welwyn Garden City, UK) is a solution for sacral anterior root stimulation [44] for voiding. Efferent fibers, supplying both the detrusor and urethra, are stimulated by implanting electrodes intradurally or extradurally, resulting in the contraction of both. However, when the stimulus is removed the urethra relaxes much faster than the detrusor and this leads to voiding. This voiding would be non-physiological but yields high voiding efficiencies. For the solution to be effective, posterior

http://dx.doi.org/10.5772/intechopen.76673

rhizotomy has been performed in subjects to increase the bladder capacity and compliance by suppressing hyperreflexive contractions of the bladder. Unfortunately, this irreversibly eliminates the reflex erection, reflex micturition and any remaining pelvic sensation. To treat hyperreflexive contractions of the bladder, Interstim system (Medtronic, MN, USA) for sacral nerve neuromodulation has been implemented by delivering a low frequency and low amplitude stimulus via an electrode set, usually placed near the S3 spinal nerve. The effect has been attributed to the inhibitory modulatory effect of somatic afferents on motor fibers to the bladder. Other examples are tibial nerve stimulation to suppress the NDO and the pelvic nerve stimulation to facilitate voiding.

Efferent and afferent pathways of pudendal nerve innervate various pelvic structures including the external anal sphincter, external urethral sphincter, genitalia and urethra. The nerve is paired and its lateral branches originate from both sides of the sacrum. Through the lower part of the left and right sciatic foramen, the nerve's lateral branches enter the gluteal region. Turning around the sacrospinous ligament near its attachment to the ischial spine, the pudendal nerve re-enters the perineum via the lesser sciatic foramen. Running across the distal edge of the ischioanal fossa, the nerve gives rise to its terminal branches in an area referred to as the Alcock's canal [45, 46]. The branching pattern and the course of the nerve have been shown to be highly variable across different individuals [47]. Pudendal nerve stimulation for LUT dysfunction in SCI patients at various positions along the course of the nerve, through different terminal branches and using various stimulation parameters has been the topic of research by many workers [48–55]. A key point is that depending on the position and frequency of the stimulus current pulses, voiding or inhibition effects may be achieved. Also, the pattern of the stimulus pulses may induce different responses. These may be attributed to pudendal afferents being involved in both voiding and storage reflexes.

6.5. A wearable device

The course of the pudendal nerve provides the opportunity of its trans-rectal stimulation using an ano-rectally worn probe with surface electrodes. If electrodes are positioned near the ano-rectal junction, it is likely to expose the compound pudendal nerve to the stimulus current. This way, it may be possible to devise a wearable solution to treat various bladder dysfunctions after SCI. Another interesting feature of this solution when used to improve continence is that the efficacy and viability of the solution may be improved using conditional stimulation in SCI [56]. The majority of cases of the NDO after SCI are accompanied by the DSD [41]. Thus, any contraction in the bladder is mirrored by contractions in the external urethral sphincter. It has been shown that electromyogram (EMG) signal recorded from the external anal sphincter may be used as a surrogate for that of the external urethral sphincter which demonstrates the onset of the hyper-reflexive contractions of the bladder [57, 58]. Due to the closeness of these sites the recording and stimulation may be performed by surface electrodes mounted on the same probe.

Intra-anal and vaginal stimulation is primarily used for an acute delivery of the stimulus current to pelvic structures and for therapeutic purposes. An early attempt to stimulate the pudendal nerve trans-rectally reported current levels were as high as 100 mA to yield a visible response [59] which is in agreement with more recent results [57]. This might impede the chronic use of the solution. Knight et al. [57] based on a patented technology [60] for conditional trans-rectal neuromodulation of the pudendal nerve and inspired by an earlier study by Brindley et al. [61] in a study on 6 SCI patients suffering from the NDO and DSD showed that by the conditional neuromodulation of the pudendal nerve, hyper-reflexive contractions of the bladder were suppressed. The control signal was set to be the EMG signal from external anal sphincter. They designed the probe more suited to chronic use by considering the anatomy of the anal canal and making the probe from silicone rubber with stainless steel surface electrodes. However, still high levels of required stimulus current were reported. **Figure 1** shows the probe used in experiments.

The study presented by Knights et al. [57] similar to previous studies was not tested for prolonged use. For chronic use, high levels of the stimulus current may result in the habituation of the nerve even under the conditional stimulation. For sensate individuals, such levels of the stimulus current may be beyond the pain threshold and at high levels of the stimulus current the targeting accuracy required may be lost. Any structure in the vicinity may also be stimulated, resulting in unforeseen responses. Also, high levels of the stimulus current increase the possibility of electrode or tissue damage. Therefore, the design of electrodes should be optimized to minimize the stimulus current. Even if the current is minimized, however, the challenge is that the course of the nerve is highly variable between different individuals as discussed earlier in this chapter. Furthermore, the pelvic structures move depending on the posture of the individual. This adds to the variations of the course of the nerve with respect to the device. Using computational model we designed [62] an electrode setting that based on modeling results may be capable of activating the pudendal nerve at much lower thresholds considering all the factors described above. Such models implement what is referred to as hybrid models in the literature. The electrical potential field generated by the stimulation electrodes is simulated in a volume conductor model then the field is applied to more realistic HH type cable models of the nerve to predict their response.

The other feature regarding this solution is the recording compartment. Various studies have investigated different features of surface EMG signal recorded from the external anal sphincter [63]. However, the design of the recording electrodes should be carefully carried out to

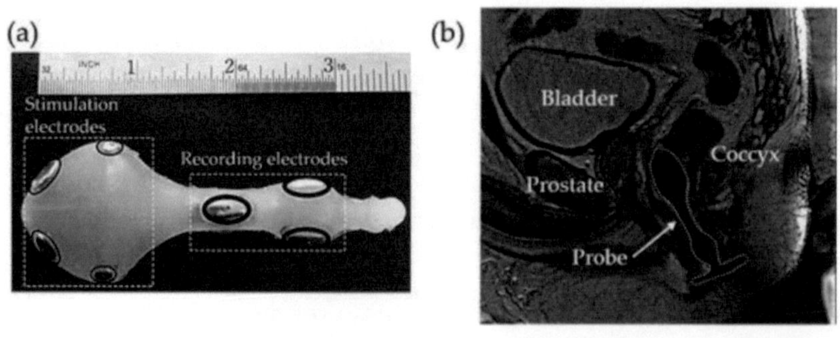

Figure 1. (a) The device used by Knight et al. [57] and (b) the MRI of the device *in situ*.

http://dx.doi.org/10.5772/intechopen.76673

ensure the recording is selective enough, sufficient EMG amplitudes are recorded and surface electrodes maintain contact for chronic use. In a proof of concept study we demonstrated [64] a design that met the aforementioned criteria.

These led to the design of a new prototype to be tested in the subsequent clinical tests. An issue to bear in mind is the design of the electrode-bearing probe in terms of variability of the anatomy in different individuals. This should be further tested in a population of the patient group. Upon successful clinical trials the electronics may be embedded in the probe to have a fully autonomous wearable solution.

7. Conclusions

Neuromodulation is an interdisciplinary field that draws from expertise in various fields and is currently predominantly used for medical purposes. Neuromodulation may be applied to various cites of central or peripheral nervous system and the applications may range from treatment of mental disorders to incontinence upon SCI. Most existing neuromodulators may be classified as wearable as they are mostly compact, portable, may utilize connectivity and may have a control interface for the patient. However, if one is to exclude invasive devices from this category only a select group utilizing surface electrodes will qualify as wearable. Unfortunately, evidence shows that non-invasive neuromodulators tend to have lower efficacies primarily as the neural tissue is not directly accessed. Future directions may involve devising design strategies that improve the efficacies of non-invasive neuromodulators or reducing the risks associated with invasive ones.

Author details

Arsam N. Shiraz[1], Brian Leaker[2*] and Andreas Demosthenous[1]

*Address all correspondence to: brian.leaker@heartlungcentre.com

1 Electronic and Electrical Engineering Department, University College London, London, United Kingdom

2 Nephro-Urology Clinical Trials (NUCT) Ltd, Queen Anne Street Medical Centre, London, United Kingdom

References

[1] Kandel ER et al. Principles of Neural Science. Vol. 4. New York: McGraw-Hill; 2000

[2] Noback CR et al. The Human Nervous System: Structure and Function. Berlin: Springer Science & Business Media; 2005

[3] Kiernan J, Rajakumar R. Barr's the Human Nervous System: An Anatomical Viewpoint. Philadelphia: Lippincott Williams & Wilkins; 2013

[4] Carpenter R, Reddi B. Neurophysiology: A Conceptual Approach. Boca Raton: CRC Press; 2012

[5] McIntyre CC, Richardson AG, Grill WM. Modeling the excitability of mammalian nerve fibers: Influence of afterpotentials on the recovery cycle. Journal of Neurophysiology. 2002; **87**(2):995-1006

[6] Hodgkin AL, Huxley AF. A quantitative description of membrane current and its application to conduction and excitation in nerve. The Journal of Physiology. 1952;**117**(4):500-544

[7] Merrill DR, Bikson M, Jefferys JG. Electrical stimulation of excitable tissue: Design of efficacious and safe protocols. Journal of Neuroscience Methods. 2005;**141**(2):171-198

[8] Cogan SF. Neural stimulation and recording electrodes. Annual Review of Biomedical Engineering. 2008;**10**:275-309

[9] Tyler DJ, Polasek KH. Electrodes for the neural interface. In: Neuromodulation. Amsterdam: Elsevier; 2009. pp. 181-213

[10] Sahyouni R et al. Interfacing with the nervous system: A review of current bioelectric technologies. Neurosurgical Review. 2017:1-15

[11] Arzuaga P. Cardiac pacemakers: Past, present and future. IEEE Instrumentation & Measurement Magazine. 2014;**17**(3):21-27

[12] Kumar K et al. Spinal cord stimulation versus conventional medical management for neuropathic pain: A multicentre randomised controlled trial in patients with failed back surgery syndrome. Pain. 2007;**132**(1-2):179-188

[13] Schwedt TJ et al. Occipital nerve stimulation for chronic headache—Long-term safety and efficacy. Cephalalgia. 2007;**27**(2):153-157

[14] Lefaucheur J-P et al. Evidence-based guidelines on the therapeutic use of repetitive transcranial magnetic stimulation (rTMS). Clinical Neurophysiology. 2014;**125**(11):2150-2206

[15] Mordillo-Mateos L et al. Effects of transcranial direct current stimulation on temperature and pain perception. Scientific Reports. 2017;**7**(1):2946

[16] De Hemptinne C et al. Therapeutic deep brain stimulation reduces cortical phase-amplitude coupling in Parkinson's disease. Nature Neuroscience. 2015;**18**(5):779

[17] Baizabal-Carvallo JF et al. The safety and efficacy of thalamic deep brain stimulation in essential tremor: 10 years and beyond. Journal of Neurology, Neurosurgery, and Psychiatry. 2014;**85**(5):567-572

[18] Cury RG et al. Thalamic deep brain stimulation for tremor in Parkinson disease, essential tremor, and dystonia. Neurology. 2017;**89**(13):1416-1423

[19] Schrock LE et al. Tourette syndrome deep brain stimulation: A review and updated recommendations. Movement Disorders. 2015;**30**(4):448-471

[20] Boccard SG, Pereira EA, Aziz TZ. Deep brain stimulation for chronic pain. Journal of Clinical Neuroscience. 2015;**22**(10):1537-1543

[21] Lehtimäki K et al. Outcome based definition of the anterior thalamic deep brain stimulation target in refractory epilepsy. Brain Stimulation: Basic, Translational, and Clinical Research in Neuromodulation. 2016;**9**(2):268-275

[22] Delorme C et al. Deep brain stimulation of the internal pallidum in Huntington's disease patients: Clinical outcome and neuronal firing patterns. Journal of Neurology. 2016; **263**(2):290-298

[23] Dougherty DD et al. A randomized sham-controlled trial of deep brain stimulation of the ventral capsule/ventral striatum for chronic treatment-resistant depression. Biological Psychiatry. 2015;**78**(4):240-248

[24] Wilson BS, Dorman MF. Stimulation for the return of hearing. In: Neuromodulation. Amsterdam: Elsevier; 2009. pp. 713-722

[25] Memon M, Rizzo JF. The development of visual prosthetic devices to restore vision to the blind. In: Neuromodulation. Amsterdam: Elsevier; 2009. pp. 723-742

[26] Menzel-Severing J et al. Implantation and explantation of an active epiretinal visual prosthesis: 2-year follow-up data from the EPIRET3 prospective clinical trial. Eye. 2012; **26**(4):501

[27] Stingl K et al. Subretinal visual implant alpha IMS–clinical trial interim report. Vision Research. 2015;**111**:149-160

[28] Eiber CD et al. Multipolar field shaping in a suprachoroidal visual prosthesis. IEEE Transactions on Neural Systems and Rehabilitation Engineering. 2017;**25**(12):2480-2487

[29] Nishida K et al. Visual sensation by electrical stimulation using a new direct optic nerve electrode device. Brain Stimulation: Basic, Translational, and Clinical Research in Neuromodulation. 2015;**8**(3):678-681

[30] Jawwad A et al. Modulating lateral geniculate nucleus neuronal firing for visual prostheses: A Kalman filter-based strategy. IEEE Transactions on Neural Systems and Rehabilitation Engineering. 2017;**25**(10):1917-1927

[31] Lewis PM et al. Restoration of vision in blind individuals using bionic devices: A review with a focus on cortical visual prostheses. Brain Research. 2015;**1595**:51-73

[32] Selkirk SM et al. Feasibility of restoring walking in multiple sclerosis with multichannel implanted electrical stimulation. American Journal of Physical Medicine & Rehabilitation. 2017;**96**(9):e170-e172

[33] Khamis S et al. Is functional electrical stimulation an alternative for orthotics in patients with cerebral palsy? A literature review. European Journal of Paediatric Neurology. 2018; **22**(1):7-16

[34] Hausmann J et al. Functional electrical stimulation through direct 4-channel nerve stimulation to improve gait in multiple sclerosis: A feasibility study. Journal of Neuroengineering and Rehabilitation. 2015;**12**(1):100

[35] Howlett OA et al. Functional electrical stimulation improves activity after stroke: A systematic review with meta-analysis. Archives of Physical Medicine and Rehabilitation. 2015;**96**(5):934-943

[36] Parissis J et al. Functional electrical stimulation of lower limb muscles as an alternative mode of exercise training in chronic heart failure: Practical considerations and proposed algorithm. European Journal of Heart Failure. 2015;**17**(12):1228-1230

[37] Ho CH et al. Functional electrical stimulation and spinal cord injury. Physical Medicine and Rehabilitation Clinics. 2014;**25**(3):631-654

[38] de Groat WC, Griffiths D, Yoshimura N. Neural control of the lower urinary tract. Comprehensive Physiology. 2015;**5**(1):327-396

[39] Fowler CJ, Griffiths D, De Groat WC. The neural control of micturition. Nature Reviews Neuroscience. 2008;**9**(6):453

[40] Maynard FM Jr et al. International standards for neurological and functional classification of spinal cord injury. Spinal Cord. 1997;**35**(5):266

[41] Corcos J, Ginsberg DD, Karsenty G. Textbook of the Neurogenic Bladder. Boca Raton: CRC Press; 2015

[42] McGee MJ, Amundsen CL, Grill WM. Electrical stimulation for the treatment of lower urinary tract dysfunction after spinal cord injury. The Journal of Spinal Cord Medicine. 2015;**38**(2):135-146

[43] Gaunt RA, Prochazka A. Control of urinary bladder function with devices: Successes and failures. Progress in Brain Research. 2006;**152**:163-194

[44] Brindley G. The first 500 patients with sacral anterior root stimulator implants: General description. Spinal Cord. 1994;**32**(12):795

[45] Shafik A et al. Surgical anatomy of the pudendal nerve and its clinical implications. Clinical Anatomy. 1995;**8**(2):110-115

[46] Mahakkanukrauh P, Surin P, Vaidhayakarn P. Anatomical study of the pudendal nerve adjacent to the sacrospinous ligament. Clinical Anatomy. 2005;**18**(3):200-205

[47] Gustafson KJ et al. Fascicular anatomy and surgical access of the human pudendal nerve. World Journal of Urology. 2005;**23**(6):411-418

[48] Lee Y-H, Creasey GH. Self-controlled dorsal penile nerve stimulation to inhibit bladder hyperreflexia in incomplete spinal cord injury: A case report. Archives of Physical Medicine and Rehabilitation. 2002;**83**(2):273-277

[49] Peters KM, Feber KM, Bennett RC. Sacral versus pudendal nerve stimulation for voiding dysfunction: A prospective, single-blinded, randomized, crossover trial. Neurourology and Urodynamics. 2005;**24**(7):643-647

[50] Spinelli M et al. A new minimally invasive procedure for pudendal nerve stimulation to treat neurogenic bladder: Description of the method and preliminary data. Neurourology and Urodynamics. 2005;**24**(4):305-309

[51] Yoo PB et al. Pudendal nerve stimulation evokes reflex bladder contractions in persons with chronic spinal cord injury. Neurourology and Urodynamics. 2007;**26**(7):1020-1023

[52] Horvath EE et al. Conditional and continuous electrical stimulation increase cystometric capacity in persons with spinal cord injury. Neurourology and Urodynamics. 2010;**29**(3):401-407

[53] Kennelly MJ et al. Electrical stimulation of the urethra evokes bladder contractions in a woman with spinal cord injury. The Journal of Spinal Cord Medicine. 2010;**33**(3):261-265

[54] Yoo PB et al. Multiple pudendal sensory pathways reflexly modulate bladder and urethral activity in patients with spinal cord injury. The Journal of Urology. 2011;**185**(2):737-743

[55] Kennelly MJ et al. Electrical stimulation of the urethra evokes bladder contractions and emptying in spinal cord injury men: Case studies. The Journal of Spinal Cord Medicine. 2011;**34**(3):315-321

[56] Kirkham A et al. The acute effects of continuous and conditional neuromodulation on the bladder in spinal cord injury. Spinal Cord. 2001;**39**(8):420

[57] Knight SL et al. Conditional neuromodulation of neurogenic detrusor overactivity using transrectal stimulation in patients with spinal cord injury: A proof of principle study. Neurourology and Urodynamics. 2018;**37**(1):385-393

[58] Wenzel BJ et al. Detection of neurogenic detrusor contractions from the activity of the external anal sphincter in cat and human. Neurourology and Urodynamics. 2006; **25**(2):140-147

[59] Vodusek DB, Light JK, Libby JM. Detrusor inhibition induced by stimulation of pudendal nerve afferents. Neurourology and Urodynamics. 1986;**5**(4):381-389

[60] Craggs M. Neuromodulation device for pelvic dysfunction. Google Patents; 2014

[61] Brindley G, Rushton D, Craggs M. The pressure exerted by the external sphincter of the urethra when its motor nerve fibres are stimulated electrically. BJU International. 1974;**46**(4):453-462

[62] Shiraz AN et al. Minimizing stimulus current in a wearable pudendal nerve stimulator using computational models. IEEE Transactions on Neural Systems and Rehabilitation Engineering. 2016;**24**(4):506-515

[63] Peng Y et al. Modern theories of pelvic floor support. Current Urology Reports. 2018; **19**(1):9

[64] Shiraz A et al. Design of sEMG assembly to detect external anal sphincter activity: A proof of concept. Physiological Measurement. 2017;**38**(11):L17

Chapter 5

Wearable Dialysis: Current State and Perspectives

Nikolai Bazaev, Nikita Zhilo, Viktor Grinval'd and Sergey Selishchev

Additional information is available at the end of the chapter

http://dx.doi.org/10.5772/intechopen.75552

Abstract

For more than four decades, scientists and engineers are trying to miniaturise dialysis machines to make them wearable. There are many reasons for that—from increased biocompatibility and cost-efficiency to longer life expectancy and higher quality of life. That can be achieved by continuous blood treatment like in natural kidneys, which softly filter blood for 168 h a week when hemodialysis does that quickly—for approximately 20 h a week, which affects the organism in a bad way. Along with that, during hemodialysis, the patient must be near the dialysis machine, in contrary to wearable apparatus that can be carried anywhere. To achieve these advantages, dialysis fluid regeneration system must be developed, and it is a problem to be solved in the next few years. In this chapter, we describe current prototypes of wearable artificial kidneys, their design principles and results of our investigations.

Keywords: wearable health, dialysis, wearable artificial kidney, personalised medicine, dialysate regeneration

1. Introduction

In case of renal failure (RF), products of metabolism remain in blood and cells, and excessive fluid is not removed from the body. There are two types of RF—acute renal failure (ARF), which is often reversible with proper treatment, and chronic kidney disease (CKD), which is usually irreversible. The causes of CKD are diverse, most often diabetes mellitus, high blood pressure, glomerulonephritis, and so on. In **Figure 1**, incidence and accidence of terminal stage of CKD in several countries is shown.

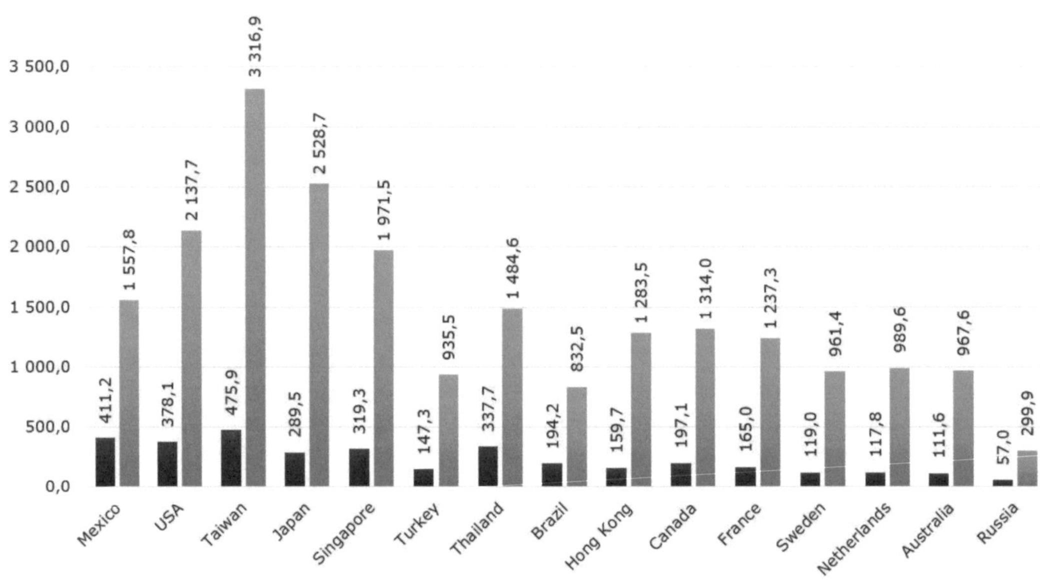

Figure 1. Incidence and prevalence of terminal stage CKD in the world by USRDS Atlas 2013 [1].

The only way of adequate treatment of CKD is kidney transplantation. Renal replacement therapy (RRT) is an artificial blood purification process that is achieved by several methods that do not cure kidney failure but keep living organism of patients for years while there will be a possibility to carry out a kidney transplantation. ARF is also treated via artificial blood purification. Artificial blood purification (or dialysis) is a process of removing metabolites and excess fluid from patient's organism. Dialysis does not replace all kidney functions (e.g., endocrine function), and in some cases, patients require additional therapy (e.g., hormonal). Clinical tasks of renal replacement therapy are to purify the body, remove excess fluid (Ultrafiltration, or UF) and maintain electrolyte balance (concentration of K^+, Na^+, Ca^{2+} ions, etc.).

Methods of dialysis blood purification are divided into two categories: extracorporeal methods (hemodialysis or HD) and intracorporeal methods (peritoneal dialysis or PD). Extracorporeal methods perform blood pumping through a dialyser, in which the metabolic products pass across membranes into dialysate solution due to diffusion and convection. In case of intracorporeal methods, peritoneal (abdominal) cavity represents itself a semi-permeable membrane across which the mass transfer is carried out due to diffusion and osmosis. For HD and PD, different dialysis solutions are used. Generally, dialysate consists of highly purified water, sodium, potassium, magnesium, calcium and chlorine ions in concentrations close to their concentrations in the blood. In case of PD, dialysis fluid must also contain an osmotic substance like glucose or icodextrin. There are other differences in pH, buffers, and so on.

Since the peritoneal membrane has a thickness of up to 1000 μm, the efficiency of peritoneal dialysis is significantly lower than the efficiency of hemodialysis because dialyser membranes

http://dx.doi.org/10.5772/intechopen.75552

are about 200 μm thick. Therefore, renal replacement therapy should be carried out continuously: abdominal cavity must be filled with a dialysis solution for long periods, where it is saturated with metabolites within a few hours, after which solution is replaced with fresh one. Continuity is the main advantage of peritoneal dialysis, as the procedure for purifying the blood proceeds more physiologically, the cardiovascular system experiences less additional stresses, and the residual function of the kidneys remains longer.

Comparison between HD and PD is presented in **Table 1**.

For more than 40 years, scientists and engineers are developing technologies that will enable to create a wearable artificial kidney (**WAK**). WAK has potential to overcome the negative sides of existing methods of RRT. There are some advantages of WAK over HD:

- increasing patient's mobility due to the reduction of its size and weight;
- adjusting blood purification to human physiology (metabolite elimination rate is closer to metabolite production rate);
- reduction of water consumption (2l of dialysate per one procedure vs. 120–150 l for hemodialysis);
- cost reduction (~€ 45,000 annually using WAK [2] vs ~€ 59,600 annually for continuous ambulatory peritoneal dialysis [3]);
- and the following advantages over continuous ambulatory peritoneal dialysis:
- lowering frequency of peritonitis due to rare dialysate exchange (once a day);
- gaining life quality due to the ability to use WAK at work or while travelling;
- reduction of water consumption (2 l of dialysate vs 8 l for 1 day of CAPD).

Dialysis type	Pros	Cons
Peritoneal dialysis	Keep residual kidney function longer	Necessity of peritoneal catheter implantation 2 weeks before the procedures start
	Home dialysis	Existing solutions degrade peritoneal membrane within several years (5+)
	Dialysis for patients with low mass index	Passive source of glucose, it is necessary to monitor the intake of carbohydrates
	Fewer amounts of dialysate are used (cheaper method)	
Hemodialysis	Dialysis takes 3–4 h for 3 days	Rigid schedule of procedures
		Spent time: about 20 h per week
		Necessity to create a vascular access
		Expensive method (up to 85 k€/year/patient)

Table 1. Comparison between HD and PD.

2. WAK design principles

WAK must comply with several requirements, including:

- safety and biocompatibility (WAK must be equipped with temperature, volume, pH, ionic solution sensors, and must have a system for bacterial contamination prevention);
- ease of use (the must be light and ergonomic);
- reliability (the device must function for a long period);
- availability (use of the device should be cheaper than traditional PTA methods);
- continuous power supply (the apparatus must provide continuous operation from the battery for at least 8 h);
- portability or implantability (the device should not significantly reduce patient mobility);
- elimination of substances normally removed by the kidneys (the apparatus should provide an adequate level and speed of removal of uremic toxins). The history of WAK development shows that HD was chosen as the major method. However, PD seems to be a more promising method for a WAK. Comparison between PD- and HD-based WAKs is presented in **Table 2**.

For the last several years, PD is chosen as the main method for WAK development in Europe. Typical schemes of WAKs are presented in **Figure 2**.

Implementation of HD as a base for WAK development is connected to the necessity of anticoagulant infusion to prevent blood clotting in the dialyser. Besides that in case of HD, tubing

Dialysis type/ characteristics	WAK	
	PD-based	HD-based
Weight and dimensions	2.5…5 kg, waist bag, shoulder bag or backpack	
Usage	Home dialysis, workplace dialysis, nocturnal dialysis, 24/7 dialysis	Short day dialysis, nocturnal home dialysis
Preparative operations	Peritoneal catheter implantation	Arteriovenous fistula formation
Eliminated substances	Small, medium and large molecules removal	Small and medium molecules removal
Infection risk	Peritonitis possible if solution changed frequently	Blood infection possible
Used with	Osmotic agent (glucose/dextrose/icodextrin/amino acids)	Anticoagulant (heparin)
Annual Cost	~€ 45,000 [3]	

Table 2. Comparison between PD- and HD-based WAKs.

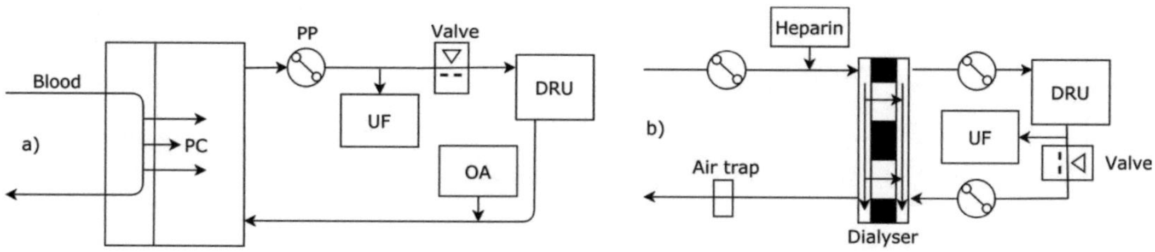

Figure 2. Typical WAK schemes based on a) peritoneal dialysis (AWAK Technologies), b) hemodialysis (nanodialysis); PP—peristaltic pump, DRU—dialysate regeneration unit, PC—peritoneal cavity, UF—ultrafiltrate, OA—osmotic agent.

sets must have a dialyzer and an air trap, while PD-based WAK can avoid using such elements. HD-based WAKs must also have means to fix blood catheters so that they do not disconnect from blood vessels due to physical activity. Therefore, the usage of HD-based WAK is possible only in clinics. Conversely, WAK on PD basis has potential to be used at home or in the office and consequently increase patients' quality of life.

Special technical aspects of PD-based WAK are connected to the necessity of keeping the initial concentration of osmotic agent in dialysis fluid during artificial blood purification.

Another aspect of WAKs is the method of ultrafiltration. Particularly, ultrafiltration in dialysis regeneration units on HD basis requires two peristaltic pumps to create transmembrane pressure between blood and dialysate compartments of the dialyser. In case of PD-based WAK, drainage of excess fluid occurs when spent dialysate exits patients' body and technically only one peristaltic pump is needed (**Figure 2**).

3. Dialysis regeneration methods

Regeneration of spent dialysate in WAK occurs in dialysis regeneration unit; its structure can be different, as it can be seen in **Figure 3**. Ideal WAK must eliminate all metabolites from spent dialysis solution, but this is difficult to implement and validate. It means that dialysis fluid that comes out of dialysis regeneration unit must be chemically equal to fresh dialysis fluid. However, at the current stage of WAK evolution, this cannot be achieved. In these conditions, we can outline substances that are crucial to being eliminated from the dialysis fluid. They are urea, creatinine, uric acid, phosphates, p-cresol, and potassium. An important aspect of dialysis regeneration is maintaining an ionic compound of dialysis fluid, including concentrations of Ca^{2+}, Mg^{+}, K^{+}, Cl^{-}.

Sorption is the most widespread method for eliminating a wide range of metabolites. This method is easy to implement, but its urea and other small molecules elimination capability is poor, and therefore it must be combined with effective urea removal method. Research and development of new sorbents can increase dialysis efficiency and possibly make WAK usage cheaper. There are several well-known methods for urea elimination.

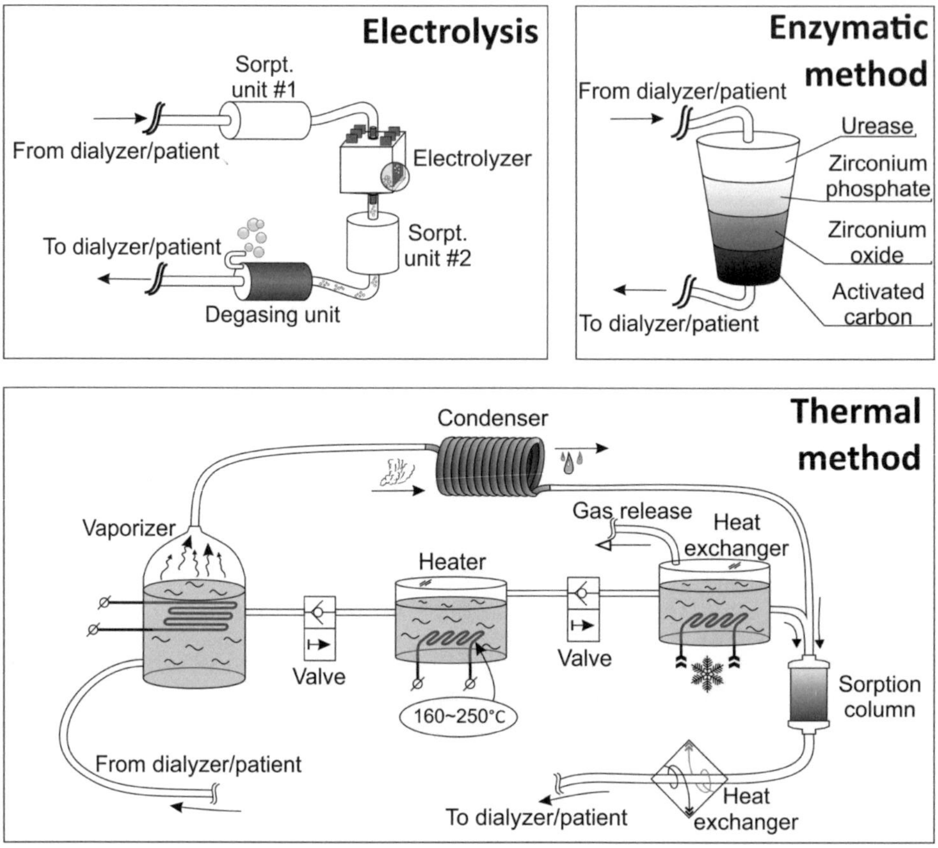

Figure 3. Types of dialysis regeneration units.

1. Enzymatic method is using animal- or plant-derived urease ferment to hydrolyse urease and reduce it to carbon dioxide and ammonia; its combination with sorption is used in many WAK prototypes. In our work, we extracted a plant-derived enzyme from jack beans. For the experimental verification, a sorption column was constructed (**Figure 4** top left). The column was tested on a model solution, which consisted of peritoneal dialysis solution (1 l), urea (~36 mmol), creatinine (~820 mmol/l), uric acid (~700 mmol/l). Model solution circulated through the column with use of a peristaltic pump with a flow speed of 100 ml/min. Metabolites' and ions concentrations were measured each hour for 8 hours. Experimental results are presented in **Figures 4** and **5**. After 8 h, this unit eliminated 7.15 g of urea, 2.08 g of creatinine and 0.4 g of uric acid.

The advantage of this method is its ease of use. Expendable materials, in this case, consist of a tubing set with sorption element that can be easily replaced by the patient. Drawbacks of the method are: complexity of storage and preparation of immobilised urease (up to 1 month at 4°C), high cost of the method (expensiveness of zirconium phosphate), presence of aluminium in the dialysis solution, the short lifetime of the sorption element (4–6 h) as well as an increase of pH, that leads to necessity of buffers infusion.

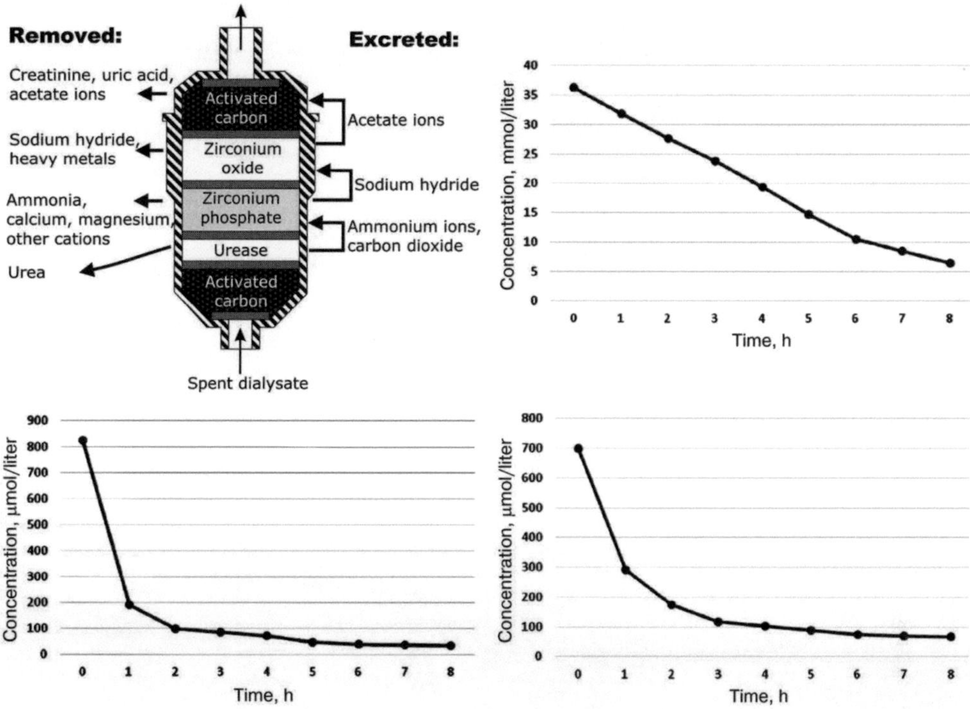

Figure 4. Top left—Sorption column structure: Top right—Urea concentration dynamics; bottom left—Creatinine concentration dynamics; top right—Uric acid concentration dynamics.

The electrochemical method uses electrolysers with special electrodes to electrolyse urea and other metabolites but produces by-products, which can be removed by activated carbons

Urea can be oxidised either directly on the anode (Eq. (1)) [4, 5] or in solution by interacting with the hypochlorite ion released at the anode (Eqs. (2)–(4)). However, toxic chlorine-containing compounds and free chlorine can accumulate in solution during electrolysis. Free chlorine is formed because of the interaction of the chlorine ion with water.

$$(NH_2)_2\,CO + H_2O \rightarrow N_2\uparrow + CO_2\uparrow + 6\,H^+ + 6\,e^- \tag{1}$$

$$2\,Cl^- \rightarrow Cl_2\uparrow + 2\,e^- \tag{2}$$

$$Cl_2 + H_2O \rightarrow HOCl + HCl \tag{3}$$

$$(NH_2)_2\,CO + 3\,OCl^- \rightarrow N_2\uparrow + 3\,CO_2\uparrow + 3\,Cl^- + 2\,H_2O \tag{4}$$

The main requirement for the electrolysis of urea is the choice of an effective and safe electrocatalyst. Studies of various electrode materials have shown that urea can be electrochemically

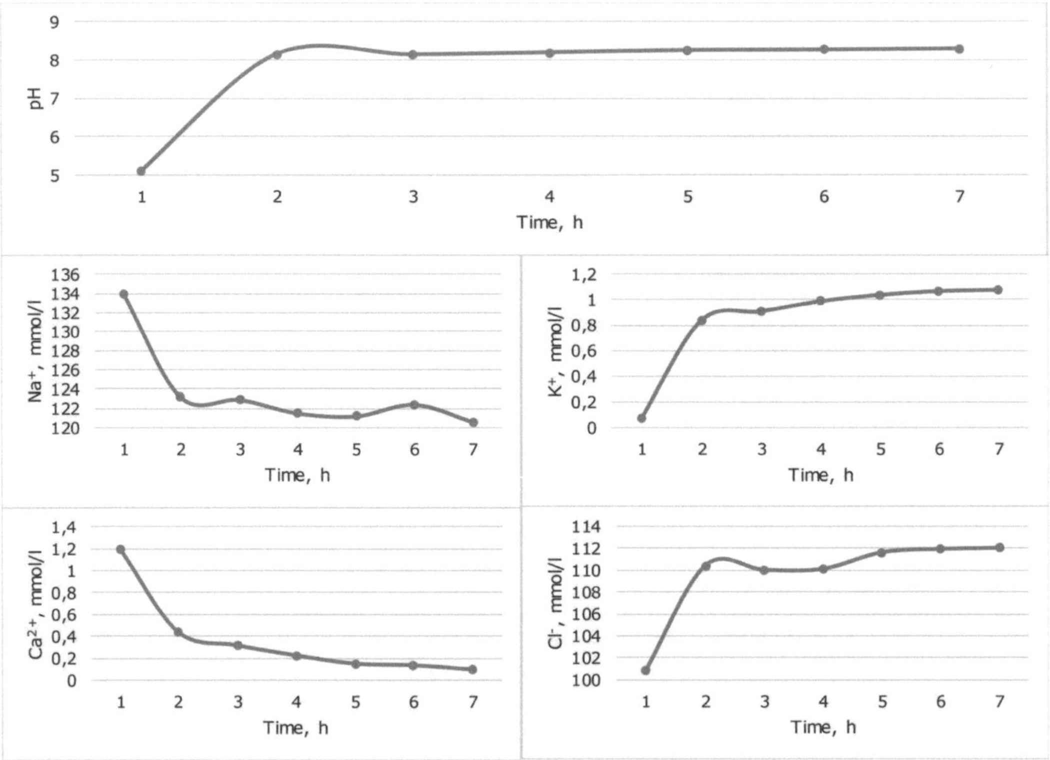

Figure 5. pH and ions' concentration dynamics during dialysate regeneration by urease.

oxidised in a neutral medium using catalysts made of noble metals such as Ru-TiO_2, Ti-Pt, Ti- (Pt-Ir) etc. However, the high cost of such materials is a significant obstacle to their wide practical application.

The possibility of electrochemical urea oxidation on electrodes made of platinum group metals was investigated. To determine the rate of urea removal, depending on the type of electrode, a series of experiments were performed with the following parameters: current density 5 mA/cm^2; initial urea concentration 30 ± 2 mmol/l; perfusion rate 100 ml/min; the distance between electrodes is 1 mm. As an object of study, next electrode types were chosen: graphite; platinum, electrochemically deposited on titanium (Ti-$Pt_{(ec)}$) or by explosion-rolling (Ti-$Pt_{(er)}$); rhodium, electrochemically deposited on titanium (Ti-$Rh_{(ec)}$); ruthenium, electrochemically deposited on titanium (Ti-$Ru_{(ec)}$), as well as electrodes from foamed coal ($C_{(foam)}$); silicon-carbon films deposited on titanium substrates by vacuum spraying, doped with molybdenum (Ti-$SiC_{(Mo)}$); silicon-diamond films deposited on titanium substrates (Ti-$SiC_{(diam)}$); platinum sprayed on titanium substrates (Ti-$Pt_{(spray)}$). The results of the experiment are shown in **Table 3**. Zero removal rate of urea is indicated for electrodes, the coating of which has dissolved before the end of the experiment.

Material	Urea elimination rate, mg/g	Anode surface area, cm^2	Specific urea elimination rate, mg/cm^2·h
Graphite	274	150	1.83
Ti-Pt$_{(ec)}$	73	150	0.49
Ti-Pt$_{(er)}$	70	150	0.47
Ti-SiC$_{(diam)}$	32	100	0.32
Ti-Rh$_{(ec)}$	30	100	0.30
C$_{(foam)}$	20	100	0.20
Ti-Ru$_{(ec)}$	0	100	0
Ti-SiC$_{(Mo)}$	0	50	0
Ti-Pt$_{(spray)}$	0	18	0

Table 3. Urea elimination rates by electrode material type.

As can be seen, the highest specific rate of urea elimination occurs on the graphite, Ti-Pt$_{(ed)}$ and Ti-Pt$_{(er)}$ electrodes. The operating time of the platinum-coated electrodes is limited due to the active transition of the coating to the solution: electrodeposited platinum has dissolved within 40 h; explosion-rolling platinum remained on the substrate for more than 200 h. Lifetime of graphite electrodes could not be determined. In connection with this, further investigations were carried out on graphite electrodes, since their coating proved to be the most stable.

For these electrodes, a number of additional experiments were carried out, the purpose of which was to study the effect of electrolysis on the ionic composition and the acid-base state of the solution. Results are presented in **Figure 6**. Electrolysis on graphite electrodes alkalises the solution and affects the ionic composition of the solution. The pH of the solution in 7 h of electrolysis is increased from 5.2 to 5.9, and an increase in the concentration of chloride ions and chlorine compounds (including sodium hypochlorite) is observed. In the course of the experiment, there was also a slight decrease in the calcium concentration, which is probably due to the formation of calcium hydroxide on the cathode surface. From the results obtained, it follows that the main advantage of the electrochemical method is the elimination of urea, but at the same time, electrolysis affects the acid-base state of the solution and its ionic composition. To eliminate this disadvantage, it is necessary to apply a post-treatment solution, namely sorption columns with activated carbon.

However, research and development of other perspective electrode materials (i.e. graphite and other carbon materials) is in progress.

2. Thermolysis is heating spent dialysate to the temperature of urea decomposition (around 150°C). This method cannot be used in WAK because of high-energy usage, and difficult to use even in the stationary setting because of many thermolysis by-products and caramelisation of glucose contained in the dialysate. In case of WAK usage, it is only usable in HD conditions.

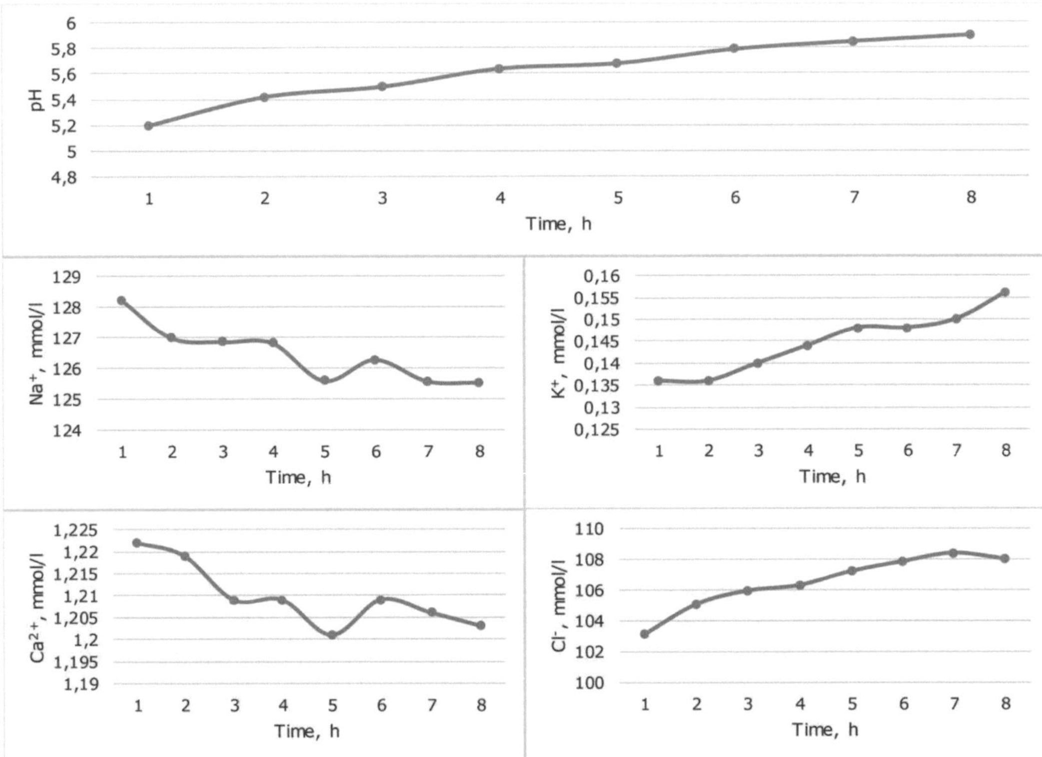

Figure 6. pH and ions' concentration dynamics during dialysate regeneration by electrolysis.

When the spent dialysate is heated to 100°C in a closed loop, organic dialysis products decompose to a gaseous state. For example, thermolysis of urea proceeds according to the following Eq. (5):

$$(NH_2)_2 CO + H_2 O \xrightarrow{\theta > 373°K} (NH_3)_2 CO_2 \xrightarrow{Carbonate} 2NH_3\uparrow + CO_2\uparrow \tag{5}$$

where Θ is the thermolysis temperature, °K.

The results of the studies [8] demonstrate that nitrogen-containing components (urea, uric acid, and creatinine) elimination rate exponentially depends on the temperature.

Thermal dialysate regenerator functioning can be represented by a sequence of steps. The first stage involves the filling of the thermal regenerator with dialysate and its subsequent heating up to a given thermolysis temperature Θ. The second stage is the regeneration of dialysate. The third stage is cooling the dialysate down and its removal from the thermal regenerator.

The results of the experimental testing of the thermal regenerator presented in [6] confirmed the possibility of using the proposed method for dialysate regeneration; urea elimination rate was 2.5–10.0 g/h in temperature range of 160–250°C, but this method was not further developed due to the significant energy consumption and the need for the thermal reactor to function at high excess pressures.

Considering all the above, it could be said that combinations of sorption + enzymatic method and sorption+ electrolysis are most perspectives to use in WAKs. Some of the current prototypes are presented below.

4. Prototypes of WAK

4.1. ViWAK PD (Vicenza University, Italy)

The device is implemented in accordance with the scheme a) of **Figure 2**. The device [7] is placed in a vest and is controlled wirelessly by a mobile device. The apparatus includes a system for spent dialysate regeneration based on the urease enzyme. The regeneration unit REDY (REcirculating DialYsis) includes urease, zirconium phosphate, zirconium oxide and activated carbon for the removal of urea, heavy metals, creatinine, uric acid and by-products of chemical reactions. The regeneration unit is designed to purify 12l of spent dialysate (about a day of continuous operation). ViWAK PD uses a double hollow catheter for intraperitoneal infusion of dialysate, which is then purified in an extracorporeal circuit. The device operates from an external battery for 10 h, while the mass of the device is about 200 g. The device is also equipped with a sterilising filter, a degasser, a pressure sensor and a rotary pump.

The apparatus does not perform ultrafiltration (removal of excess fluid from the body), which reduces the applicability of the device in patients with zero residual renal function.

4.2. WAK (University of California, Los Angeles, USA)

Portable hemodialysis apparatus [8] implements HD according to scheme b) of **Figure 2**. The device consists of two sections:

1. A blood transport section in which the patient's blood moves to the dialyzer along the arterial line and then returns to the patient's cardiovascular system;
2. The dialysate transport section, where the dialysis solution enters the mass transfer device and then moves through the regeneration system, at the same time it is cleared of accumulated toxins and saturated with sodium bicarbonate. The apparatus also has pumps for controlling the flow of anticoagulant and for performing ultrafiltration.

The device is made in the form of a belt and weighs 5 kg. The device consists of four pumps, powered by external batteries (provide continuous operation for 8 h), and regulate the removal and addition of fluids in the circuits through the blood and dialysate. The dialysate is continuously regenerated by passing through three containers with sorbents containing urease, activated carbon, zirconium oxide and zirconium phosphate (REDY system).

Initial testing of the device was performed on eight patients, and an average urea removal rate of 1.6 mmol/h, creatinine 1.2 mmol/h was obtained.

4.3. SORB (Fresenius Medical Care, Waltham, USA)

The device [9] realises the PD method according to scheme a) of **Figure 2**. The device is a belt with a mass of 2 kg, in which the dialysate moves through a series of sorption containers, along hollow fibres, on the outside of which there is a sorbent absorbing phosphate, organic substances and ammonium from the volume of dialysate, and urea decomposition by enzymatic method (using urease) with subsequent removal of the reaction products.

Among the shortcomings of this device can be identified the lack of ultrafiltration, as well as the absorption of calcium and magnesium from the dialysate, which requires the use of an infusion pump to return them to the dialysate.

4.4. WABPU (ZITC, Moscow, Russia)

The device [10] implements the PD method according to scheme c) of **Figure 2**. The device is a 3.5 kg backpack containing a hydraulic circuit that realises recirculation and dialysate regeneration and an electrical circuit that realises the control of the procedure, the system as a whole, and communication with the smartphone displaying user interface. The functional diagram of the device is shown in **Figure 7**.

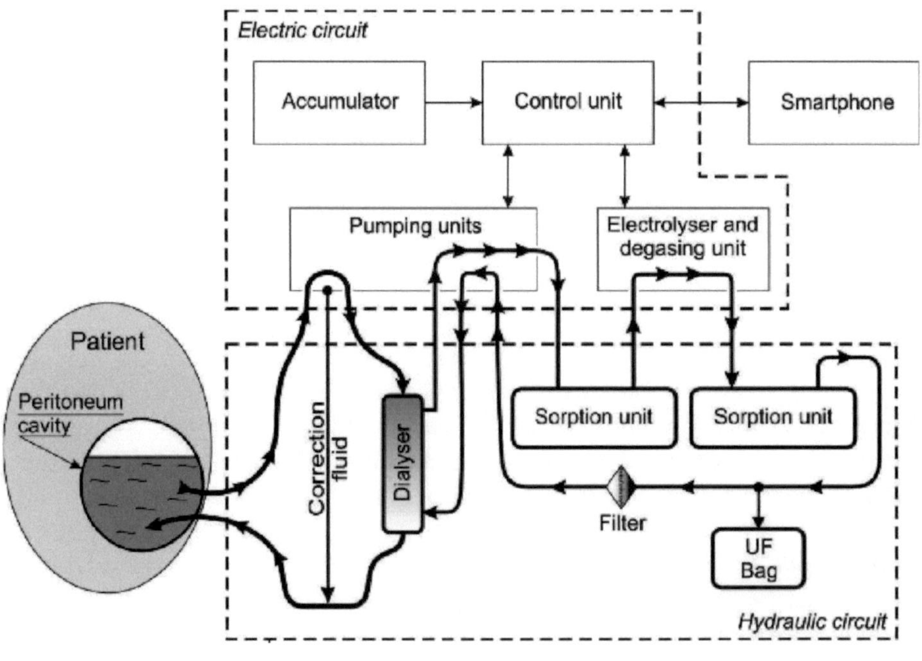

Figure 7. Functional diagram of WABPU: UV—ultrafiltrate [12].

Regeneration of the dialysate solution is carried out by sorption of metabolites by activated charcoal and electro-oxidation of urea in an electrolytic cell. The combination of the material of electrodes and sorbents allows maintaining the pH of the dialysate at a constant level (7.1 … 7.3), but slightly changes the ion composition, which is solved by the selection of specialised sorbents and ion exchange resins. The device includes a control unit, a battery (provides continuous operation for 8 h), pump modules, a dialysate regeneration unit, a trunk and a smartphone. The regeneration unit includes an electrolyzer, a degasser and two sorption columns with activated carbon. The device was successfully tested on an animal model. Now the device is in preparation for clinical trials.

4.5. WAKD (nephron + project, Netherlands)

The apparatus [11] uses a combination of sorption and electrolysis of the spent dialysate [12]. The concept of the device was initially based on the method of hemodialysis, but in the last 2 years, it was changed in favour of peritoneal dialysis. The device is made in the form of a block, located on the belt. Management is carried out using a smartphone. The device assumes the presence of a special pump for introducing a glucose solution to maintain the concentration of the osmotic agent in the dialysate during dialysis to maintain the patient's fluid balance. The hemodialysis-based apparatus underwent preclinical tests on animals. WAKD weighs 3.2 kg, but the developers claim about its possible miniaturisation to 2.5 kg.

A comparison of the existing WAK prototypes devices is presented in **Table 4**.

Name	Type	Regeneration method	Characteristics	Current status
ViWAK [7]	PD	Sorption + urease	0.2 kg, 17 × 8 × 3 cm, 10 h of battery life	The prototype was not further developed
The WAK [8]	HD	Sorption + urease	5 kg, on-waist wearable, 8 h of battery life, urea removal rate 1.6 mmol/h, creatinine 1.2 mmol/h	Pre-clinical trials are under way
EO NAIP [10]	PD	Sorption + electrolysis	3.5 kg, backpack, urea removal rate 1.2 g/h, creatinine 0.3 g/h	Preclinical tests passed
SORB [9]	PD	Sorption + urease	2 kg, on-waist wearable	The prototype was not further developed
AWAK [12]	PD	Sorption + urease	1 kg, shoulder bag or vest, 16 h of battery life, expendables exchange every 7 or 12 h	Pre-clinical trials are under way
WAK-MAN [13]	HD	Sorption	Vest, 24 h of battery life,	The prototype was not further developed
WAKD [11]	HD and PD	Sorption + electrolysis	3.2 kg, on-waist wearable, up to 30 ml of urea clearance	Pre-clinical trials are under way

Table 4. Current WAK prototypes.

Table 4 shows that the current state of development of equipment for artificial blood purification allows you to count on their clinical use in the next 2–5 years.

Wearable equipment for artificial blood purification will overcome the shortcomings of existing apparatus and methods of dialysis and is one of the most promising areas in the field of biomedical engineering of artificial organs.

5. General problems of WAK development

Osmotic agents are added to the peritoneal dialysis solution to remove excess fluid from the body. They make the solution hyperosmolar in comparison with extracellular fluid, which leads to the removal of the liquid through ultrafiltration. The ideal osmotic agent must be cheap, biocompatible, with a sufficiently small molecular mass for changing the viscosity of the RPD, but with a sufficiently large molecular mass to not be absorbed into the bloodstream. The list of potential osmotic agents examined includes glucose, glycerol, xylitol, sorbitol, fructose, mannitol, gelatine, glucose polymers, polypeptides and dextrans. Most of them turned out to be unsuitable due to side effects. Relative success was seen with the use of glycerine, amino acids and glucose polymers, but only glucose, glucose polymers and amino acids are used in everyday clinical practice.

The main disadvantage of low molecular weight osmotic agents is their rapid absorption, resulting in a loss of ultrafiltration speed and metabolic disturbances. Large, loosely absorbed osmotic agents theoretically allow the ultrafiltration to last longer. The driving force, in this case—the colloid osmotic pressure, is determined to a greater extent by the number of osmotically active macromolecules than by the osmotic gradient. This means that osmosis can be maintained even with the use of peritoneal dialysis solution with osmolarity close in value to the osmolarity of extracellular fluid.

6. Conclusion

Development of WAK is held all over the world. The state of the art is that some of them went through animal trials and undergo clinical trials. Most probably first commercial WAKs will penetrate the market in 2–3 years. However, there is still very much work to be done because the tendency is in achieving the performance of native kidney. In the first stage, WAKs must be cheaper than HD and PD that will be the major factor to spread them and that will be a very good option to raise the number of patients with CKD who receive RRT.

Conflict of Interest

This work was partly financed by the Ministry of Education and Science of the Russian Federation (project ID RFMEFI57917X0152, Contract № 14.579.21.0152, 26.09.2017).

Author details

Nikolai Bazaev[1]*, Nikita Zhilo[2], Viktor Grinval'd[1] and Sergey Selishchev[3]

*Address all correspondence to: bazaev-na@yandex.ru

1 Zelenograd Innovative-Technological Centre, I.M Sechenov First Moscow State Medical University, Moscow, Russia

2 Zelenograd Innovative-Technological Centre, National Research University of Electronic Technology (MIET), Moscow, Russia

3 Zelenograd Innovative-Technological Centre, Moscow, Russia

References

[1] United States Renal Data System. 2013 Atlas of CKD & ESRD [Internet]. 2013. Available from: https://www.usrds.org/atlas13.aspx [Accessed: 20-01-2018]

[2] Nanodialysis. Nanodialysis Miniature Dialysis System brochure [Internet]. Available from: http://www.nanodialysis.nl/media/Nanodialysis_Miniature_Dialysis_System_Brochure_2015.pdf [Accessed 27-03-2017]

[3] Eriksson JK, Neovius M, Jacobson SH, Elinder CG, Hylander B. Healthcare costs in chronic kidney disease and renal replacement therapy: A population-based cohort study in Sweden. BMJ Open. 2016;**6**(10):e012062

[4] Koster K, Wendt H, Gallus J, Krisam G, Lehmann HD. Regeneration of hemofiltrate by anodic oxidation of urea. Artificial Organs. 1983;**7**:163-168

[5] Grinval'd V, Leshchinskii GM, Rodin VV, Strelkov SI, Yakovleva AA. Development and testing of a unit for electrochemical oxidation of products of hemodialysis. Biomedical Engineering. 2003;**37**:67-72

[6] Grinvald VM, Zalko GA, Mikhailov YuN. Method for cleaning dialysis solution in the apparatus "artificial kidney". RU Patent No. 2008927. Mar. 1. 2011

[7] Ronco C, Fecondini L. The vicenza wearable artificial kidney for peritoneal dialysis (ViWAK PD). Blood Purification. 2007;**25**:P.383-388

[8] Gura V, Rambod E. Wearable Continuous Renal Replacement Therapy Device. US Patent No. 7,896,829 B2. Mar. 3. 1991

[9] Ofsthun NJ, Stennett AK. An integrated membrane/sorbent PD approach to a wearable artificial kidney. IFMBE Proceedings. 2009;**25**:729-732

[10] Bazaev NA, Grinvald VM, Selishchev SV, Kalinov AV, Kozachuk AV, Kosatkin VV, Tyunder FF, Federyakin DV. Experimental research of wearable artificial kidney (In

Russ.). Russian Journal of Transplantology and Artificial Organs. 2017;**19**(3):46-52. DOI:10.15825/1995-1191-2017-3-46-52

[11] Sorbent system for blood purification. Nanodialysis. [Internet]. Available from: http://www.nanodialysis.nl/sorbents/ [Accessed: 01-20-2018]

[12] Lee DB, Roberts MA. Peritoneal-based automated wearable artificial kidney. Clinical and Experimental Nephrology. 2008;**12**:171-180

[13] Ronco C, Davenport A, Gura V. The future of the artificial kidney: Moving towards wearable and miniaturized devices. Nefrología. 2011;**31**:9-16

Section 2

New Materials

Chapter 6

Smart Materials for Wearable Healthcare Devices

Han Jin, Qinghui Jin and Jiawen Jian

Additional information is available at the end of the chapter

http://dx.doi.org/10.5772/intechopen.76604

Abstract

Wearable devices seem to have great potential that could result in a revolutionary non-clinical approach to health monitoring and diagnosing disease. With continued innovation and intensive attention to the materials and fabrication technologies, development of these healthcare devices is progressively encouraged. This chapter gives a concise review of some of the main concepts and approaches related to recent advances and developments in the scope of wearable devices from the perspective of emerging materials. A complementary section of the review linking these advanced materials with wearable device technologies is particularly specified. Some of the strong and weak points in development of each wearable material/device are clearly highlighted and criticized.

Keywords: wearable device, smart materials, healthcare, diagnostic biomarkers, physiological index

1. Introduction

Wearable devices are apparatuses that can be worn directly on the skin in different parts of the body. These devices have gained considerable attention owing to their ease of collection of crucial information in real-time regarding a wearer's health, both continuously and non-invasively [1–7]. In contrast to the traditional non-invasive methodologies (e.g., X-ray, type-B ultrasonic, nuclear magnetic resonance (NMR)), non-invasive diagnosis implemented by wearable devices creates new opportunities for remote and continuous healthcare monitoring in non-clinical settings [5, 8], with the ability to detect developing diseases at intervals between routine examinations. The use of wearable healthcare devices also encourages people to take greater interest in their own healthcare in a more convenient and cheaper way, thereby improving their compliance.

Wearable devices are becoming smaller and more mobile with time, opening new alternatives to traditional ways that providers have interacted with patients, carried out tests, collected data and delivered treatments [9–11]. Wearables come in many forms; there are smart wristbands [12], watches [13], shirts [14], shoes [15], headbands [16], eyeglasses [17] and necklaces [18], among others. Most of them contain sensors that gather raw data that is fed into a database or software application for analysis. Analysis typically triggers a response that would, for example, alert a physician to contact a patient who is experiencing abnormal symptoms, or might send an appreciative message when an individual achieves a fitness or diet goal [11].

Recent insights into wearable devices have resulted in rapid development of various kinds monitoring different parameters, such as pressure/strain, body vital signs (e.g. heartbeat rate, respiration rate and temperature) and biomarkers that can be found either in body fluids (e.g. saliva, sweat and tears) or in skin odor and breath [19–23]. Among these reported wearable healthcare settings, skin-based wearable devices have considerable advantage in allowing simultaneous monitoring of multi-physiological indexes and biomarkers [19, 21, 22, 24]. By continuous or frequent detection of the level of physical markers (e.g. pressure pulses and body temperature) using different sensing techniques, wearable devices can provide one of the most comprehensive feedbacks on human health states [7, 8, 19, 22, 24–32]. Perspiration contains a diversity of chemical species (e.g. minerals, trace elements, lactic acid, urea and volatile organic compounds) that can be detected and monitored in sweat, which can assist in determining the health status of a person. A good example of the potential of sweat diagnostics is the detection of cystic fibrosis [33, 34], currently in clinical use [35]. However, small sample volumes, dilution of analytes and variability in the measurement of concentration are some of the challenges that are encountered using sweat for diagnosis. **Table 1** summarizes the different physiological indexes and diagnostic biomarkers using wearable devices. As inferred in **Table 1** for gathering broad information of the health status, the skin-based wearable devices should integrate multiple sensing features. Consequently, the performance of a wearable device depends heavily on the proper sensing characteristics of its integrated sensing parts.

Although many challenging issues are of concern, developing technologies proposed in the last few years significantly enhance the practicability of applying these wearable devices in clinic use and academic research. One of the main reasons for the advance of wearable devices is covertly the rapid emergence of new materials with interesting properties [19, 23, 24]. As a result, modern wearable devices are becoming cheaper, smaller and more reliable. Several reviews on these aspects have already been published [38–40]. The aim here is to present the current state of non-invasive skin-based wearable devices from the perspective of materials science and engineering. In view of these emerging skin-based wearable devices that are expected to meet diverse healthcare applications, smart materials used for these purpose should be compliant with the following basic criteria:

- **Highly sensitivity to subtle changes in the physiological index:** subtle change in the physiological index, for example, body temperature, may be an early indicator of health problems. For the purpose of catching very slight changes in the human body, stimuli-responsive materials used in skin-based wearables should be very sensitive. For instance,

Sample source	Physiological parameter	Sensing devices	Refs.
Temperature	• T_{body}[a]	• Resistometric sensor • FET[b]	[7, 8]
Odor	• VOCs[c]	• Resistometric • Colorimetric sensor	[19]
Pressure/strain	• Heart/respiration pulse • Motion posture	• Resistometric sensor • Capacitive sensor • Piezoelectric sensor	[24]
Sweat	• Metal ions • Lactate • Uric acid • pH • Ammonium • Chloride • Skin humidity	• Potentiometric sensor • Amperometric sensor • Voltammetric sensor • Resistometric sensor • Colorimetric sensor	[27, 29–32, 36, 37]

[a]T_{body} = Body temperature
[b]FET = Field effect transistor
[c]VOC = Volatile organic compound

Table 1. Typical non-invasive skin-based wearable healthcare devices.

thermal-responsive materials used in body temperature sensors should sense down to 0.1°C variation, obviating missing any sign of health risk.

- **Adequate detection limit:** stimuli and/or analytes obtained from human skin are generally at low level regime, for example, pressure of pulsing artery is typically within several hundreds of Pa, which gives the benchmark for designing wearable pulsing sensors. Consequently, an adequate detection limit is particularly required by the responsive materials for these stimuli and/or analytes. Since the level of analytes coming from human skin is usually lower than that of analytes obtained directly from tissue or body fluid (excluding sweat), detection limit of these smart materials dedicated in the design of skin-based wearables is even lower compared with those wearable counterparts developed for other uses (e.g. implantable devices, breath sample analyzer, etc.).
- **Compatible with human skin:** materials that are featured as flexible, stretchable and less irritating are especially to be considered in these skin-based wearables, largely due to their

unique biocompatibility with soft and moist skin. Flexible and stretchable elastomers are more suitable for fabricating strain-responsive devices.

Owing to the continuous increase in the essential requirements of and interest in, the skin-based wearables, an update on the relevant materials is presented in this review, with an emphasis on novel materials and sensing techniques, with regard to their advantages, limitations, challenges and future trends. A comprehensive description of the unique technique in accessing health conditions by monitoring skin biomarkers and physiological parameters has also been also particularly in focus.

2. Smart materials for wearable healthcare devices

2.1. Wearable temperature sensors

To meet the requirements of non-invasive and quantitative monitoring of body temperature, wearable temperature devices are required to be flexible, biocompatible, lightweight and highly sensitive within temperatures range 35–42°C [41, 42]. Several temperature-sensing devices have been invented in the past, but flexible wearable temperature-sensing devices (**Table 2**) are mainly limited to thermally resistance (resistometric) and thermally sensitive field-effect transistor (FET) transduction mechanisms. While the operation mode previously relied on resistance changes of the sensing elements according to the temperature change, the latter relies on detecting FET transfer characteristics as a result of thermal effects [7, 41–48, 50, 53].

The attractive advantages of low-cost, simple configuration and ease of mass production made thermal-resistance-based temperature devices the most popular as skin-mounted sensors [43–46, 48]. These devices relay thermal expansion and/or variation in temperature coefficient of resistance (TCR) induced by changing temperature [46]. An acceptable thermal expansion coefficient and TCR of metal and metal oxides have become available and have been used in the past decades [45, 48, 54]. A representative metal-based sensing element for tracking skin temperature is gold foil or nanoparticles that possess a desirable TCR [45, 54]. For TCR-based devices, gold has good ductility and can be deposited on a range of substrates, including glass, SiO_2 and flexible polymer [45, 54]. This allows the design of ultrathin [55, 56] and flexible wearable temperature devices suitable for a range of complex measurement under different conditions.

An example of a successful wearable temperature sensor is the epidermal device designed for continuous thermal monitoring of human skin (**Figure 1a**). With integrating 4 × 4 TCR (Cr/Au) sensor array on a thin elastomeric substrate, this ultrathin conformal device gives reliable performance and minor deviations [45, 61]. However, the sensing range for most of metal-based sensing elements is limited to relatively high temperatures (normally >42°C) because of their low thermal-response [46]. Inadequate sensitivity of metal-based sensing elements also brings the need for sophisticated measurement electronics for increasing signal-to-noise ratio. Hence, to increase the thermal response range and improve the sensitivity of these devices, and increase efficiently the signal-to-noise ratio, alternatives have been devised, one being

Sensing materials	Working principle	Sensitivity	Range of measured temperature	Refs.
Si nanoribbon	Resistometric	N.M[a]	37–58 (°C)/310.15–331.15 (K)	[41]
MWCNT/PVBC-Et_3N[b]	Resistometric	−0.004 K^{-1}	20–40 (°C)/293.15–313.15 (K)	[43]
MWCNT/SEB[c]	Resistometric	−0.028 K^{-1}	20–50 (°C)/293.15–323.15 (K)	[44]
Cr/Au	Resistometric	N.M.	23–31 (°C)/296.15–304.15 (K)	[45]
GNWS/PDMS[d]	Resistometric	0.214°C^{-1}	35–45 (°C)/308.15–318.15 (K)	[46]
Graphite/acrylate copolymers	Resistometric	N.M.	37–58 (°C)/310.15–331.15 (K)	[47]
Ni-filled PEO/PE[e]	Resistometric	0.3 V°C^{-1}	35–42 (°C)/308.15–315.15 (K)	[48]
SWCNT/self-healing polymer	Resistometric	N.M.	20–80 (°C)/353.15 (K)	[49]
R-GO/PU[f]	FET	0.09°C^{-1}	30–80 (°C)/303.15–353.15 (K)	[7]
R-GO	FET	6.7 nSK^{-1}	30–80 (°C)/303.15–353.15 (K)	[50]
P(VDF-TrFE)/$BaTiO_3$[g]	FET	N.M.	35–40 (°C)/308.15–313.15 (K)	[51]
Ag/pentacene	FET	N.M.	15–70 (°C)/288.15–343.15 (K)	[52]

[a]N.M. = Not mentioned.
[b]MWCNT/PVBC-Et_3N = Multi-wall carbon nanotubes/poly(vinylbenzyl chloride) derivative with triethylamine.
[c]MWCNT/SEB = Multi-wall carbon nanotubes/poly(styrene-b-(ethylene-co-butylene)-b-styrene).
[d]GNWS/PDMS = Graphene nanowalls/polydimethylsiloxane.
[e]Ni-filled PEO/PE = Ni-filled polyethylene oxide/polyethylene.
[f]R-GO/PU = Reduced graphene oxide/polyurethane.
[g](VDF-TrFE)/$BaTiO_3$ = Poly(vinylidenefluoride-trifluoro-ethylene)/$BaTiO_3$.

Table 2. Performance of typical examples in measuring epidermal temperature.

based on carbon nanotubes (CNTs) where resistance is strictly dependent on temperature. By dispersing proportional amounts of CNTs into specific polymer such as poly(styrene-b-(ethylene-co-butylene)-b-styrene; SEBS), flexible sensing composites had very high sensitivity to changes in ambient temperature from 20 to 40°C [43, 44].

Compared to thermal resistance-based devices, FET-based temperature devices have greater sensitivity, and are easily incorporated into integrated circuits of convenience in signal amplification [7, 42, 50]. By applying thermal-responsive materials as an active channel, electrode or gate dielectric layer into the FET structure, promising results in determining skin temperature have been achieved. Few examples of FET-based temperature sensors include graphene-based sensing elements, metal-polymer hybrids, inorganic-polymer hybrids [7, 42, 47, 50, 51, 53]. Owing to the appropriate and adjustable thermal activation energy of the charger carrier for these thermal-responsive materials, the detection limit of the FET-based temperature devices can reach milli-Kelvin [47]. The inherent characteristics of FET-based devices also help in the precise sensing signals, with a deviation of <0.1°C [47].

Inadequate thermal-response, unsatisfactory stability and significant interference from the ambient environment are all challenging issues for wearable temperature devices [43, 46, 50].

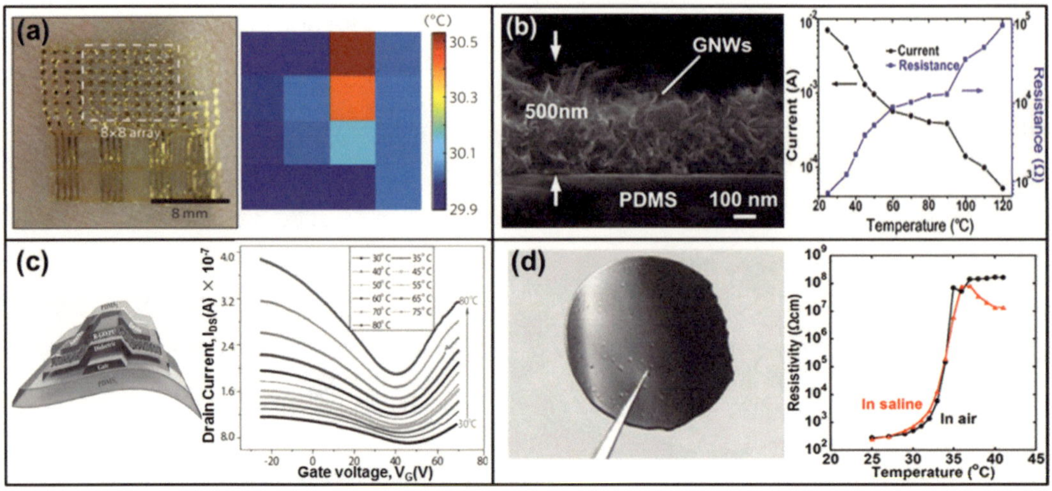

Figure 1. Schema of temperature sensors. (a) Image and the sensing performance of a 4 × 4 TCR sensor array after application to the skin using a water-soluble adhesive tape based on polyvinyl alcohol. (b) Cross-section SEM image of GNWs on PDMS and its "current and resistance versus temperature" curves. (c) Schematic of the TS-gated device and its response to temperature. (d) Photograph of a film of copolymer with graphite filler, with the temperature dependence of the resistivity of the temperature sensor. Reconstructed from Refs.[57–60].

Latterly, graphene nanowalls (GNWs) in combination with polydimethylsiloxane (PDMS) have been used as the sensing chip in a thermal-resistive sensor (**Figure 1b**) [46]. In comparison to conventional temperature sensors, this device has a sensitivity over 108.6 Ω $°C^{-1}$ between 35 and 45°C, which is sufficient to monitor human body temperature. Further evidence proves that the high response of the sensor can be attributed to the excellent TCR (0.214$°C^{-1}$) of GNWs and large expansion coefficient of PDMS [46]. Stability of these temperature devices has been many times reported, who found that the transition of morphology and phase of sensing elements during cooling-heating cycle is the main reason for irreproducibility of a thermal response [44, 48]. Adsorption of water vapor and/or ambient gas species is another crucial reason that generates hysteresis and causes significant interference [50]. Consequently, thermally stable materials are screened and extra encapsulation layers are added to decrease hysteresis, thereby improving stability and enhancing reproducibility (**Figure 1c**) [44, 48, 50]. Taking these materials from the lab to the technological/industrial benchmark would require advanced printing techniques. A graphite-polymer-based ultrathin thermometer of 15 μm thickness that could be accomplished by such simple print and press method is a good example for fabrication of large-area stretchable and transparent wearable temperature devices in the future. These fabricated thermometers should be able to measure temperature precisely between 25 and 50°C (**Figure 1d**) [47].

2.2. Wearable strain sensors

Wearable strain sensors could be useful for the detection and monitoring of movement-based signals, such as heartbeat and respiration rate. Generally, wearable strain devices should be lightweight, reliable, flexible and stretchable to match the mechanical properties of human skin. Sensitivity of these strain sensors must also be in line with their diverse healthcare applications. For instance, mean value of feet pressure for adults mostly ranges from 140 to 868 kPa [62], whereas systolic pressure (artery pulse) for healthy persons is within 0.67–5.32 kPa (5–40 mmHg) [63]. This suggests that high sensitivity in a low-pressure regime (perhaps <10 kPa [64]) is required for strain sensor trying to detect arterial pulse. To date, many working models for strain devices have been developed [65]. Among them are flexible and stretchable devices based on piezoresistive, piezocapacitive and piezoelectric principles seem to be closer to practical implementation, due to their relatively simple sensory configuration, uncomplicated read-out systems and acceptable dynamic performance [24, 66–68].

2.2.1. *Piezoresistive-based strain sensors*

Piezoresistive-based strain devices are typically composed of electrically conductive sensing films coupled with flexible substrates. When the structure of the device is deformed, changes in the microstructure of sensing films lead to changes in electrical resistance as a function of applied strain. Metal-based foil and graphitic-based sensing networks (e.g. InGaZn and graphene) are widely used in pressure/strain sensing [73–75]. In conductive networks, electrons can pass through overlapped nanomaterials within the percolation network. Stretching or bending in the conductive network results in disconnection between overlapped areas and thereby loss electrical connection; consequently, there is an increase of the electrical resistance of sensing film. Another known strain-responsive mechanism is crack propagation, that is, cracks originate and propagate in brittle thin films coated on the top of soft polymer layers upon stretching [20]. Generally speaking, the sensitivity (gauge factor) of the conventional metal foil-based strain device is in the range of 2–5 (a.u.) and semiconductor-based strain sensors might be 100 or greater. The value of sensitivity for flexible strain devices can be in the range of 1–100 s, depending on the sensing mechanisms, materials and micro/nanostructures [20].

To meet the demand of practical healthcare applications, a proof-of-concept wearable pressure device based on ultrathin Au nanowires (NWs) has been developed (**Figure 2a**) [76]. The Au NWs directly coated on a tissue paper can detect pressure forces as low as 13 Pa with high sensitivity (41.1 kPa^{-1}). Furthermore, the devices also provide fast response times (<17 ms) and high stability (450,000 loading–unloading cycles) [76]. A similar strategy has also been used in the design of piezoresistive strain devices comprised of other sensing elements, for example, graphene [74]. Furthermore, these prototypes give excellent performance in real-time monitoring of human blood pulses and detection of music motion to confirm their usefulness in healthcare applications [74, 76].

Strain sensors based on the piezoresistive conductive polymer (CP) and their composites have also been receiving considerable attention due to their excellent flexibility and biocompatibility [77]. These CP composites are usually based on polymeric matrix filled with stainless steel fibers (SSFs) and carbon nanotubes (CNTs) [57]. CP and their composites generally have robust piezoresitive behavior for a long period time under the temperature ranging from −50 to +200°C [57]. Note that their piezoresistivity is closely dependent on the conductive phase content. When this is low, piezoresistivity monotonically increases; in contrast, the piezoresistivity is monotonically decreasing at high conductive phase content [58]. Such flexible piezoresistive CP or CP composites are attractive candidates for wearable devices, for instance,

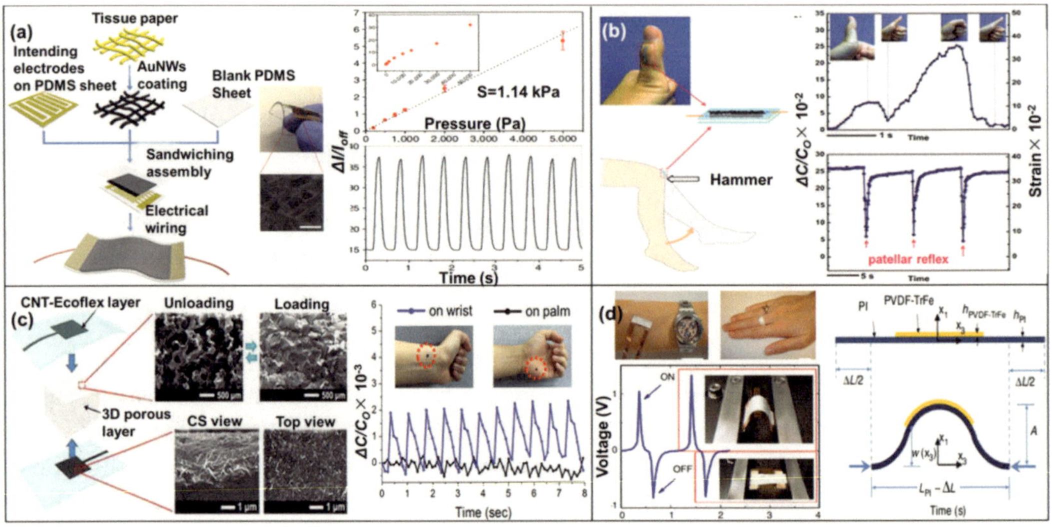

Figure 2. Strain-responsive devices based on emerging materials and fabricating technology. (a) AuNWs-coated tissue paper and (b) piezocapacitive-based strain device composed with AgNWs. (c) Three-dimensional (3D) microporous dielectric elastomer for a piezocapacitive strain device. (d) Piezoelectric devices based on aligned arrays of nanofibers of PVDF. Reconstructed from Refs. [69–72].

through coating poly(3,4-ethyl-enedioxythiophene–poly(styrenesulfonate) (PEDOT:PSS)/ polyurethane composite polymer on the spring-like compressible micro-pyramid-structured substrate, a stretchable strain sensor with high sensitivity has been developed [59]. The sensor had a sensitivity of 10.3 kPa^{-1} when stretched by 40%, which is sufficiently sensitive to detect human pulse pressure (23 Pa) [59]. Furthermore, its suitability in non-invasive healthcare applications is validated by showing excellent piezoresistivity and a fast response rate in measuring the primary features of the waveform in the human pulse.

It should be noted that beyond those conventional strain-responsive materials, stretchable silicon has received considerable attention from researchers because of its superior response to strain [60]. This silicon can be reversibly stretched and compressed to a large extent without damage, a new finding that might provide opportunities to design tunable structures that are compatible with mechanical strain [60]. Miniaturization and simplicity of configuration of the devices, as well as decreasing power consumption, is another challenging issue. Hence, a high-resolution unpixelated smart device was proposed for sensing stimuli. The sensing film of the device is constructed from two parallel GNP strips with antiparallel sensitivity gradients [78]. This design enables the device not only to detect the loaded strain, but also to discriminate the spatial position of the strain [78]. The gradient design of the sensing film can remarkably decrease the complexity and power consumption of a future strain-responsive device.

2.2.2. Piezocapacitive-based strain devices

Piezocapacitive-based strain devices use a highly compliant dielectric layer sandwiched between a pair of stretchable electrodes. Applied strain changes the distance between two electrodes, which results in a change of capacitance [24]. For instance, a piezocapacitive strain sensor composed of silver nanowire (AgNWs), Ecoflex, liquid metal and PDMS was built for finger flexing and knee motions among other body movement tracking (**Figure 2b**) [79]. Quick response (<40 ms) and good pressure mapping function was discovered during the examine process; and the designed piezocapacitive strain device has satisfactory wearability when mounted on the human body.

Besides these conventional strain devices, a transparent flexible pieocapactive pressure/strain sensor has been developed by Lipomi et al. [80]. The pieocapactive sensor, made by spraying transparent carbon nanotube on PDMS, gave a desirable performance in detecting pressure down to ~50 kPa [80]. Note that this transparent piezocapacitive skin-like strain sensor is extremely compliant mechanically, physically robust and easily fabricated, although it is less sensitive than its other counterparts.

In contrast to those piezoresistive counterparts, piezocapacitive strain devices have high linearity, stretchability and low hysteresis, but suffer from low sensitivity. Thus, a 3D microporous dielectric elastomer with giant piezocapacitive effect was used to construct a pressure device (**Figure 2c**) [81]. Due to the presence of micropores within the elastomeric dielectric layer, the resulting piezocapacitive pressure device is highly deformable by the minimal amount of pressure, leading to a marked increase in sensitivity. The gradual closure of micropores under compression also increased the effective dielectric constant, thereby enhancing sensitivity. The 3D microporous dielectric layer with serially stacked springs of elastomer bridges can cover a much wider range of pressure than those of previously reported micro-/nanostructured sensing materials. This proof-of-concept can help in monitoring both ultralow and high levels of human activity. Noteworthily, materials with unique microstructure will enhance their piezocapacitive performance. Compared with those piezocapacitive materials with flat surface, the pressure sensor comprised of pyramidal dielectric elastomer significantly enhanced blood pulse monitoring and force tracking [82]. Similar improvements have also been investigated for piezoresistive elastomer with hollow-sphere [83], thickness-gradient [84] and pyramidal microstructures [85], implying an efficient approach to improve the performance of strain sensors by well-designed strain-responsive materials.

2.2.3. Piezoelectric-based strain sensors

Piezoelectric-based strain sensors detect changes in pressure or strain; in particular, sensing signal (voltage) is only generated by dynamic changes of applied strain or force. Piezoelectric materials, especially certain crystals (e.g. ZnO and $LiNbO_3$) and some ceramics (e.g. $BaTiO_3$ and $Pb(Zr11_xTi_x)O_3$ (PZT)), generate a voltage potential when the crystal lattice is deformed [86, 87]. The sensitivity of piezoelectric materials depends on their crystal structure and the voltage generated is directly proportional to the applied strain. These sensors have very good high-frequency responses, which makes them ideal for measuring dynamic forces [88]. To match the mechanical properties of human skin, recent piezoelectric materials used in wearable devices are mechanically flexible [87, 89, 90]. These flexible piezoelectric materials can be based on either rigid piezoelectric materials embedded in a flexible substrate or a piezoelectric polymer film.

Compared with piezoresistive- and piezocapacitive-based strain sensors, strain devices comprised of piezoelectric materials have greater dynamic durability. The nature of good high-frequency response to an applied strain makes this device more suitable for measuring vibrations. For example, piezoelectric-based strain sensors could be installed in a smart watch for monitoring heartbeat rate. However, the dynamic piezoelectric sensing mechanism disallows them from measuring static forces. A classic piezoelectric material commonly used in flexible strain devices is poly(vinylidene fluoride; PVDF), a piezoelectric polymer that has considerable flexibility [90]. Its flexibility and high piezoelectric factor make PVDF an attractive material for flexible strain devices. For example, Persano et al. (**Figure 2d**) reported on aligned arrays of nanofibers of PVDF [91]. Free-standing 3D architectures of aligned arrangements of PVDF fibers have been designed (**Figure 2d**); furthermore, the polymer chains inside the fibers adopt strongly preferential orientations [91]. This microstructure enables ultra-high sensitivity in measuring pressure, even at exceptionally low values (0.1 Pa); in particular, exceptional piezoelectric characteristics offered by these devices indicate an application in monitoring human motion.

2.3. Wearable devices for recording electrical conductivity through the skin

Wearable devices that detect tiny electrical changes on the skin could be helpful for monitoring the electrical activity of heart to locate an arrhythmia (abnormal heartbeat). These results can help one or one's doctor to decide whether you need medicine, a pacemaker, an implantable cardioverter defibrillator (ICD), cardiac neural ablation or some other surgery. For wearable and accurate monitoring of electrophysiological signals in a variety of everyday conditions, reliable biopotential-capturing electrodes that are reusable with long-term stability required, and they should also not cause skin irritations or allergic reactions.

To date, disposable silver/silver chloride (Ag/AgCl)-gelled electrodes are most commonly used, but these electrodes have limited storage time (<1 year) and are not reusable [96]. The best possible alternatives are flexible dry electrodes, which are usually made of polymeric- and textile-based materials [69, 96–98]. For instance, after patterning a metal layer on a thin PDMS substrate, a biocompatible and flexible PDMS-based (polymetric) electrocardiogram (ECG) recording device can be obtained (**Figure 3a**) [97]. In the case of textile-based electrodes, these are normally fabricated by knitting, weaving, embroidering and non-weaving methods. **Figure 3b** shows photographic images and comparison of the response of the textile-based dry electrode and (Ag/AgCl)-gelled electrode in wearable ECG recording devices. These flexible dry electrodes in general are more skin compatible compared with those of gelled electrodes (**Figure 3**) [69, 96–98]. However, disturbances to ECG signals caused by body movement can often appear. Despite larger disturbances due to movement, these flexible dry electrodes cause relatively little irritation or itches on human skin compared with (Ag/AgCl)-gelled electrodes. After wearing the polymeric dry electrodes for 7 days, no significant changes on skin were seen (**Figure 3a**); however, the skin under the (Ag/AgCl)-gelled electrode turned red and became itchy in two subjects. In conclusion, these newly proposed dry electrodes have comparatively good fidelity and are comfortable regarding body contact as well as being acceptable in long-term stability, indicating broad applicability to the ubiquitous field of biosignal monitoring.

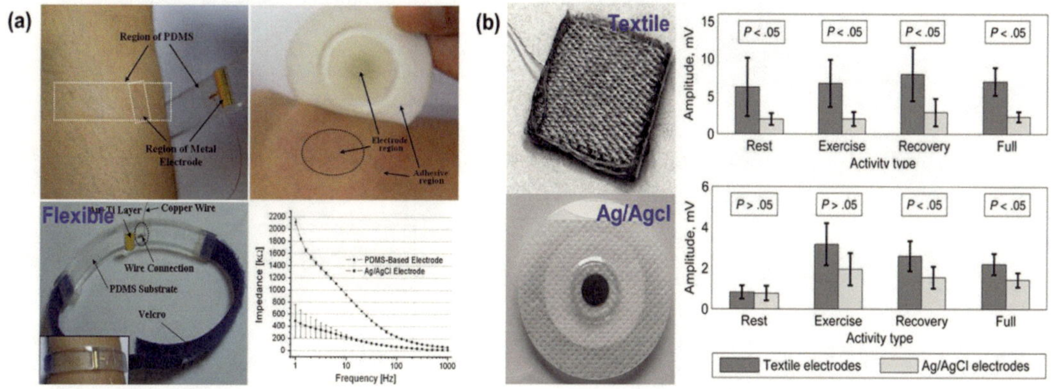

Figure 3. Typical examples of flexible dry electrodes for ECG signal recording. Photographs of (a) polymeric- and (b) textile-based electrodes, and their response to biosignals compared with those of commercial Ag/AgCl electrodes. Reconstructed from Refs. [92–95].

2.4. Wearable sensors for analyzing sweat metabolites

Levels of sweat metabolites (such as lactate and uric acid) and electrolytes (metal ions, such as sodium and potassium), as well as skin humidity, are useful physiological parameter indirectly reflecting health status of an individual, and can potentially be used to probe body conditions by non-invasive monitoring [21, 23, 33, 34, 37, 99]. **Table 1** summarizes the use of sweat body fluids as physiological biomarkers [7, 8, 27, 33]. The concentration of uric acid and urea in sweat is much higher in uremic patients [100]. Levels of skin humidity and pH of sweat, as well as its electrolytes, are important indicators of dehydration and electrolyte loss during exercise [101]. Accordingly, sweat-based wearable devices are required for monitoring these non-invasive health states.

To date, sweat-based healthcare devices are mainly based on an electrochemical sensing principle, owing to the low-cost, high-performance and excellent portability of electrochemical devices [21, 23]. **Figure 4a** shows the configuration of a generally reported electrochemical sweat-based device [102]. Three electrodes are printed on the flexible substrate and, as might be necessary, a fourth electrode can be present, that is, a working electrode (WE), a counter electrode (CE, sometimes as anode), a reference electrode (RE), and a cathode. WE and CE are normally composed of conductive materials, such as carbon black, CNTs, Pt black, modified Ag or Au, etc. [21, 102, 103]. Ag/AgCl is usually used as the RE. Conformal contact between the electrode surface and the skin is necessary to ensure efficient functioning of the devices. Accordingly, substrates for these devices should be made of fabric (e.g. wool, cotton or nylon) or flexible plastics [21]. Unique technology is required for fabricating firm electrodes/substrate configuration.

Figure 4b–d shows three typical approaches [21, 23, 37, 70, 103–105] for attachment of electrodes to a flexible substrate or skin [21, 23]. The first approach relies on screen-printed electrodes [23]. This approach has been massively employed in production due to its low-cost, industrial-scale fabrication of robust electrodes on different substrates. The geometries and thickness of the electrodes

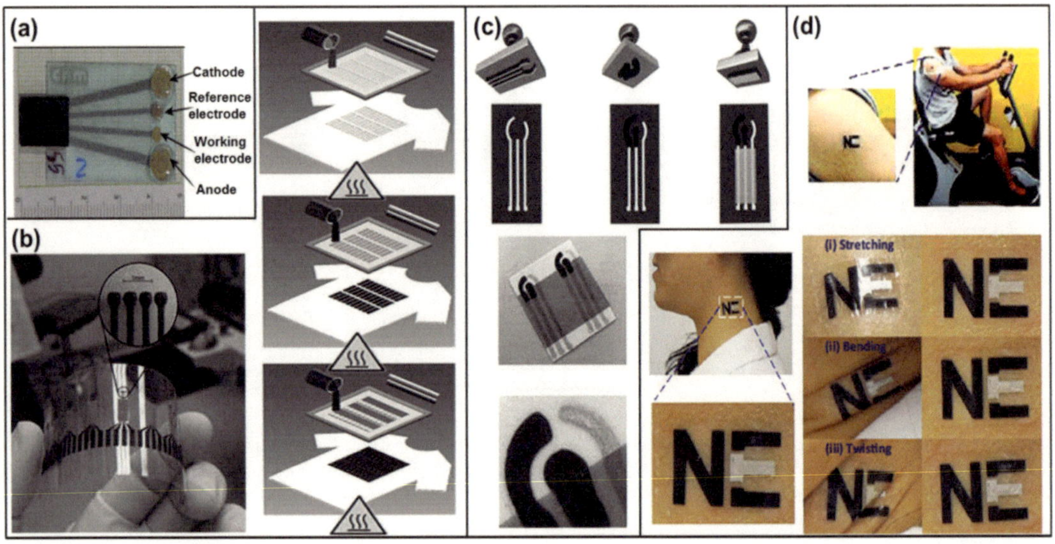

Figure 4. General fabrication technology of sweat-based wearable devices. (a) Primary configuration of the sweat-based electrochemical wearable devices. (b) Steps involved in screen printing thick-film electrochemical devices. Left of **Figure 4b**: Representative design of flexible device fabricated by a screen-printing technique. (c) Preparation progress of stamp transfer electrodes. (d) Typical example of a tattoo-based wearable device. Reconstructed from Refs. [21, 23].

can be readily adjusted by changes of the screen mask to meet the design requirements of the wearable device. However, preparation of wearable devices by screen printing involves several complicating issues; for instance, the influence of manufacturing conditions and substrate properties, as well as the composition of the ink, need to be taken into consideration, since these factors significantly determine the performance of the devices. Generally, the natural properties of substrates must be compatible with the printing process and the specific operational environment. Some necessary modification, for example, doping of the noble catalyst, may be required for the inks to obtain specific functionality. The annealing temperature also requires optimization to avoid undesired deformation of the electrode and substrate. The left part of **Figure 4b** depicts the screen-printed microelectrodes on polyethylene naphthalate (PEN) for dopamine detection, suggesting the convenience of depositing responsive materials by this approach.

The second approach used for attachment of electrodes to a flexible substrate or skin relies on stamp transferred electrodes [23]. The pattern-transfer technique is very useful in fabricating sensing materials on a non-planar substrate, particularly on uniformly non-planar substrates. **Figure 4c** shows details of the stamp transfer process. It is known that, through pattern-transfer technique, responsive materials are highly compatible with irregular substrates possessing diverse surface morphologies (e.g. skin), without compromising the structural integrity of the pattern. Stamp transferred electrodes notably give better performance than those fabricated by screen-printing technique [23]. This fabricating technique provides a novel method to form the ink-based printable materials on non-planar and oversized surfaces incompatible with standard screen-printing methods.

The third approach used for attachment of electrodes to a flexible substrate or skin relies on epidermal suspended electrodes [21]. The rationale behind this approach is to deal with the major problem for traditional wearable device by restricted intimate contact with the skin at limited regions – something that can be solved if the sensing electrodes can be directly attached on the skin. Inspired by tattooing, the strategy of using elastomeric stamps or the tattoo technique to from electrodes directly on human epidermis has been proposed. The strategy includes steps of screen printing of conductive and insulating inks on commercial, temporary tattoo-base paper, to form the electrodes and/or devices, and then flip and apply the fabricated tattoo or printed electrodes to the skin. To adapt for mechanical stress and overcome the potential deformations due to bodily movements, it was suggested that finely dispersed carbon fibers should be incorporated into the tattoo-based devices. Numerous results showed the success and convenience of applying this new technique in preparing electrodes and/or wearable devices on the skin. Among these examples, a classic prototype reported by Jia et al. [71] was noteworthy. In the prototype (**Figure 4d**), an enzymatic tattoo amperometric biosensor was built for continuously monitoring lactate levels in the perspiration as a biomarker of physical stress. This flexible, printed, temporary-transfer tattoo, electrochemical biosensor adhered well to the skin and had acceptable selectivity to lactate, with linearity up to 20 mmol/L. In particular, the device had commendable resilience against continuous the mechanical deformation expected from epidermal wear.

Simultaneous screening of target biomarkers is helpful in ensuring the accuracy of measurements. Consequently, multifunctional wearable devices combining sensing and therapy have been particularly a focus of attention. Through providing sufficient information of heater, temperature, humidity, glucose and pH, as well as several chemicals in sweat, these multifunctional wearable devices (e.g. graphene-based electrochemical device) give a comprehensive profile of health status [27, 106]. On account of these attractive results, it is expected that, through combination of advanced materials science and these innovative preparation techniques, sweat-based wearable devices could possibly alert traditional disease prevention.

Spontaneously measuring chemical and ECG parameters will provide more comprehensive information about health status; however, most of the reported wearable healthcare sensors can only measure chemical or ECG parameters. To obtain fuller information of the human body, these monofunctional sensors need to be integrated into a sensor matrix, which increases the complexity of fabrication. An alternative strategy to resolve this issue is to design sensors that can be used for monitoring chemical and ECG parameters spontaneously. Based on this strategy, a wearable chemical-ECG hybrid biosensing system is proposed for real-time health and fitness monitoring [107]. The proposed system should give a desirable performance in simultaneously sensing lactate (0–28 mmol/L) and monitoring ECG parameters [107]. Such proof-of-concept opens a new way in designing flexible smart sweat-based healthcare devices.

2.5. Wearable sensors for detection of volatile biomarkers

Numerous experimental data have verified the robust correlation between volatile organic compounds (VOCs; organic compounds that have a high vapor pressure under ordinary room temperature conditions) and health status [72, 92–95, 108–143], indicating that changes in VOC profiles (amount and/or species) of human skin can be used for early diagnosis and warning of potential health risks [109, 110, 112]. The ideal approach to track VOCs exuded

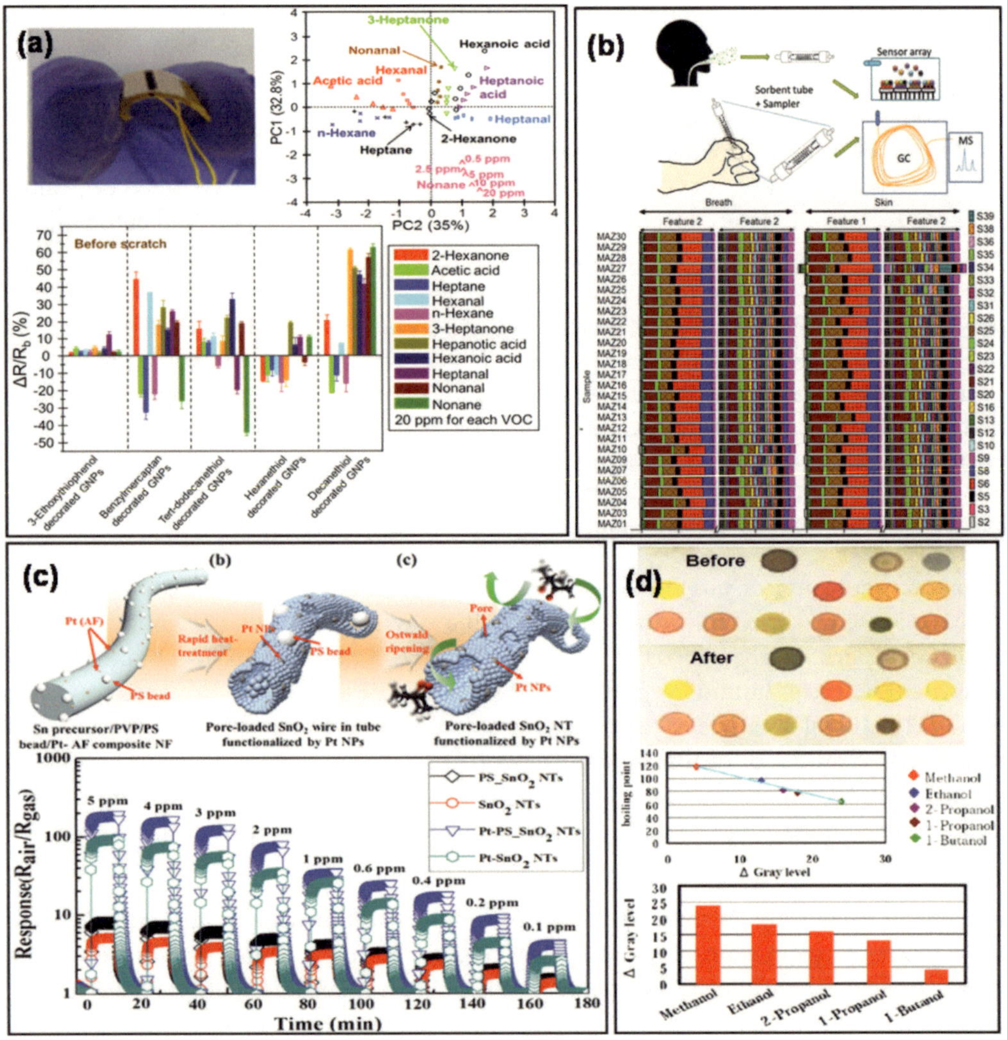

Figure 5. Typical examples of the practical and potential sensing elements for skin VOCs detection. (a) Functional group (at different chain-length) modified GNPs-based flexible device. (b) Combined volatolomics monitoring by nanomaterial-based devices. (c) Fiber-like SnO_2 for tracking trace amounts of VOCs. (d) A colorimetric VOCs device composed of different odor-responsive pigments. Reconstructed from Refs. 71, 78, 128 and [153, 163].

from human skin is expected to be non-invasive, real-time and operatable in non-clinical settings.These wearable devices can be managed either autonomically or remotely by a general practitioner.The concentration of these VOCs is normally at tens of ppb, suggesting high sensitivity and low LODs is required for these skin VOCs tracking sensors.

2.5.1. Metal nanoparticle-based resistometric sensors

In view of confirmation of the usability of metal nanoparticles [110, 113, 144, 145] in trace VOC detection, and other physical and environmental parameter monitoring (e.g. pressure, relative humidity and temperature) [146], Segev et al. [78] developed a smart patch based on gold nanoparticle (GNP) film which gave a satisfactory performance. Based on this strategy, they developed series of analogous GNP-based devices for tracking VOC biomarkers extracted from human skin [54, 111]. **Figure 5a** shows 2 typical examples of a GNP-based device and its performance relative to several kinds of VOCs found in human skin/breath. Interestingly, an acceptable discrimination capability was observed for the device after combination with specific algorithms [54, 111]. To prove the feasibility of the GNP-based devices in practical healthcare application, Broza et al. [93] used the approach developed to monitor VOCs of skin samples (**Figure 5b**); they included a heterogeneous group of 30 volunteers with no acute or active disease. Although their main goal was to associate VOCs from skin and breath of specific volunteers, the second goal was to achieve a chemical profiling based on the VOC samples from either skin or breath. This method successfully created a unique chemical barcoding enabling classification and clustering of the volunteers, for example, clustering based on age-related VOCs.

2.5.2. Potential candidates for tracking skin VOCs

Several materials appear to be sensitive to VOCs from human breath and odor, as well as air pollutants. Although there is no direct evidence of the use of these materials to track VOCs from the skin, they have no great potential for the near future. Thus, a general description of these candidates is introduced in the following sub-sections.

2.5.2.1. Metal oxides

Gas-sensitive metal oxides can be divided into transition- and non-transition metal oxides [147]. It is generally believed that the former are more gas-sensitive than the non-transition metal oxides because of their more active chemical structure. Primary factors that influence sensing behavior of metal oxides include chemical components, surface area (surface-to-volume ratio), microstructures of sensing layers, humidity and temperature [148].

Generally, very few unitary metal oxides possess favorable properties for sensing VOCs, especially in tracing gas species at ppb level. For this level of sensitivity, recent studies have focused mainly on composites, that is, binary, ternary, quaternary or even more complex metal oxides (such as noble metal decorated metal oxides) [147, 148]. Details of the enhancement mechanism have been published [147–149]. Besides the chemical composition, materials with high surface-to-volume ratio (high surface area) provide more reaction sites. Fabrication of nanofibers by electrospinning (**Figure 5c**), generation of thin film by sputtering, as well as creating porous structures with help of pore forming agents, are the most popular ways of obtaining sensing candidates with a high surface area [150–153]. It has now been shown that, when the grain size is less than twice the thickness of surface charge layers, the grain is fully involved in the space-charge layer; consequently, gas sensitivity of the oxides can be significantly increased by adjusting grain size [149, 151]. Synthesis of oxides with unique shape and morphology is a very desirable method of enhancing the sensitivity of metal oxide gas sensors, as the shape and morphology of the oxides determines the crystallographic facets exposed on the surface of a nanocrystal, and will therefore determine the number of atoms located at the edges or corners [151]. By controlling the shape and

Materials	Synthesis routes	Morphology	BET surface area ($m^2.g^{-1}$)	Response to 100 ppm ethanol	Temperature (°C)	Response time (s)	Refs.
SnO_2	Thermal evaporation	Nanowire	N.M.	2.1	400	N.M	[154]
	Coprecipitation	Nonporous	19.5	13	300	13	[155]
	Electrospinning	Fiber	N.M.	15	340	10	[152]
	Hydrothermal	Nanosheets	31	35	270	13	[156]
	Coprecipitation	Microcubes	42.7	58	280	N.M.	[155, 157]
SnO_2/ZnO	Thermal evaporation and hydrothermal	hierarchical	N.M.	6	400	N.M	[154]
Ag-doped TiO_2/SnO_2	Wet impregnation and nanocasting	Hierarchical flower	49	112.5	275	3.5	[158]

Table 3. Comparison of ethanol sensing performance, based on different SnO_2-based sensing materials.

morphology using different synthetic routes, the reactivity and selectivity of a nanocatalyst can be manually tailored.

One of the main challenges in metal oxide sensors is that water adsorption appreciably lowers the sensitivity of metal oxide gas sensors, since water adsorption on the metal oxide surface prevents donation of electrons to the sensing layers. Furthermore, prolonged exposure to humid environments leads to gradual formation of stable chemisorbed OH^- on the surface, resulting in progressive degradation. However, surface hydroxyls start to desorb at ~400°C so that the hydroxyl ions can be removed by heating to temperatures >400°C [151, 159]. Thus, an optimal operating temperature will essentially eliminate humidity interference. As the balance of adsorption/desorption is markedly affected by temperature, optimization of operating temperature improves the magnitude of response/recovery speed and response. **Table 3** lists several typical examples covering the above-mentioned strategies using SnO_2 as the sensing element [147, 150, 152, 154–158]. It is clear that shape/morphology, surface area and operating temperature influence the performance.

2.5.2.2. Conducting polymers

Conducting polymers (CPs) and their derivatives have been used as the active layers of gas sensors over the last century [160]. Compared to metal oxides sensors, devices made of conducting polymers have high sensitivities and short response durations at room temperature [161]. Conducting polymers are easily synthesized and their molecular chain structure can be conveniently modified by copolymerization or structural derivations [148, 161]. Furthermore, conducting polymers have good mechanical properties, which allow easy fabrication of sensor devices, and therefore considerable attention has been paid to these sensors.

Some of the earliest studies indicated that the sensing mechanism of CPs to VOCs involves swelling, namely the adsorbed VOC leads to the separation of polymer chains and increases the hopping distance of charge career for charge transport. As a result, the electrical resistance (response signal) increases [148, 160, 161]. In some cases, polar VOC molecules behave as electron acceptors or donors contributing a charge carrier to the polymer, and subsequently participate as dopants that increase or decrease carrier concentration [148, 160]. Hybridization of CPs with inorganic and graphitic nanoscale building blocks, such as metal oxides, CNTs or graphene, are more advantageous in polymer processing [148]. These hybridized CPs can also provide tunable sensing performance to VOCs through hybridization of CPs with optimal nanoscale building blocks, furthering the design of advanced sensing devices for specific monitoring of VOC biomarkers.

2.5.2.3. Responsive dyes

Theoretically, if a VOC molecule interacts with a responsive dye, a visual response can be obtained by a change in color. Based on this working principle, colorimetric VOCs devices have been designed based on two fundamental requirements: (1) each chemically responsive dye must contain a center that strongly interacts with the analyte; (2) each interaction center must be strongly coupled to an intense chromophore. Metalloporphyrins (e.g. Cu(II), Zn(II), Mn(II), Co(III), Cr(III), Sn(IV)-based porphyrins, etc.) and some pH sensitive dyes (e.g. methyl red, bromophenol blue, chlorophenol red, etc.) are among the primary compositions for sensors based on responsive dyes [162]. Similarly, colorimetric devices showed their merits in identifying odorous biomarkers [163]. Following the interaction of VOCs with an array of responsive dyes, a unique color map was generated for each VOC and different color patterns could be seen at different concentrations of VOCs (**Figure 5d**) [163]. Combined with a specific algorithm, excellent discrimination was obtained. LODs of the colorimetric devices made of responsive dyes have generally ranged from hundreds of ppb to several ppm. Due to the strong interaction between VOC molecules and responsive dyes, this kind of device can often be very sensitive to VOC biomarkers. In summary, these disposable responsive dye-based colorimetric devices have given interesting results for consideration in the future of skin-based wearable devices.

3. Future perspectives

The rapid development of material science and engineering essentially stimulates progress in the development of wearable healthcare devices. Introduction of innovative sensing techniques has recently reformed our approach to disease diagnosis. It is undoubtedly reasonable to expect that real-time monitoring devices can give a much greater comprehensive assessment of the wearer's health status and improve his or her quality of life. Indeed, many more organizations are paying attention to this field since continuous personalized health monitoring using wearable healthcare devices that can be expected to provide low-cost solutions especially for remote monitoring, and will be particularly useful in indicating efficient treatment where early diagnosis can be made. To meet the market demands, extensive progress must be made in this field, which relies on the collaboration between material sciences and electronics, as well as sensing signal acquisition and processing. The requirement for highly active sensing materials has spurred tremendous interests in developing monodisperse, single-component small-size NPs and encouraged researchers to put greater effort in searching for creative strategies in designing and synthesizing sensing elements with novel properties. Considering this goal, a better understanding of sensing mechanism can provide deep insight into the connection between sensing characteristics and materials composition/structure/morphology that will be instructive to scientists and engineers for exploring high-performance healthcare devices.

Although encouraging results have been achieved with a wide variety of wearable sensors, major obstacles remain that prevent the routine use of wearable devices by the general population. First, there is a lack of "bioaffinity" protocols [21], namely the relationship between these biomarkers and physiological index or a specific disease. Although some reports discuss this issue, the intrinsic connections remain ambiguous. Second, the precision, sensitivity, long-term stability and biocompatibility of devices must improve to meet the standard requirements from diagnostic devices, as defined by regulatory agencies. Third, there is a need to effectively reduce power consumption of these devices. In most situations, a sensing matrix should be built into these wearable devices for recording comprehensively body information, leading to considerable power consumption. Even though fabrication of self-powered devices suggests the way to solve this issue, power generation efficiency of this kind of devices remains questionable.

Although the importance of smart materials have been validated in designing high-performance, multifunctional, compact skin-based wearables, updates on the materials sciences still lags behind progress being made in the relevant fields. A typical example is that the biomaterials are less often reported in wearable healthcare devices, whereas this type of candidate will probably improve the current status. To date, biomaterials have been widely directed at their therapeutic applications; in particular they have shown unique features in the application of implantable healthcare devices [164, 165]. Owing to their natural biocompatibility and biodegradability, it is expected that skin-based wearable devices made of biomaterials will achieve even more unexpected results. One possible promotion is to replace the frequently used polymeric elastomer with more biocompatible biomaterials based substrates, for example, protein films, which will eliminate completely any irritation caused by the wearables. Their biodegradability will also be especially useful in designing environmental friendly products. As well as the more usual biomaterials, multifunctional materials also need more attention. As discussed in the section on multifunctional devices, these multifunctional materials that are spontaneously energy harvesting/self-healing and stimuli-responsive, or multi stimuli-responsive, will probably decrease the size, cost and the complexity of fabricating these healthcare devices. Although until now few relevant multifunctional candidates have been successfully devised, we speculate that wearables with smart multifunctional materials will lead to further innovative developments.

Overall, with advances in sensor sophistication and miniaturization, and battery solutions as well as materials sciences, much is expected to be packed into a patch. As with any wearable solution, there may need to be trade-offs in the number of data-points collected, the period of use and the processing power. Generally speaking, the closer the connection between the wearable device and the signal it is collecting, the greater will be the accuracy of the data. Because patches provide direct contact with the skin, they can detect very subtle changes in the patient's physiological condition and give comprehensive profile of one's health status.

Author details

Han Jin*, Qinghui Jin and Jiawen Jian

*Address all correspondence to: jinhan@nbu.edu.cn

Environmental Monitoring and Sensing Technology Laboratory, School of Information Science and Engineering, Ningbo University, Ningbo, P.R. China

References

[1] Stoppa M, Chiolerio A. Wearable electronics and smart textiles: A critical review. Sensors. 2014;**14**:11957-11992

[2] Tamsin M. Wearable biosensor technologies. International Journal of Scientific Research. 2015;**13**:697-703

[3] Zeng W, Shu L, Li Q, Chen S, Wang F, Tao XM. Fiber-based wearable electronics: A review of materials, fabrication, devices, and applications. Advanced Materials. 2014; **26**:5310-5336

[4] Ajami S, Teimouri F. Features and application of wearable biosensors in medical care. Journal of Research in Medical Sciences. 2015;**20**:1208-1215

[5] Appelboom G, Camacho E, Abraham ME, Bruce SS, Dumont EL, Zacharia BE, et al. Smart wearable body sensors for patient self-assessment and monitoring. Archives of Public Health. 2014;**72**:1-9

[6] Esteban M, Castaño A. Non-invasive matrices in human biomonitoring: A review. Environment International. 2009;**35**:438-449

[7] Trung TQ, Ramasundaram S, Hwang BU, Lee NE. An all-elastomeric transparent and stretchable temperature sensor for body-attachable wearable electronics. Advanced Materials. 2016;**28**:502-509

[8] Kantoch E, Augustyniak P. Human activity surveillance based on wearable body sensor network. Computing in Cardiology. 2012;**39**:325-328

[9] Comina G, Suska A, Filippini D. Autonomous chemical sensing Interface for universal cell phone readout. Angewandte Chemie, International Edition. 2015;**54**:8708-8712

[10] Munos B, Baker PC, Bot BM, Crouthamel M, Vries GD, Ferguson I, et al. Mobile health: The power of wearables, sensors, and apps to transform clinical trials. Annals of the New York Academy of Sciences. 2016;**1375**:3-18

[11] Patel S, Park H, Bonato P, Chan L, Rodgers M. A review of wearable sensors and systems with application in rehabilitation. Journal of Neuroengineering and Rehabilitation. 2012;**9**:1-17

[12] Nelson EC, Verhagen T, Noordzij ML. Health empowerment through activity trackers: An empirical smart wristband study. Computers in Human Behavior. 2016;**62**:364-374

[13] Chuah HW, Rauschnabel PA, Krey N, Bang N, Ramayah T, Lade S. Wearable technologies: The role of usefulness and visibility in Smartwatch adoption. Computers in Human Behavior. 2016;**65**:276-284

[14] Finlay DD, Nugent CD, Donnelly MP, Mccullagh PJ, Black ND. Optimal electrocardiographic lead systems: Practical scenarios in smart clothing and wearable health systems. IEEE Transactions on Information Technology in Biomedicine. 2008;**12**:433-441

[15] Jung PG, Oh S, Lim G, Kong K. A mobile motion capture system based on inertial sensors and smart shoes. Journal of Dynamic Systems Measurement and Control. 2013; **136**:692-697

[16] Kim SH, Ryoo DW, Bae C. U-healthcare system using smart headband. Proceedings of 30th Annual International IEEE Engineering in Medicine and Biology Society Conference, Vancouver, British Columbia, Canada. 2008; 1557-1560

[17] Constant N, Douglas-Prawl O, Johnson S, Mankodiya K. Pulse-glasses: An unobtrusive, wearable HR monitor with internet-of-things functionality. 12th IEEE International Conference on Wearable & Implantable Body Sensor Networks. 2015

[18] Khandwalla RM, Birkeland K, Zimmer R, Banet M, Pede S, Kedan I. Predicting heart failure events with home monitoring: Use of a novel wearable necklace to measure stroke volume, cardiac output and thoracic impedance. Journal of the American College of Cardiology. 2016;**67**:1296

[19] Konvalina G, Haick H. Sensors for breath testing: From nanomaterials to comprehensive disease detection. Accounts of Chemical Research. 2013;**47**:66-76

[20] Amjadi M, Kyung KU, Park I, Sitti M. Stretchable, skin-mountable, and wearable strain sensors and their potential applications: A review. Advanced Functional Materials. 2016;**26**:1678-1698

[21] Bandodkar AJ, Wang J. Non-invasive wearable electrochemical sensors: A review. Trends in Biotechnology. 2014;**32**:363-371

[22] Vashist SK. Non-invasive glucose monitoring technology in diabetes management: A review. Analytica Chimica Acta. 2012;**750**:16-27

[23] Windmiller JR, Wang J. Wearable electrochemical sensors and biosensors : A review. Electroanalysis. 2012;**24**:1-18

[24] Chortos A, Bao ZN. Skin-inspired electronic devices. Materials Today. 2014;**17**:321-331

[25] Oliver NS, Toumazou C, Cass AEG, Johnston DG. Glucose sensors: A review of current and emerging technology. Diabetes Medication. 2009;**26**:197-210

[26] So CF, Choi KS, Wong TKS, Chung JWY. Recent advances in noninvasive glucose monitoring. Journal of Medical Devices. 2012;**5**:45-52

[27] Gao R, Emaminejad S, Nyein HYY, Challa S, Chen K, Peck A, et al. Fully integrated wearable sensor arrays for multiplexed in situ perspiration analysis. Nature. 2016;**529**:509-514

[28] Windmiller JR, Bandodkar AJ, Parkhomovsky S, Wang J. Stamp transfer electrodes for electrochemical sensing on non-planar and oversized surfaces. Analyst. 2012;**137**:1570-1575

[29] Bandodkar AJ, Hung VWS, Jia WZ, Ramírez GV, Windmiller JR, Martinez AG, et al. Tattoo-based potentiometric ion-selective sensors for epidermal pH monitoring. Analyst. 2013;**138**:123-128

[30] Guinovart T, Bandodkar AJ, Windmiller JR, Andrade FJ, Wang J. A potentiometric tattoo sensor for monitoring ammonium in sweat. Analyst. 2013;**138**:7031-7038

[31] Ruiz JG, Mas R, Haro C, Cabruja E, Camero R, Lomillo MAA, et al. Early determination of cystic fibrosis by electrochemical chloride quantification in sweat. Biosensors and Bioelectronics. 2009;**24**:1788-1791

[32] Solovei D, Žák J, Majzlíková P, Sedláček J, Hubálek J. Chemical sensor platform for non-invasive monitoring of activity and dehydration. Sensors. 2015;**15**:1479-1495

[33] Paliwal S, Hwang BH, Tsai KY, Mitragotri S. Diagnostic opportunities based on skin biomarkers. European Journal of Pharmaceutical Sciences. 2013;**50**:546-556

[34] Mishra A, Greaves R, Massie J. The relevance of sweat testing for the diagnosis of cystic fibrosis in the genomic era. Clinical Biochemist Reviews. 2005;**26**:135-153

[35] Vermeulen F, Lebecque P, Boeck KD, Leal T. Biological variability of the sweat chloride in diagnostics sweat tests: A retrospective analysis. Journal of Cystic Fibrosis. 2016;**15**:S30-S35

[36] Koh A, Kang D, Xue YG, Lee S, Pielak RM, Kim J, et al. A soft, wearable microfluidic device for the capture, storage, and colorimetric sensing of sweat. Science Translational Medicine. 2016;**8**:1-13

[37] Schazmann B, Morris D, Slater C, Beirne S, Fay C, Reuveny R, et al. A wearable electrochemical sensor for the real-time measurement of sweat sodium concentration. Analytical Methods. 2010;**2**:342-348

[38] Sanders JP, Loveday A, Pearson N, Edwardson C, Yates T, Biddle SJ, et al. Devices for self-monitoring sedentary time or physical activity: A scoping review. Journal of Medical Internet Research. 2016;**18**:e90

[39] Page T. A forecast of the adoption of wearable technology. International Journal of Technology Diffusion. 2015;**6**:12-29

[40] Paul BK, Panat R, Mastrangelo C, Kim D, Johnson D. Manufacturing of smart goods: Current state, future potential, and research recommendations. Journal of Micro and Nano-Manufacturing. 2016;**4**:044001-044011

[41] Kim J, Lee M, Shim HJ, Ghaffari R, Cho HR, Son D, et al. Stretchable silicon Nanoribbon electronics for skin prosthesis. Nature Communications. 2014;**5**:5747

[42] Kaltenbrunner M, Sekitani T, Reeder J, Yokota T, Kuribara K, Tokuhara T, et al. An ultra-lightweight Design for Imperceptible Plastic Electronics. Nature. 2013;**499**:458-463

[43] Giuliani A, Placidi M, Francesco FD, Pucci A. A new polystyrene-based Ionomer/ MWCNT nanocomposite for wearable skin temperature sensors. Reactive and Functional Polymers. 2014;**76**:57-62

[44] Matzeu G, Pucci A, Savi S, Romanelli M, Francesco FD. A temperature sensor based on a MWCNT/SEBS nanocomposite. Sensors and Actuators A. 2012;**178**:94-99

[45] Webb RC, Bonifas AP, Behnaz A, Zhang YH, Yu KJ, Cheng HY, et al. Ultrathin conformal devices for precise and continuous thermal characterization of human skin. Nature Materials. 2013;**12**:938-944

[46] Yang J, Wei DP, Tang LL, Song JZ, Luo W, Chu J, et al. Wearable temperature sensor based on graphene nanowalls. RSC Advances. 2015;**5**:25609-25615

[47] Yokota T, Inoue Y, Terakawa Y, Reeder J, Kaltenbrunner M, Ware T, et al. Ultraflexible, large-area, physiological temperature sensors for multipoint measurements. Proceedings of the National Academy of Sciences of the United States of America. 2015;**112**:14533-14538

[48] Jeon J, Lee HBR, Bao ZN. Flexible wireless temperature sensors based on Ni microparticle-filled binary polymer composites. Advanced Materials. 2013;**25**:850-855

[49] Yang H, Qi DP, Liu ZY, Chandran BK, Wang T, Yu JC, et al. Soft thermal sensor with mechanical adaptability. Advanced Materials. 2016;**28**:9175-9181

[50] Trung TQ, Tien NYT, Kim D, Jung JH, Yoon OJ, Lee NE. High thermal responsiveness of a reduced graphene oxide field-effect transistor. Advanced Materials. 2012;**24**:5254-5260

[51] Tien NT, Jeon D, Kim DI, Trung DD, Jang JS, Hwang BH, et al. A flexible bimodal sensor Array for simultaneous sensing of pressure and temperature. Advanced Materials. 2014;**26**:796-804

[52] Ren XC, Chan PKL, Lu JB, Huang BL, Leung DCW. High dynamic range organic temperature sensor. Advanced Materials. 2013;**25**:1291-1295

[53] Hattori Y, Falgout L, Lee E, Jung JH, Poon E, Lee E, et al. Multifunctional skin-like electronics for quantitative clinical monitoring of cutaneous wound healing. Advanced Healthcare Materials. 2014;**3**:1597-1607

[54] Huynh TP, Haick H. Self-healing, fully functional, and multiparametric flexible sensing platform. Advanced Materials. 2015;**28**:138-143

[55] Haick H, Ambrico M, Ghabboun J, Ligonzo T, Cahen D. Contacting organic molecules by metal evaporation. Physical Chemistry Chemical Physics. 2004;**6**:4538-4541

[56] Haick H, Ghabboun J, Cahen D. Pd versus au as evaporated metal contacts to molecules. Applied Physics Letters. 2005;**86**:042113/1-042113/3

[57] Wang LH, Li YL. A review for conductive polymer piezoresistive composites and a development of a compliant pressure transducer. IEEE Transactions on Instrumentation and Measurement. 2013;**6**:495-502

[58] Zhang MH, Chen JK, Lu CS. Influence of strain rate on the piezoresistive behavior of conductive polyamide composites. Composites Science and Technology. 2016;**133**:1-6

[59] Choong CL, Shim MB, Lee BS, Jeon S, Ko DS, Kang TH, et al. Highly stretchable resistive pressure sensors using a conductive elastomeric composite on a micropyramid Array. Advanced Materials. 2014;**26**:3451-3458

[60] Khang DY, Jiang HQ, Huang Y, Rogers JA. A stretchable form of single-crystal silicon for high-performance electronics on rubber substrates. Science. 2006;**311**:208-212

[61] Webb RC, Pielak RM, Bastien P, Ayers J, Niittynen J, Kurniawan J, et al. Thermal transport characteristics of human skin measured in vivo using ultrathin conformal arrays of thermal sensors and actuators. PLoS One. 2015;**10**:e0118131

[62] Cavanagh PR, Rodgers MM, Iiboshi A. Pressure distribution under symptom-free feet during barefoot standing. Foot & Ankle International. 1987;**7**:262-276

[63] Wald H, Guernsey M, Scott FH. Some effects of alteration of posture on arterial blood pressure. American Heart Journal. 1937;**14**:319-330

[64] Schwartz G, Tee BC-K, Mei J, Appleton AL, Kim DH, Wang H, et al. Flexible polymer transistors with high pressure sensitivity for application in electronic skin and health monitoring. Nature Communications. 2013;**4**:1859

[65] Tiwana MI, Redmond SJ, Lovell NH. A review of tactile sensing technologies with applications in biomedical engineering. Sensors and Actuators. 2012;**179**:17-31

[66] Amjadi M, Kyung KU, Park I, Sitti M. Stretchable, skin-mountable, and wearable strain sensors and their potential applications: A review. Advanced Functional Materials. 2016;**26**:1678-1698

[67] Tai YL, Mulle M, Ventura IA, Lubineau G. A highly sensitive, low-cost, wearable pressure sensor based on conductive hydrogel spheres. Nanoscale. 2013;**7**:14766-14773

[68] Boutry CM, Nguyen A, Lawal QO, Chortos A, Rondeaugagné S, Bao Z. A sensitive and biodegradable pressure sensor Array for cardiovascular monitoring. Advanced Materials. 2015;**43**:6954-6961

[69] Park S, Jayraman S. Smart textile-based wearable biomedical systems: A transition plan for research to reality. IEEE Transactions on Information Technology in Biomedicine. 2010;**14**:86-92

[70] Windmiller JR, Bandodkar AJ, Ramirez FH, Parkhomovsky S, Martinez AG, Wang B. Electrochemical sensing based on printable temporary transfer tattoos. Chemical Communications. 2012;**48**:6794-6796

[71] Jia WZ, Bandodkar AJ, Ramírez GV, Windmiller JR, Yang ZJ, Ramírez JL, et al. Electrochemical tattoo biosensors for real-time noninvasive lactate monitoring in human perspiration. Analytical Chemistry. 2013;**85**:6553-6560

[72] Amal H, Leja M, Funka K, Skapars R, Sivins A, Ancans G, et al. Detection of precancerous gastric lesions and gastric cancer through exhaled breath. Gut. 2016;**65**:400-407

[73] Park IJ, Jeong CY, Cho IT, Lee JH, Cho ES, Kwon SJ, et al. Fabrication of amorphous InGaZnO thin-film transistor-driven flexible thermal and pressure sensors. Semiconductor Science and Technology. 2012;**27**:105019-105024

[74] Wang Y, Wang L, Yang TT, Li X, Zang XB, Zhu M, et al. Wearable and highly sensitive graphene strain sensors for human motion monitoring. Advanced Functional Materials. 2014;**24**:4666-4670

[75] Li YQ, Samad YA, Liao K. From cotton to wearable pressure sensor. Journal of Materials Chemistry A. 2014;**3**:2181-2187

[76] Gong S, Schwalb W, Wang YW, Chen Y, Tang Y, Si J, et al. A wearable and highly sensitive pressure sensor with ultrathin gold nanowires. Nature Communications. 2014;**5**:163-180

[77] Stassi S, Cauda V, Canavese G, Pirri CF. Flexible tactile sensing based on piezoresistive composites: A review. Sensors. 2014;**14**:5296-5332

[78] Segev-Bar M, Konvalina G, Haick H. High-resolution unpixelated smart patches with antiparallel thickness gradients of nanoparticles. Advanced Materials. 2015;**27**:1779-1784

[79] Yao SS, Zhu Y. Wearable multifunctional sensors using printed stretchable conductors made of silver nanowires. Nanoscale. 2014;**6**:2345-2352

[80] Lipomi DJ, Vosgueritchian M, Tee BC-K, Hellstrom SL, Lee JA, Fox CH, et al. Skin-like pressure and strain sensors based on transparent elastic films of carbon nanotubes. Nature Nanotechnology. 2011;**6**:788-792

[81] Kwon D, Lee TI, Shim J, Ryu S, Kim MS, Kim S, et al. Highly sensitive, flexible, and wearable pressure sensor based on a giant piezocapacitive effect of three-dimensional microporous elastomeric dielectric layer. ACS Applied Materials & Interfaces. 2016;**8**:16922-16931

[82] Tee BC-K, Chortos A, Dunn RR, Schwartz G, Eason E, Bao Z. Tunable flexible pressure sensors using microstructured elastomer geometries for intuitive electronics. Advanced Functional Materials. 2014;**24**(34):5427

[83] Pan LJ, Chortos A, Yu GH, Wang YQ, Isaacson S, Allen R, et al. An ultra-sensitive resistive pressure sensor based on hollow-sphere microstructure induced elasticity in conducting polymer film. Nature Communications. 2013;**5**:3002

[84] Liu HY, Qi DP, Guo PZ, Liu Y, Zhu BW, Yang H, et al. Thickness-gradient films for high gauge factor stretchable strain sensors. Advanced Materials. 2015;**27**:6230-6237

[85] Zhu BW, Niu ZQ, Wang H, Leow WR, Wang H, Li YG, et al. Microstructured graphene arrays for highly sensitive flexible tactile sensors. Small. 2014;**10**:3625-3631

[86] Anton SR, Sodano HA. A review of power harvesting using piezoelectric materials. Smart Materials and Structures. 2007;**16**:R1-R21

[87] Sun KT, Zhang HQ, Hu LZ, Yu DQ, Qiao SS, Sun JC, et al. Growth of Ultralong ZnO microwire and its application in isolatable and flexible piezoelectric strain sensor. Physica Status Solidi A. 2010;**207**:488-492

[88] Dagdeviren C, Shi Y, Joe P, Ghaffari R, Balooch G, Usgaonkar K, et al. Conformal piezoelectric systems for clinical and experimental characterization of soft tissue biomechanics. Nature Materials. 2015;**14**:728-736

[89] Gullapalli H, Vemuru VSM, Kumar A, Mendez AB, Vajtai R, Terrones M, et al. Flexible piezoelectric ZnO-paper nanocomposite strain sensor. Small. 2010;**6**:1641-1646

[90] Sharma T, Je SS, Gill B, Zhang JXJ. Patterning piezoelectric thin film PVDF-TrFE based pressure sensor for catheter application. Sensors and Actuators A. 2012;**177**:87-92

[91] Persano L, Dagdeviren C, Su YW, Zhang YH, Girardo S, Pisignano D, et al. High performance piezoelectric devices based on aligned arrays of nanofibers of poly (vinylidenefluoride-co-trifluoroethylene). Nature Communications. 2013;**4**:67-88

[92] Amal H, Shi D-Y, Ionescu R, Zhang W, Hua Q-L, Pan Y-Y, et al. Assessment of ovarian cancer conditions from exhaled breath. International Journal of Cancer. 2015;**136**: E614-EE22

[93] Amann A, Mochalski P, Ruzsanyi V, Broza YY, Haick H. Assessment of the exhalation kinetics of volatile cancer biomarkers based on their physicochemical properties. Journal of Breath Research. 2014;**8**:016003

[94] Bachar N, Mintz L, Zilberman Y, Ionescu R, Feng X, Mullen K, et al. Polycyclic aromatic hydrocarbon for the detection of nonpolar analytes under counteracting humidity conditions. ACS Applied Materials & Interfaces. 2012;**4**:4960-4965

[95] Barash O, Peled N, Tisch U, Bunn PA, Hirsch FR, Haick H. Classification of the lung cancer histology by gold nanoparticle sensors. Nanomedicine (New York, NY, US). 2012; **8**:580-589

[96] Marozas V, Petrenas A, Daukantas S, Lukosevicius A. A comparison of conductive textile-based and silver/silver chloride gel electrodes in exercise electrocardiogram recordings. Journal of Electrocardiology. 2011;**44**:189-194

[97] Baek JY, An JH, Choi JM, Park KS, Lee SH. Flexible polymeric dry electrodes for the long-term monitoring of ECG. Sensors and Actuators A: Physical. 2008;**143**:423-429

[98] Xu PJ, Zhang H, Tao XM. Textile-structured electrodes for electrocardiogram. Tex Prog. 2008;**40**:183-213

[99] Cizza G, Marques AH, Eskandari F, Christie IC, Torvik S, Silverman MN, et al. Elevated Neuroimmune biomarkers in sweat patches and plasma of premenopausal women with major depressive disorder in remission: The POWER study. Biological Psychiatry. 2008;**64**:907-911

[100] Tamer YYA, Hadi EA, Badrani IEIA. Sweat urea, uric acid and Creatinine concentrations in uraemic patients. Urological Research. 1997;**25**:337-340

[101] Costill DL. Water and electrolyte requirements during exercise. Clinics in Sports Medicine. 1984;**3**:639-648

[102] Guinovart T, Parrilla M, Crespo GA, Riusa FX, Andrade FJ. Potentiometric sensors using cotton yarns, carbon nanotubes and polymeric membranes. Analyst. 2013;**138**:5208-5215

[103] Weber J, Kumar A, Kumar A, Bhansali S. Novel lactate and pH biosensor for skin and sweat analysis based on single walled carbon nanotubes. Sensors and Actuators B. 2006;**17**:308-313

[104] Kim J, Jeerapan I, Imani S, Cho TN, Bandodkar A, Cinti S, et al. Noninvasive alcohol monitoring using a wearable tattoo-based Iontophoretic-biosensing system. ACS Sensors. 2016;**1**:1011-1019

[105] Yang YL, Chuang MC, Lou SL, Wang J. Thick-film textile-based amperometric sensors and biosensors. Analyst. 2012;**135**:1230-1234

[106] Lee H, Choi1 TK, Lee YB, Cho HR, Ghaffari R, Wang L, et al. A graphene-based electrochemical device with thermoresponsive microneedles for diabetes monitoring and therapy. Nature Nanotechnology. 2016;**11**:566-572

[107] Imani S, Bandodkar AJ, Mohan AMV, Kumar R, Yu SF, Wang J, et al. A wearable chemical-electrophysiological hybrid biosensing system for real-time health and fitness monitoring. Nature Communications. 2016;**7**:11650

[108] Broza YY, Mochalski P, Ruzsanyi V, Amann A, Haick H. Hybrid volatolomics and disease detection. Angewandte Chemie, International Edition. 2015;**54**:11036-11048

[109] Broza YY, Zuri L, Haick H. Combined volatolomics for monitoring of human body chemistry. Scientific Reports. 2014;**4**:4611-4617

[110] Haick H, Broza YY, Mochalski P, Ruzsanyi V, Amann A. Assessment, origin, and implementation of breath volatile cancer markers. Chemical Society Reviews. 2014;**43**: 1423-1449

[111] Jin H, Huynh TP, Haick H. Self-healable sensors based nanoparticles for detecting physiological markers via skin and breath: Toward disease prevention via wearable devices. Nano Letters. 2016;**16**:4194-4202

[112] Nakhleh MK, Amal H, Jeries R, Broza YY, Aboud M, Gharra A, Ivgi H, Khatib S, Badarneh S, Har-Shai L, Glass-Marmor L, Lejbkowicz I, Miller A, Badarny S, Winer R, Finberg J, Cohen-Kaminsky S, Perros F, Montani D, Girerd B, Garcia G, Simonneau G, Nakhoul F, Baram S, Salim R, Hakim M, Gruber M, Ronen O, Marshak T, Doweck I, Nativ O, Bahouth Z, Shi DY, Zhang W, Hua QL, Pan YY, Tao L, Liu H, Karban A, Koifman E, Rainis T, Skapars R, Sivins A, Ancans G, Liepniece-Karele I, Kikuste I, Lasina I, Tolmanis I, Johnson D, Millstone SZ, Fulton J, Wells JW, Wilf LH, Humbert M, Leja M, Peled N, Haick H. Diagnosis and classification of 17 diseases from 1404 subjects via pattern analysis of exhaled molecules. ACS Nano. 2017;**11**:112-125

[113] Peng G, Tisch U, Adams O, Hakim M, Shehada N, Broza YY, et al. Diagnosing lung cancer in exhaled breath using au nanoparticles. Nature Nanotechnology. 2009;**4**:669-673

[114] Amal H, Leja M, Funka K, Lasina I, Skapars R, Sivins A, et al. Breath testing as potential colorectal cancer screening tool. International Journal of Cancer. 2016;**138**:229-236

[115] Barash O, Zhang W, Halpern JM, Hua QL, Pan YY, Kayal H, et al. Differentiation between genetic mutations of breast cancer by breath volatolomics. Oncotarget. 2015;**6**: 44864-44876

[116] Cohen-Kaminsky S, Nakhleh M, Perros F, Montani D, Girerd B, Garcia G, et al. A proof of concept for the detection and classification of pulmonary arterial hypertension through breath analysis with a sensor Array. American Journal of Respiratory and Critical Care Medicine. 2013;**188**:756-759

[117] Davies MP, Barash O, Jeries R, Peled N, Ilouze M, Hyde R, et al. Unique volatolomic signatures of TP53 and KRAS in lung cells. British Journal of Cancer. 2014;**111**:1213-1221

[118] Hakim M, Billan S, Tisch U, Peng G, Dvrokind I, Marom O, et al. Diagnosis of head and neck cancer from exhaled breath. British Journal of Cancer. 2011;**104**:1649-1655

[119] Hakim M, Broza YY, Barash O, Peled N, Phillips M, Amann A, et al. Volatile organic compounds of lung cancer and possible biochemical pathways. Chemical Reviews. 2012;**112**:5949-5966

[120] Homede E, Abo Jabal M, Ionescu R, Haick H. Printing ultrasensitive artificially intelligent sensors array with a single self-propelled droplet containing nanoparticles. Advanced Functional Materials. 2016;**26**:6359-6370

[121] Ionescu R, Broza Y, Shaltieli H, Sadeh D, Zilberman Y, Feng X, et al. Detection of multiple sclerosis from exhaled breath using bilayers of polycyclic aromatic hydrocarbons and Single-Wall carbon nanotubes. ACS Chemical Neuroscience. 2011;**2**:687-693

[122] Karban A, Nakhleh MK, Cancilla JC, Vishinkin R, Rainis T, Koifman E, et al. Programmed nanoparticles for tailoring the detection of inflammatory bowel diseases and irritable bowel syndrome disease via Breathprint. Advanced Healthcare Materials. 2016;**5**:2339-2344

[123] Konvalina G, Haick H. Sensors for breath testing: From nanomaterials to comprehensive disease detection. Accounts of Chemical Research. 2014;**47**:66-76

[124] Marom O, Nakhoul F, Tisch U, Shiban A, Abassi Z, Haick H. Gold nanoparticle sensors for detecting chronic kidney disease and disease progression. Nanomedicine (London, UK). 2012;**7**:639-650

[125] Nakhleh M, Broza YY, Haick H. Monolayer-capped gold nanoparticles for disease detection from breath. Nanomedicine (London, UK). 2014;**9**:1991-2002

[126] Nakhleh MK, Amal H, Awad H, Gharra A, Abu-Saleh N, Jeries R, et al. Sensor arrays based on nanoparticles for early detection of kidney injury by breath samples. Nanomedicine (New York, NY, US). 2014;**10**:1767-1776

[127] Nakhleh MK, Badarny S, Winer R, Jeries R, Finberg J, Haick H. Distinguishing idiopathic Parkinson's disease from other Parkinsonian syndromes by breath test. Parkinsonism & Related Disorders. 2015;**21**:150-153

[128] Nakhleh MK, Baram S, Jeries R, Salim R, Haick H, Hakim M. Artificially intelligent nanoarray for the detection of preeclampsia under real-world clinical conditions. Advanced Materials Technology. 2016;**1**:1600132-1600136

[129] Nakhleh MK, Jeries R, Gharra A, Binder A, Broza YY, Pascoe M, et al. Detecting active pulmonary tuberculosis by breath test using nanomaterial-based sensors. The European Respiratory Journal. 2014;**43**(5):1522

[130] Nardi-Agmon I, Abud-Hawa M, Liran O, Gai-Mor N, Ilouze M, Onn A, et al. Exhaled breath analysis for monitoring response to treatment in advanced lung cancer. Journal of Thoracic Oncology. 2016;**11**:827-837

[131] Peled N, Barash O, Tisch U, Ionescu R, Broza YY, Ilouze M, et al. Volatile fingerprints of cancer specific genetic mutations. Nanomedicine (New York, NY, US). 2013;**9**:758-766

[132] Peled N, Hakim M, Bunn Jr PA, Miller YE, Kennedy TC, Mattei J, et al. Non-invasive breath analysis of pulmonary nodules. Journal of Thoracic Oncology. 2012;**7**:1528-1533

[133] Peng G, Hakim M, Broza YY, Billan S, Abdah-Bortnyak R, Kuten A, et al. Detection of lung, breast, colorectal, and prostate cancers from exhaled breath using a single Array of nanosensors. British Journal of Cancer. 2010;**103**:542-551

[134] Peng G, Tisch U, Adams O, Hakim M, Shehada N, Broza YY, et al. Diagnosing lung cancer in exhaled breath using gold nanoparticles. Nature Nanotechnology. 2009;**4**:669-673

[135] Shehada N, Brönstrup G, Funka K, Christiansen S, Leja M, Haick H. Ultrasensitive silicon nanowire for real-world gas sensing: Noninvasive diagnosis of cancer from breath volatolome. Nano Letters. 2014;**15**:1288-1295

[136] Shehada N, Cancilla JC, Torrecilla JS, Pariente ES, Bronstrup G, Christiansen S, et al. Silicon nanowire sensors enable diagnosis of patients via exhaled breath. ACS Nano. 2016;**10**:7047-7057

[137] Shuster G, Gallimidi Z, Reiss AH, Dovgolevsky E, Billan S, Abdah-Bortnyak R, et al. Classification of breast cancer precursors through exhaled breath. Breast Cancer Research and Treatment. 2011;**126**:791-796

[138] Tisch U, Haick H. Nanomaterials for cross-reactive sensor arrays. MRS Bulletin. 2010;**35**:797-803

[139] Vishinkin R, Haick H. Nanoscale sensor technologies for disease detection via volatolomics. Small. 2015;**11**:6142-6164

[140] Wang B, Cancilla JC, Torrecilla JS, Haick H. Artificial sensing intelligence with silicon nanowires for ultraselective detection in the gas phase. Nano Letters. 2014;**14**:933-938

[141] Wang B, Huynh T-P, Wu W, Hayek N, Do TT, Cancilla JC, et al. Highly sensitive ambipolar field effect transistor-based diketopyrrolopyrrole copolymer for selective detection and discrimination of xylene isomers. Advanced Materials. 2016;**28**:4012-4018

[142] Broza Y, Kremer R, Tisch U, Gevorkyan A, Abdah-Bortnyak R, Shiban A, et al. A nanomaterial-based breath test for short-term follow-up after lung tumor resection. Nanomedicine (New York, NY, US). 2013;**9**:15-21

[143] Tisch U, Schlesinger I, Ionescu R, Nassar M, Axelrod N, Robertman D, et al. Detection of Alzheimer's and Parkinson's diseases from exhaled breath using nanomaterial-based sensors. Nanomedicine (London, UK). 2013;**8**:43-56

[144] Dovgolevsky E, Konvalina G, Tisch U, Haick H. Monolayer-capped cubic platinum nanoparticles for sensing nonpolar analytes in highly humid atmospheres. Journal of Physical Chemistry C. 2010;**114**:14042-14049

[145] Dovgolevsky E, Tisch U, Haick H. Chemically sensitive resistors based on monolayer-capped cubic nanoparticles: Towards configurable nanoporous sensors. Small. 2009;**5**:1158-1161

[146] Zaretski AV, Root SE, Savchenko A, Molokanova E, Printz AD, Jibril L, et al. Metallic nanoislands on graphene as highly sensitive transducers of mechanical, biological, and optical signals. Nano Letters. 2016;**16**:1375-1380

[147] Korotcenkov G. Metal oxides for solid-state gas sensors: What determines our choice? Materials Science and Engineering B. 2007;**139**:1-23

[148] Hangarter CM, Chartuprayoon N, Hernández SC, Choa YH, Myung NV. Hybridized conducting polymer chemiresistive Nano-sensors. Nano Today. 2013;**8**:39-55

[149] Yamazoe N, Sakai G, Shimanoe K. Oxide semiconductor gas sensors. Catalysis Surveys from Asia. 2003;**7**:63-75

[150] Li ZP, Zhao QQ, Fan WL, Zhan JH. Porous SnO_2 nanospheres as sensitive gas sensors for volatile organic compounds detection. Nanoscale. 2011;**3**:1646-1652

[151] Wang CX, Yin LW, Zhang LY, Xiang D, Gao R. Metal oxide gas sensors: Sensitivity and influencing factors. Sesnors. 2010;**10**:2088-2106

[152] Zhang Y, Li JP, An GM, He XL. Highly porous SnO_2 fibers by electrospinning and oxygen plasma etching and its ethanol-sensing properties. Sensors and Actuators B. 2010;**144**:43-48

[153] Jang JS, Choi JM, Kim SJ, Kim ID. Rational design of highly porous SnO_2 nanotubes functionalized with biomimetic nanocatalysts for direct observation of simulated diabetes. Advanced Functional Materials. 2016;**26**:4740-4748

[154] Khoang ND, Trung DD, Duy NV, Hoa ND, Hieu NV. Design of SnO_2/ZnO hierarchical nanostructures for enhanced ethanol gas-sensing performance. Sensors and Actuators B. 2012;**174**:594-601

[155] Huang JR, Wang LY, Gu CP, Wang ZJ, Sun YF, Shim JJ. Preparation of porous SnO_2 microcubes and their enhanced gas-sensing property. Sensors and Actuators B. 2015;**207**:782-790

[156] Sun P, Mei XD, Cai QY, Ma J, Sun JC, Liang XH, et al. Synthesis and gas sensing properties of hierarchical SnO_2 nanostructures. Sensors and Actuators B. 2013;**187**:301-307

[157] Huang JR, Xu XJ, Gu CP, Wang WZ, Geng BY, Sun YF, et al. Effective VOCs gas sensor based on porous SnO_2 microcubes prepared via spontaneous phase segregation. Sensors and Actuators B. 2012;**173**:599-606

[158] Tomer VK, Duhan S. Ordered mesoporous Ag-doped TiO/SnO_2 nanocomposite based highly sensitive and selective VOCs sensor. Journal of Materials Chemistry A. 2016;**4**:1033-1043

[159] Shankar P, Rayappa JBB. Gas sensing mechanism of metal oxides: The role of ambient atmosphere, type of semiconductor and gases – A review. Science Letters. 2015;**4**:1-18

[160] Bai H, Shi GQ. Gas sensors based on conducting polymers. Sesnors. 2007;**7**:267-307

[161] Kolmakov A, Moskovits M. Chemical sensing and catalysis by one-dimensional metal-oxide nanostructures. Annual Review of Materials Research. 2004;**35**:151-180

[162] Askim JR, Mahmoudi M, Suslick KS. Optical sensor arrays for chemical sensing: The optoelectronic nose. Chemical Society Reviews. 2013;**42**:8649-8682

[163] Han F, Inoue T, Wasai T, Kurauchi Y. K. Ohga, Odor sensor based on the information of colour. Journal of Engineering and Applied Sciences. 2006;**1**:1-6

[164] Geelhood SJ, Horbett TA, Ward WK, Wood MD, Quinn MJ. Passivating protein coatings for implantable glucose sensors: Evaluation of protein retention. Journal of Biomedical Materials Research. Part B, Applied Biomaterials. 2007;**81**:251-260

[165] Sokolov A, Hellerud BC, Lambris JD, Johannessen EA, Mollnes TE. Activation of polymorphonuclear leukocytes by candidate biomaterials for an implantable glucose sensor. Journal of Diabetes Science and Technology. 2011;**5**:1490-1498

Recent Progress in Nanostructured Zinc Oxide Grown on Fabric for Wearable Thermoelectric Power Generator with UV Shielding

Pandiyarasan Veluswamy, Suhasini Sathiyamoorthy, Hiroya Ikeda, Manikandan Elayaperumal and Malik Maaza

Additional information is available at the end of the chapter

http://dx.doi.org/10.5772/intechopen.76672

Abstract

Traditional materials for thermoelectric such as bismuth telluride have been studied and utilized commercially for the last half century, but recent advancements in materials selection are one of the principal function of the active thermoelectric device as it determines the reliability of the fabrication regarding technical and economic aspects. Recently, many researcher's efforts have been made to utilize oxide nanomaterials for wearable thermoelectric power generator (WTPG) applications which may provide environmental stable, mechanical flexibility, and light weight with low cost of manufacturing. In precise, fabric containing oxide metals have shown great promise as P–/N-type materials with improved transport and UV shielding properties. On the other hand, we have focused on ZnO nanostructures as a high-efficiency WTPG material because they are non-toxic to skin, inexpensive and easy to obtain and possess attractive electronic properties, which means that they are available for clothing with low-cost fabrication. To our observation, we are chaptering about the thermoelectric properties of ZnO and their composite nanostructures coated cotton fabric via the solvothermal method for the first time.

Keywords: ZnO, wearable device, UV shielding, nanocomposites, solvothermal method

1. Introduction

Turning textile fibers into progressively challenging and innovative product have seen manifold development in the last decade. The most promising characteristics of fibers drive the development of a wide range of fibrous products enabling digital components and electronics to be embedded in them known as smart clothing or smart textiles or smart fabrics. Adopting the new technology of smart fabrics is advantageous since they are relatively lightweight, comfortable, soft, and biodegradable. The move towards function-focus fibrous structures in pavement and embankment reinforcement may seem more appropriate, it implies technological applications such as stain resistance, antimicrobial, superhydrophobic/ superhydrophilic, antistatic, sensors, power generators, electromagnetic/ultraviolet interference shielding, wrinkles resistant, and shrink-proof abilities.

Among modern perspective, the wearable devices used to monitor a variety of health and environmental measures are becoming popular, and those devices requires capable of functioning autonomously for extended periods without replacing or recharge of batteries, but it is not practically impossible or ineffective. Since time immemorial, technology focuses to decrease the power consumption of devices so that the battery lasts for years. It has been done long ago, e.g., in watches. Low power consumption devices are developing gradually and that today offers provide opportunities for yesterday's dreams [1–6].

The main drawbacks are to supply power stable and reliable to commercialize wearable devices, the efforts are being made to explore alternative energy sources for recharging of batteries is performed on a regular or occasional basis using the auxiliary power source of energy harvesting modules such as photovoltaic cells or thermoelectric devices, which are also known as renewable energies. The renewable energy sources, being clean energy sources, noiseless, have an extreme advantage over conventional energy resources from an environmental point of view. The photovoltaic cells are variable power supply, dependent on sunlight, which could lead to an energy shortage if too much of a region's power comes from photovoltaic cells. Therefore, photovoltaic cells fail to provide long-term autonomy sufficient power for portable devices. Fortuitously, the thermoelectric devices have attracted much attention due to their ability of direct conversion of heat to electricity. The performance of thermoelectric devices is recycling wasted heat energy, lower production cost, scalability, long-lived power source, no side effects or harm, free from gas emission, easy to dispose and reliable source of energy [7–9] (**Figure 1**).

1.1. Motivation

The body heat from the human is one of the source to harvest energy, which produces electricity from tiny energies in the environment that is called thermoelectric (TE) power generator as shown in **Figure 2**. In this application, the power generator was designed in the form of textile, so-called wearable power generator (WPG). It is also made available in the form of curtain, tent, and umbrellas and therefore, it is useful not only for a daily life applications but for special applications such as natural disaster. TE materials have prominent aspects like reliability, environmental benignity, and easy incorporation into existing technologies. They can be used as long-life power sources and provide a long-lasting solution to the ever-growing demand for implantable medical devices. This increasing demands for lightweight, high flexible, stretchable, and washable presents critical challenges for the progress of WPG.

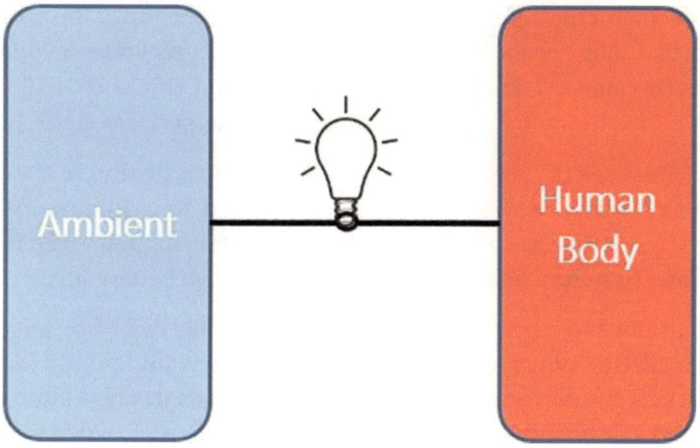

Figure 1. Commercial evolution of wearable thermoelectric devices.

Figure 2. Wearable power generation on human body.

1.2. History of wearable power generators

In 1821, the Thematron's technology developed watches at Centre Electronique Horloger (CEH) in Neuchatel, Switzerland. Basically, three different materials were used antimony, bismuth, and tellurium. The generators composed of two metal sheets: red one on the watch back side (in contact with the wrist's skin) and a blue one at the top of the case (in contact with ambient air); it is based on "Seebeck effect" principle, which generates a voltage that supplies energy to the quartz caliber. They had very low conversion efficiency, and it could not hold the electrical load, thus the distribution was eventually stopped.

With not much achievement, several efforts followed, when finally, in the year 1988, Seiko developed another thermic watch; which was connected to a lithium-ion battery with a 10 months' power reserve. It has power saving mode, and it stopped when the watch was not being worn on hand to reduce energy consumption and save battery life.

In 2010, Skinny Player was designed by Chinese engineers Chih-Wei Wang and Shou-His Fu; the player with a preloaded album onto it and to listen music you would simply stick the player onto your body; By sticking the player, it would gain power from its flexible battery charging device; it is eco-friendly and sustainable way; major issues with the player is when intended to exercise, the body is hot and usually sweats, however it was stuck on a sweating body for a long time.

A tremendous achievement in the same year by Orange Power Wellies; Boots that converts heat from your feet into electrical power to charge your mobile phone; the battery gets charged by 12 hours of walking but still its remarkable significant that the amount of energy can be harvested from the normal human activity; in order to decrease the charging period we should try more active duties like dancing or running; then the device contains pairs of p-type and n-type semiconductor materials forming a thermopile.

In 2011, the first popular TE mobile developed by Nokia E-Cu designer Patrick Hilandom; the mobile would be able to heat-conductive charging system, being in your pocket and it creating a charger-free mobile phone future; the phone body is made up of copper on which are engraved heat sinks in the form of dried and cracked earth; basically, mobile phone chargers produce 51,000 tons of emission of energy in the environment, in addition to greenhouse gases.

In 2013, Fujifilm has developed the flexible polymer TE conversion module; it can able to generate milliwatt to microwatt power; it would be possible to operate the medium and low-temperature waste heat that surrounds us all in our daily lives; the power is generated by affixing a film substrate coated with organic materials on to a heating source, and the possible power can be generated in the temperature difference ambient and human body temperature of even 1°C.

In 2016, developed by Sam Shames in Embr Labs, Wristify is a heating and cooling bracelet that lets you make yourself comfortable by sending hot or cold pulses to a patch of skin on the wrist; it is more comfortable without changing your core temperature by focusing on the temperature at the skin, and sudden changing of the skin temperature can also notice immediately make you more comfortable.

1.3. Thermoelectric materials

The performance of WPG is essential to develop high-efficient flexible TE materials. Theoretically, the efficiency of TE power generator rises monotonously with increasing dimensionless figure of merit, zT which is given by $zT = (S2\sigma)/\kappa\, T$, where S, σ, κ, and T denote the thermopower (Seebeck coefficient), electrical conductivity, thermal conductivity, and temperature respectively. Henceforth, with the aim of improving the efficiency, an increase in Seebeck coefficient and a decrease in thermal conductivity are required. Traditional materials for TE such as bismuth telluride have been studied and utilized commercially for the last half century, but recent advancements in materials selection are one of the principal function of active TE device as it determines the reliability of the fabrication regarding industrial and economic aspects [10].

1.4. Nanostructuring an efficient approach

As a resolution of this purpose, the nanostructured materials are expected to enhance the Seebeck coefficient because of a carrier confinement effect and to lower the thermal conductivity because of an increase in the boundary scattering in phonon transport. The selection of the material has a significant role in the fabrication of high-performing TE materials. Owing to its flexible nature, conducting polymers (CPs) are favorable materials for the practical TE applications. Because of their high flexibility, environmental stability, and facile synthesis, they have the potential for use on human skin. However, most of the CPs such as polyaniline, polypyrrole, and poly(3,4-ethylenedioxythiophene) polystyrene sulfonate (PEDOT: PSS) are expensive and require complex treatments to achieve good electrical conductivity. Hence efforts have been made to find an alternative for fabricating flexible TE materials and composite materials have recently been attracting more and more attention since they possess many advantages including high thermopower, easy process-ability, and cost-effectiveness. Recently, many researcher's efforts have been made to utilize oxide nanomaterials for WPG applications that may provide environmental stable, mechanical flexibility, and lightweight with low cost of manufacturing [11–13].

1.5. Conflicting wearable power generator

Human skin as one of its charge-collectors, converts muscle movements into enough power for small electronics which requires optimization of a variety of conflicting properties. At the same time, it should enhance the performance of the material and increase the zT for which characteristics such as high electrical conductivity, lower the thermal conductivity and a high thermopower (Seebeck coefficient) are required. The three parameters are strongly interdependent each other to makes the enhancement of zT highly challenging. Sometimes the nanomaterials coated fiber bond is not the strongest, so it wears down and wears off through use and another challenge, which is related conductive fibers is erosion over time. Various techniques have been considered owing to maximum performance in TE materials; a few examples of manipulation are described in the following sections [14–17].

1.5.1. Optimization of power generation on human body

Human body releases the thermal energy of about 2.4 to 4.8 watts, but more can be achieved by placing at different parts of the body. The ideal positions are indicated in **Figure 3**. Similarly, the body heat trapping power may be enhanced by physical activity involved as shown in **Table 1**. Note, however, that energy harvested from the user may require considerable settings in position of the body. Most of the times it provides better expected results [18–22].

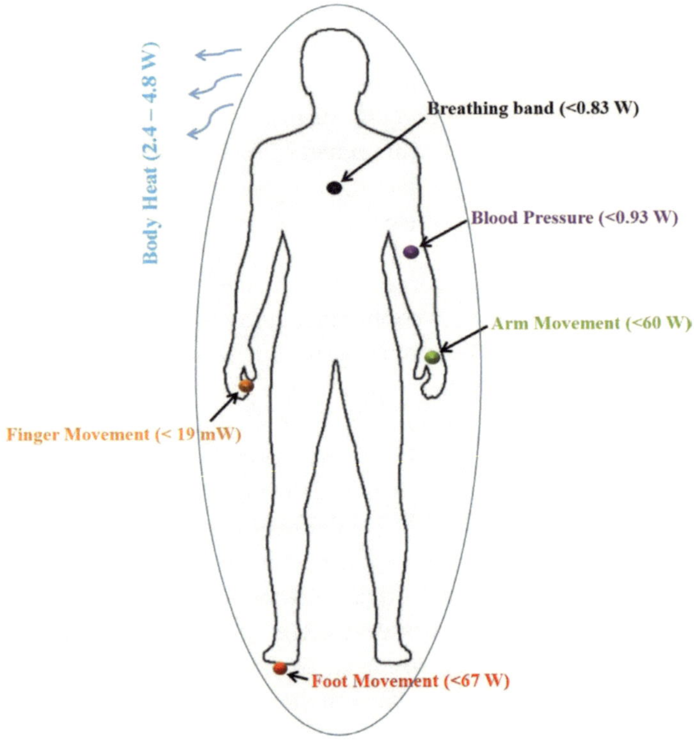

Figure 3. WPG recovery from body in different places.

1.5.2. Optimization of coating techniques

On the other hand, we have focused on ZnO and rGO nanostructure composites as a high-efficiency WPG material because they are non-toxic to skin, inexpensive, available easily and possess attractive electronic properties, making them available for engineered clothing with low-cost fabrication. Up-to-date, several coating methods have been used to grow ZnO and rGO films for tuning its size and morphology, such as sol–gel method, solvothermal synthesis, chemical precipitation, microwave method, sonochemical route, chemical vapor deposition, vapor-phase method, sputtering, and electrochemical. Among these techniques, the solvothermal method is a promising method to synthesize nanostructures with high purity and isometric crystallization. As well as it improves a surface's strength and durability. Furthermore, a variety of nanostructures such as nanorods, nanoneedles, nanotube, nanosheets, nanoflakes,

Activity	Sleeping	Sitting	Standing	Conversation	Eating	Driving	House keeping	Swimming	Running
Watts	81	93	116	128	128	163	175	582	1048
Kilo cal/ hr	70	80	110	110	110	140	150	500	900

Table 1. Power enhancement by movement of body.

nanodiscs, and nanoflowers can be obtained by solvothermal method. In precise, fabrics containing ZnO and rGO have shown great promise as N-type and P-type materials with improved transport and UV shielding properties [23–26].

1.5.3. Optimization of composites

The thermopower can be enhanced in the presence of chemical composition and crystallinity of thermoelectric materials at nano-inclusion via engineering of multicomponent nanomaterials (nanocomposites), it has proven which can reduce the lattice thermal conductivity by promoting the phonon scattering at gain boundaries and simultaneously it enhances the power factor due to electron filtering at grain boundaries [27–30].

2. Materials fabrication

2.1. Scouring process

Scouring is an important treatment stage of cotton fibers for reducing the amount of impurities such as oils, wax, gums, fatty materials, natural coloring. The absence of impurities produce the fabric more hydrophilic for achieving uniform nanomaterials coating as shown in **Figure 4** before and after scouring process. The complete scouring process as shown below [31];

The development of nanomaterials is closely linked to the following synthesis process.

2.2. Sol-gel synthesis

Sol gel approach began with inorganic ceramic materials at early mid 1800's and especially fabrication of metal oxides. In a typical sol gel reaction process [32], the reactants involve conversion of monomers into a colloidal solution then it reacts as the precursor for an integrated network. The deposition of sol gel solution produces the coatings on the cotton fabric by dipping and followed by heat treatment to form an amorphous or crystalline coatings as shown in **Figure 5** (a – d).

2.3. Microwave approach

Since in 2000, the Microwave-assisted approach is known to transform electromagnetic energy into heat for the ultra-fast synthesis of chemical reactions using microwave irradiation [33]. It has homogeneous, rapid with deep internal heating as resultant produce high yield and lower quantities of side products, purifications of yield are easier, reactions are more reproducible, and the reaction temperature is accurate but it not reproducible with conventional heating. The cotton fibers microwave assisted synthesis has established with producing nanostructures to influence both side distribution of fibers effectively as shown in **Figure 5** (g–h).

2.4. Coprecipitation method

In this process, the desired component is precipitated under the aqueous region to synthesize high purity and crystallinity without significant defects. As the novel method performed at room

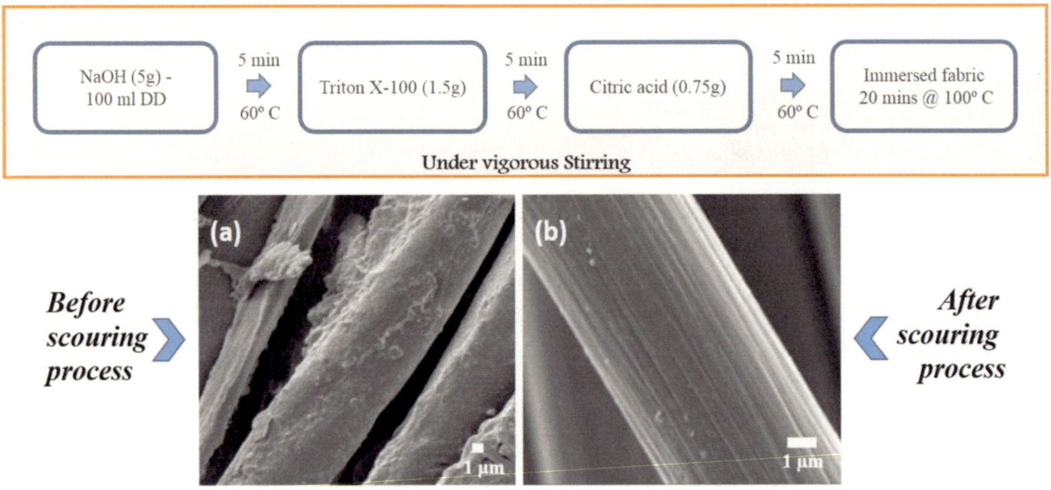

Figure 4. SEM image of cotton fabric (a) before and (b) after of scouring process.

Figure 5. Coating method for cotton fibers (a–d) sol gel approach and (g–h) microwave approach.

temperature and suitable for quantifiable production because of its high yield, excellent repeatability, and low cost. Particularly, the metal hydroxides are precipitated from their precursor element because of low solubility, but the inclusion of alkaline solution can overcome it or raising the pH. For example, **Figure 6** (a–d) shows the coated cotton fabric nanostructures is done by the co-precipitation approach with the addition of amine groups and metal precursors [34].

2.5. Chemical bath deposition

Since from 1933, it was known that for solution growth method, and it adopts the simplest form of the concentrated solution which induces a solid phase to exsolve substrate mounting for thin films. Those yields are stable, adherent, uniform coatings with good reproducibility by a simple process [35]. To grow a denser nanostructure without seed layers are also possible but the precursor species tend to be very dilute within the solution during the deposition process and tailoring of chemical bath deposition is not possible without a clear understanding or control of mechanism. We could observe the coated nanostructures on cotton fiber as shown in **Figure 6** (e – h).

Figure 6. Cotton fabric deposited with nanostructures via (a) co-precipitation method and (b) chemical bath method.

2.6. Sonochemical synthesis

It is the effect of sound wave propagating in forming acoustic cavitation in liquids, resulting high temperature and pressure in a microscopic region of the sonicated process. The most notable effects are consequences of primary and secondary radical reactions it increases chemical activity through the formation of new species [36, 37]. The effects in processes may modification of surface morphology and particle size at high velocity interparticle collisions as shown in **Figure 7** (a–d).

2.7. Hydrothermal synthesis

In principle, hydrothermal technique refers to heterogeneous reactions in aqueous media above 100°C temperature and 1 bar pressure, these resulting powder needs no more high-temperature calcination and thus can avoid the nanoparticles from re-clustering (Oswald repining). The growth of nanomaterials is performed in an apparatus consisting of a steel

Figure 7. Different techniques involved for coating on cotton fibers (a–d) sonochemical, (e–g) hydrothermal, and (i–l) solvothermal method.

vessel with Teflon container innermost that sustained the pressure. The morphology and size of the nanoparticles could be tuned easily by operating temperature, pressure and time duration. Using hydrothermal synthesis for depositing metal oxide on the cotton fabric [38, 39] was shown in **Figure 7** (e–h).

2.8. Solvothermal synthesis

The phenomena of the solvothermal method are sol gel process followed by the hydrothermal approach, and they have involved two steps process for the crystallization process: crystal nucleation formation by sol gel process and subsequent growth rates by the hydrothermal approach. Particle size and morphology can by controlling by processing variables such as temperature, pH, precursor concentration and additives or surfactants. Thus, solvothermal synthesis allows for precise control over the size, shape distribution and crystallinity of metal oxide nanostructures on cotton fabric [40–43] as shown in **Figure 7** (i–l).

3. Wearable device measurements

3.1. Electrical measurements

A current–voltage (IV) characteristics is a relationship, exemplary representation of graph between the electric current through a material and the corresponding potential difference obtained. In the precise IV measurement, current/ voltage response can be measured while applying voltage/ current to obtain current versus voltage and resistance characteristics and power dissipated or generated can be determined from the IV curve by the resistive element [44–48]. **Figure 8** (a) shows the JASCO, CEP-25BX with which IV_ characteristics were monitored in this study.

3.2. Thermopower measurements

In general, the thermoelectromotive force determined by using probes that are directly connected to the samples, and the temperature difference is measured through thermocouples. Consequently, for a good measurement of thermopower (Seebeck coefficient), it is necessary to measure the thermoelectromotive force and temperature difference instantaneously at the same place, to make sure the probes in excellent thermal and electrical connection with the samples, and to have a high precision measurement device. There are two primary methods used to measure thermopower which is the integral and differential methods. The integral method typically used for analysis with large temperature difference, where one end of the sample is maintained at a constant temperature T1 while the temperature of other end is varied in the temperature range of interest (T2 = T1 +ΔT).

The fitting technique is used to estimate the thermopower on complete measured data set of measured thermoelectromotive force V (T1. T2) as shown in **Figure 8** (b). The integral method succeeds in minimizing the influence of voltage offset. Since the applied temperature difference is large which will produce a large thermoelectromotive force. However, it is hard to maintain the temperature at one end of the sample due to the flow of heat from the high temperature applied at the other end. Moreover, it is also difficult to obtain the proper fit for evaluating the thermopower from the measured data. In additionally, the differential method can operate from a small temperature difference is applied across the sample and it is maintained at the average temperature (T = (T1 + T2)/ΔT). Then the thermopower is evaluated from the linear fit of thermoelectromotive force and temperature difference for several sets of data as shown in **Figure 8** (c). Hence, this evaluation of thermopower is valid for constant thermopower at average temperature. This technique is also effectively eliminating the offset voltages that arise from thermocouples in similarities and no equilibrium contact interfaces. In this study, the thermopower of samples is measured to be constant in the average temperature range, and a schematic representation of experimental setup shown in **Figure 8** (d).

There are two Cu plates were positioned side by side with a gap of ~1 mm. The sample for measuring placed across the gap, associate with both plates. A resistive heater was placed underneath the plate, and by controlling the heater current, a temperature difference could be produced in a plane parallel to the sample surface. A couple of probes and two K-type thermocouples were directly connected to the sample surface. The time evolution of the thermoelectromotive force was measured by a nanovoltmeter (Keithley 2182A) and simultaneously with the temperatures at the high- and low-temperature regions by a digital meter (Keithley 2700) equipped with a switching module (Keithley 7700) [49–52].

3.3. Ultraviolet ray shielding measurements

The human body have largest organ is skin, since in the beginning people desired to use sunscreen products to get beautiful and to prevent the skin burns from sun tan, without the risk towards change to burns. Now, it is necessary for all people to use sunscreen products because for the protection against ultraviolet (UV) radiation. The UV radiation is broadly divided into three distinct bands in order of decreasing wavelength and increasing energy. UVA (320–400 nm) has a longer wavelength; it penetrates skin through both the epidermis and dermis, UVB (290–320 nm) has an intermediate wavelength; these rays can be blocked by sunscreen, automobile glass, and windows but UVA is not filtered. UVC (200–290 nm) has a

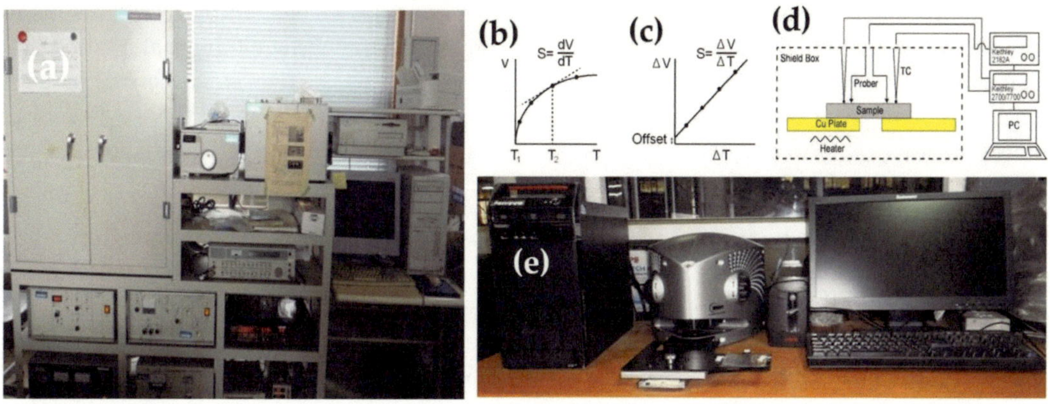

Figure 8. (a) JASCO, CEP-25X, illustration of integral (b) and differential (c) technique for measuring thermopower (Seebeck coefficient), (d) schematic representation of system for measuring the thermopower and (e) UV shielding measurement system.

shorter wavelength; it is effectively filtered by the atmosphere (Ozone layer) and is therefore not a forethought to be a considerable factor. Those imbalances can result in formation of wrinkles, hair loss, rashes, life-threatening cancers, and disorders in immune regulation.

UV protection factor (UPF) applications due to increasing global awareness on green/eco textile products and their market demand. Research has been intensified in the area of development of sustainable UV protective textile using plant extracts and other natural polymeric materials or biocompatible. The Ultraviolet ray shielding measurement was analyzed by UV scattering and absorption spectroscopy by Lapsphere UV1000F as shown in **Figure 8** (e). In the Australia/New Zealand standard AC/NZS 439:1996 method, Eq. (1) is used to determine,

$$UPF = \frac{\int_{290}^{400} E_{\lambda} \times S_{\lambda} \times d\lambda}{\int_{290}^{400} E_{\lambda} \times S_{\lambda} \times T_{\lambda} \times d\lambda} \tag{1}$$

where E_{λ} is the relative erythemal spectrum effectiveness, S_{λ} is the solar UV spectral irradiance, T_{λ} is the spectral transmittance of the specimen (incoming light that passes through the sample) and λ represents the wavelength (nm). Laundering durability of the treated fabric was measured in agreement with AATCC test method 61-2006 [53–55].

4. Fabrication of hierarchical ZnO nanostructures on cotton fabric

Various growth techniques of ZnO have been developed for tuning its size and morphology, such as sol–gel method, solvothermal synthesis, chemical precipitation, microwave method, sonochemical route, chemical vapor deposition, and vapor-phase method [56, 57]. Among these techniques, the solvothermal method is a promising method to synthesize ZnO nanostructures with high purity and isometric ZnO crystallization. Furthermore, a variety of nanostructures such as nanorods, nanoneedles, nanotube, nanosheets, nanoflakes, nanodiscs, and nanoflowers [58–61] can be obtained by solvothermal method. In applying the power generator to the textile, curtain, and so forth, the flexible TE material is required to have three principal functions of the superhydrophobic surface, ultraviolet (UV) shielding and high TE efficiency. Consequently, in this paper, we fabricated ZnO nanostructures on cotton fabric by the solvothermal method and characterized their UV shielding and thermoelectric properties. A facile solvothermal method was adapted to grow the mixed nanostructures like rods and sheets without any surfactant or amine additive. Due to the two-step process, the controlled growth and optimization of the ZnO mixed nanostructures, its functional properties were investigated [62].

In **Figure 9** shown the fabrication process of the ZnO nanostructures coating on cotton fabric by solvothermal method. To obtain the uniform and denser of ZnO coating on the cotton fabric surface, where the glass frame designed and manufactured for the small-scale screening of ZnO seed and growth conditions. The cotton fabric is treated by scouring process to eliminate the wax, fatty and other impurity materials. Pale white color of cotton fabric obtained after scouring process, surface of the fibers is smooth and it increased the surface area due to removal of unwanted materials. Two step process have been involved like seed creation and growth process, the end product apparently found that ZnO nanostructures with high density and high aspect ratios coatings obtained. The morphology images of **Figure 10** (a1 & a2) show high density and

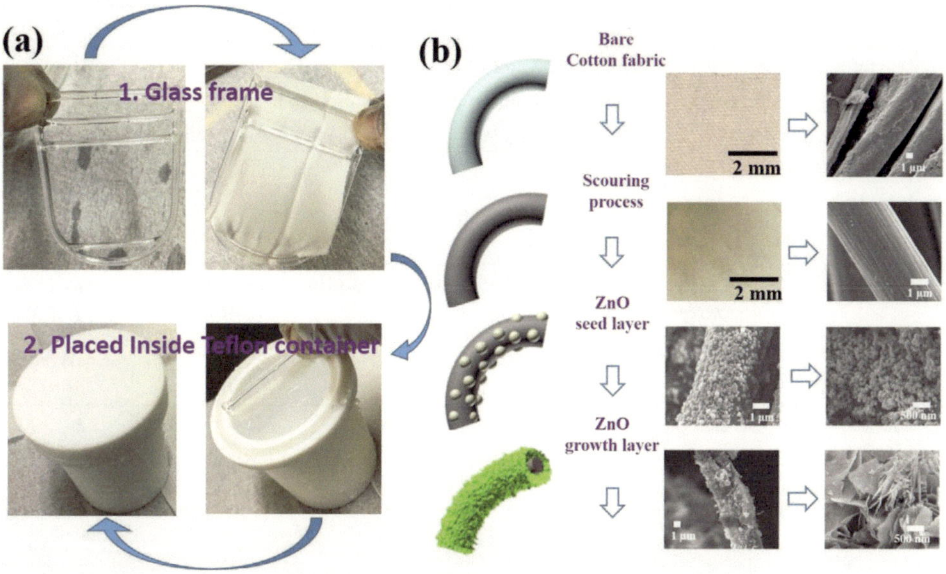

Figure 9. (a) Growth technique (b) fabrication process of ZnO nanostructures coated on cotton fibers.

uniformly aligned nanorods without any preferred orientation, which means ZnO nanorods grow preferably on c-axis direction towards the surface of each fibrous. Thus, the nanorods grew along the [002] direction. The nanosheets are shown **Figure 10** (b1 & b2), which represents the overlapped layer by layer as seen in image. Moreover, thickness of nanosheets is around to be 4 nm. **Figure 10** (c1 & c2) reveals the crumbled nanosheets with shrunken edges on nanorods.

These structural changes happening due to the concentration of zinc nitrate: hexamine as given in **Table 2**. When the growth solution of nanoparticles the OH^- anions and Zn^{2+} cations were released, which leads to the immediate precipitation of $Zn(OH)_2$ units. Successively ZnO nanostructures were made by the solvothermal of $Zn(OH)_2$ units at extended reaction temperature and time. In self-assembling of nanoparticles takes place in the growth solution, hexamine act as a capping agent or stabilizer and building block. In the meantime the presence of ZnO seed

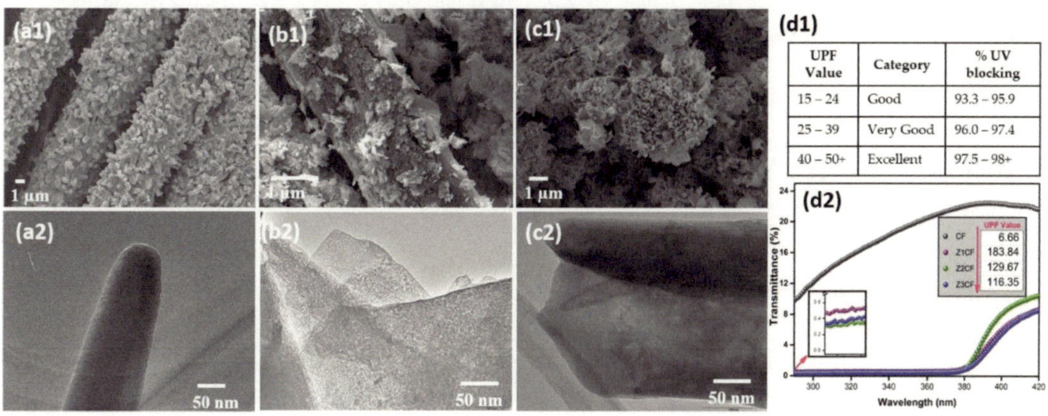

Figure 10. (a–c) morphology image of ZnO growth layer fibers and (d) UV shielding spectra.

S. no	Materials		Nanostructures
1	Zinc nitrate 1 molar	Hexamine 2 molar	Nanorods
2	Zinc nitrate 1 molar	Hexamine 1 molar	Nanosheets
3	Zinc nitrate 2 molar	Hexamine 1 molar	Nanosheets crumpled Nanorods

Table 2. Possible ZnO nanostructures.

Sample	Conduction type	Carrier concentration n (10^{19} cm^{-3})	Resistivity ρ (Ω-cm)	Mobility μ (cm^{-2}/V-s)	Seebeck coefficient S ($\mu V/K$)	Power factor PF ($\mu W/m.K^2$)
Z1	n	0.05	0.15	81.1	45	13
Z2	n	2.1	0.02	13.7	6	1.4
Z3	n	1.5	0.04	11.5	28	22

Table 3. Thermoelectric properties at room temperature.

layer can minimize the energy barrier and the lattice divergence at heterogeneous nucleation site of the growth layer. When amine group presence of high concentration, it encourages the growth oriented along the c-axis. Whereas if the amine group in equal concentration, c-axis oriented growth direction leads to longitudinal and lateral direction. In other case, subsequent growth formation which would coalescence of crumpled with shrinkage edges of nanosheets surrounded on nanorods. Ultraviolet (UV) ray protection values and description are shown in **Figure 10** (d1). UV transmittance curve of the before and after coating of ZnO on cotton fabric is shown in **Figure 10** (d2). The before coating of ZnO showed 6.66 which is poor protection against UV radiation, but coated nature which is greatly improved the blocking of UV radiation.

Thermoelectric properties at room temperature are shown in **Table 3**, when oxygen vacancy occurs the Seebeck coefficient indicates n-type conduction. The Z1 resistivity and mobility are higher in nanorods compare with Z2 & Z3 nanostructures which is related to the poorer connection between grains. In Z2 as a thin film it reduce electrical resistivity and also the boundary scattering of charge carrier. However Z3, when the nanorods surrounded by nanosheets possess the intergranular electron transport is expected to be easier than nanorod, which explains the lower value of the electric resistivity qualitatively.

5. Incorporation of ZnO and their composite nanostructures on cotton fabric

Material selection has a significant role in the fabrication of high TE performing materials. The conducting polymers are promising to its flexible nature and favorable for the TE applications. However, most of the conducting polymers are expensive and required complex synthesis procedure to achieve high electrical conductivity [63–65]. So the efforts have to be made for finding an alternative for fabricating flexible materials of TE applications and composites have recently been appealing more consideration such as thermopower, easy process ability, cost effective, and long life span [66–68]. Similar to the previous section here also two step solvothermal method for ZnO and their composite nanostructures on cotton fabric have been employed. The fabrication process of the ZnO nanocomposite coated cotton fabric represented in the **Figure 11**.

As we discussed the amine group concentration in previous session, similarly higher concentration of hexamine longer nanorods are obtained as shown in **Figure 12** (Z1) and when hexamine decreases the nanorods dramatically decreases as shown in **Figure 12** (Z2). However, Sb composite ZnO influences the orientation of nanorods towards the surface of the fibers. **Figure 12** (SZ1 & SZ2), high density Sb/ZnO composite formed on the fiber surface with intertwining into a cluster. **Figure 12** (AZ1 & AZ2) exhibits the absents of nanorods and initiate the 3D network architectures growth of nanosheets assembled with lots of interspaces/ intermesh with each other's [69, 70].

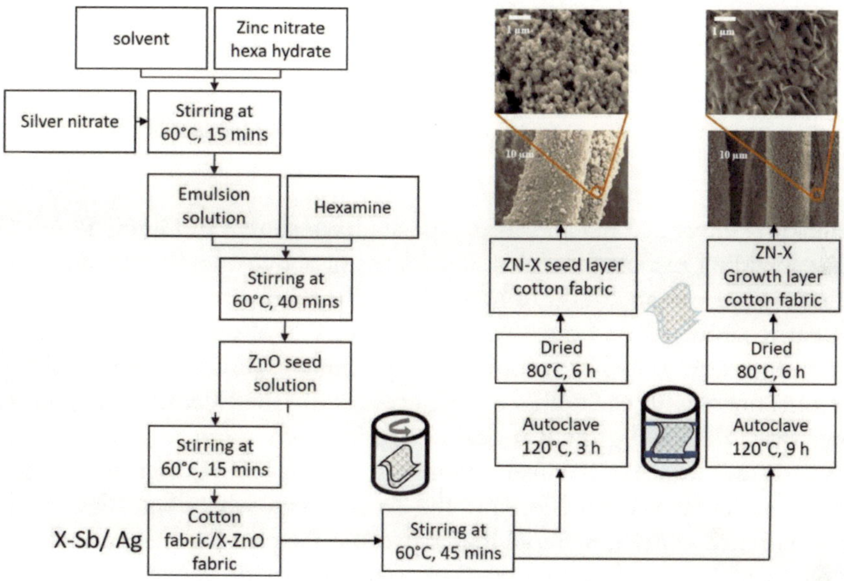

Figure 11. Fabrication process of the nanocomposite coated cotton fabric.

Figure 12. Morphology image of ZnO and its composite growth layer fibers.

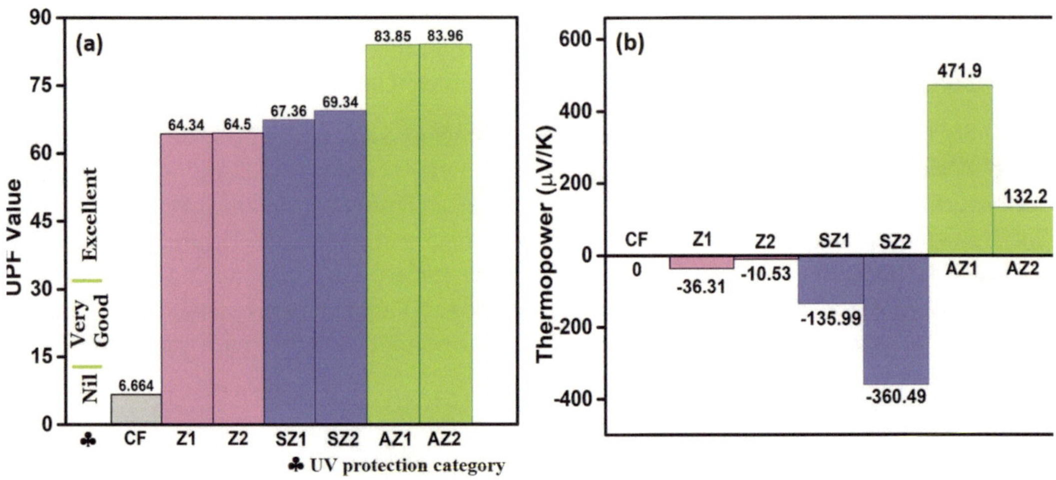

Figure 13. (a) UV shielding effect and (b) Thermopower of before/after coating of cotton fabric.

The UV shielding effect of ZnO and its composite have shown in **Figure 13** (a), bare fabric shown poor shielding response of 6.664 and coated of above 64 which means it is excellent for UV shielding performance. **Figure 13** (b) shows the thermopower of coated cotton fabric. The ZnO coated fabric depict from −36.31 to −10.53 μV/K of longer to shorter nanorods as it higher concentration to low concentration of hexamine. Thus, the longer nanorods have possibility of connect with each other's which contribute the electron transport easier than in the other case. The ZnO growth process disrupted, while the Sb composite introduced growth towards the surface of the fiber although the thermopower increases from −135.99 to −360.49 μV/K due to Sb incorporation into the Zn site. Similarly, when Ag composite led to favor the recombined with Zn material that may cause the positive effect on the charge separation efficiency of 471.9 to 13.2.2 μV/K. It plays an important role in charge transport and intergranular crystal structure.

6. Conclusion

In this chapter, we focused on wearable thermoelectric power generator since the performance of the device is recycling wasted heat energy, lower production cost, scalability, long-lived power source, no side effects or harm, free from gas emission, easy to dispose of and reliable source of energy. Herein, we have adopted the solvothermal method for the coating ZnO and its composite with various nanostructures such as nanorods, nanosheets, nanospheres, and nanowalls. The seed creation and growth condition, the concentration of precursors, growth time, have been systematically studied. The as-synthesized fabric sample of functional properties was analyzed by field emission scanning electron microscopy, transmission electron microscopy, UV shielding properties and thermoelectric properties. Besides, two different composite (Sb and Ag) were used to study the efficiency of thermopower.

Author details

Pandiyarasan Veluswamy[1,2]*, Suhasini Sathiyamoorthy[3], Hiroya Ikeda[2], Manikandan Elayaperumal[4,5] and Malik Maaza[4]

*Address all correspondence to: pandiyarasan@yahoo.co.in

1 Department of Electronics and Communication Engineering, Indian Institute of Information Technology Design and Manufacturing, Chennai, India

2 Research Institute of Electronics, Shizuoka University, Hamamatsu, Japan

3 Department of Electronics and Communication Engineering, SRM University, Chennai, India

4 UNESCO-UNISA Africa Chair in Nanosciences/Nanotechnology Laboratories, College of Graduate Studies University of South Africa (UNISA), Pretoria, South Africa

5 Department of Physics, Thiruvalluvar University College for Arts and Science (TUCAS), Thiruvalluvar University, Vellore, India

References

[1] Gilsoo C, Seungsin L, Jayoung C. Review and reappraisal of smart clothing. International Journal of Human-Computer Interaction. 2009;**25**(6):582-617. DOI: 10.1080/10447310902997744

[2] Shuhui L, Jianying H, Zhong C, Guoqiang, Yuekun L. A review on special wettability textiles: Theoretical models, fabrication technologies and multifunctional applications. Journal of Materials Chemistry A. 2017;**7**:31-55. DOI: 10.1039/c6ta07984a

[3] Gokulraja TS, Kadarkaraithangam J, Vimal KM, Pandiyarasan V, Ikeda H, Karthikeyan K. Excellent floating and load bearing properties of superhydrophobic ZnO/copper Stearate nanocoating. Chemical Engineering Journal. 2017;**320**:468-477. DOI: 10.1016/j.cej.2017.03.052

[4] Pandiyarasan V, Archana J, Pavithra A, Ashwin V, Navaneethan M, Hayakawa Y, Ikeda H. Hydrothermal growth of reduced graphene oxide on cotton fabric for enhanced ultraviolet protection applications. Materials Letters. 2017;**188**:123-126. DOI: 10.1016/j.matlet.2016.11.047

[5] Sennoga T, Jie Z, Yuying Y, Bo L. A comprehensive review of thermoelectric technology: Materials, applications, modelling and performance improvement. Renewable and Sustainable Energy Reviews. 2016;**64**:698-726. DOI: 10.1016/j.rser.2016.07.034

[6] Zheng XF, Liu CX, Yan YY, Wang Q. A review of thermoelectrics research – Research developments and potentials for sustainable and renewable energy applications. Renewable and Sustainable Energy Reviews. 2014;**32**:486-503. DOI: 10.1016/j.rser.2013.12.053

[7] Weber J, Potje-Kamloth K, Haase F, Detemple P, Volklein F, Doll T. Coin-size coiled-up polymer foil thermoelectric power generator for wearable electronics. Sensors and Actuators A, DOI: 10.1016/j.sna.2006. 2006;**132, 04**:325, 054-333

[8] Francisco S, Amin N, Daryoosh V, Mehmet CO. Designing thermoelectric generators for self-powered wearable electronics. Energy Environmental Science. 2016;**9**:2099-2113. DOI: 10.1039/C6EE00456C

[9] Vladimir L, Ruud JMV. Wearable electronics self-powered by using human body heat: The state of the art and the perspective. Renewable and Sustainable Energy. 2009;**1**: 062701. DOI: 10.1063/1.3255465

[10] Aranguren P, Astrain D, Rodriguez A, Martinez A. Experimental investigation of the applicability of a thermoelectric generator to recover waste heat from a combustion chamber. Applied Energy. 2015;**152**:121-130. DOI: 10.1016/j.apenergy.2015.04.077

[11] Sungmook J, Jongsu L, Taeghwan H, Minbaek L, Dae-Hyeong K. Fabric-based integrated energy devices for wearable activity monitors. Advance Materials. 2014;**26**:6329-6334. DOI: 10.1002/adma. 201402439

[12] Veena M, Omer O, Susasn T, David W. Flexible technologies for self-powered wearable health and environmental sensing. Proceeding of the IEEE. 2015;**108**:665-681. DOI: 10.1109/JPROC.2015.2412493

[13] Francioso L, Pascali CD, Farella I, Martucci C, Creti P, Siciliano P, Perrone A. Flexible thermoelectric generator for ambient assisted living wearable biometric sensors. Journal of Power Sources. 2011;**196**:3239-3243. DOI: 10.1016/j.jpowsour.2010.11.081

[14] Wei W, Peining C, Sisi H, Xuemei S, Huisheng P. Smart electronic textiles. Angewandte Chemie International Edition. 2016;**55**:6140-6169. DOI: 10.1002/anie.201507333

[15] Tony V, Courtney H, Joseph R, Nicholas K, Chao H, Pasindu G, David E, Rutvik J, Yanliang Z. High-performance and flexible thermoelectric films by screen printing solution-processed nanoplate crystals. Scientific Reports. 2016;**6**:33135. DOI: 10.1038/srep33135

[16] James D, Henry G, Steven M, Michael M, Shanshan Y, Feiyan L, Eric B, Brinnae B, Bongmook L, Veena M, Young Z, Omer O, Jason S, John M, David P and Alper B: Low power wearable systems for continuous monitoring of environment and health for chronic respiratory disease. IEEE Journal of Biomedical and Health Informatics. 2016:**20**:1251-1264. DOI: 10.1109/JBHI.2016.2573286

[17] Deepa M, Zuoqian W, Paul KW, James WE. Printed flexible thermoelectric generators for use on low levels of waste heat. Applied Energy. 2015;**1556**:587-592. DOI: 10.1016/j.apenergy.2015.07.066

[18] Ziyang W, Vlandimir L, Paolo F, Chris VH. Realization of a wearable miniaturized thermoelectric generator for human body applications. Sensors and Actuators A. 2009;**156**:95-102. DOI: 10.1016/j.sna.2009.02.028

[19] Wei H, Gan Z, Xingsing Z, Jie J, Guiqiang L, Xudong Z. Recent development and applications of thermoelectric generator and cooler. Applied Energy. 2015;**143**:1-25. DOI: 10.1016/j.apenergy.2014.12.075

[20] Zhisong L, Huihui Z, Cuiping M, Chang ML. Silk fabric-based wearable thermoelectric generator for energy harvesting from the human body. Applied Energy. 2016;**164**:57-63. DOI: 10.1016/j.apenergy.2015.11.038

[21] Lymberis A, Paradiso R. Smart fabrics and interactive textile enabling wearable personal applications: R & D state of the art and future challenges. IEEE EMBS Conference. 2008;**2008**:5270-5273. DOI: 10.1109/IEMBS.2008.4650403

[22] Seiichi D, Norifumi I, Yusuke I, Shoichiro I, Keisuke S, Muneaki O, Kenji S, Kentaro K. Stack-type thermoelectric power generating module with flexible section and using phase changes of low-boiling-point medium. Energy Conversion and Management. 2016;**127**:103-111. DOI: 10.1016/j.enconman.2016.08.040

[23] Robert JS, Steven JW, Karuna SK. Theoretical limits of thermoelectric power generation from exhaust gases. Applied Energy. 2014;**133**:80-88. DOI: 10.1016/j.apenergy.2014.07.075

[24] Patyk A. Thermoelectric generators for efficiency improvement of power generation by motor generators – Environmental and economic perspectives. Applied Energy. 2013;**102**:1448-1457. DOI: 10.1016/j.apenergy.2012.09.007

[25] Minijeong H, Jonghwa P, Youngoh L, Hyunhyub K. Triboelectric generators and sensors for self-powered wearable electronics. ACS Nano. 2015;**9**:3421-3427. DOI: 10.1021/acsnano.5b01478

[26] Tongcai W, Weiling L, Wei W, Shan-Tung T. Waste heat recovery through plate heat exchanger based thermoelectric generator system. Applied Energy. 2014;**136**:860-865. DOI: 10.1016/j.apenergy.2014.07.083

[27] Chaoxing W, Tae Whan K, Fushan L, Tailiang G. Wearable electricity generators fabricated utilizing transparent electronic textiles based on polyester/Ag Nanowires/Graphene core-shell nanocomposites. ACS Nano. 2016;**10**:6449-6457. DOI: 10.1021/acsnano.5b08137

[28] Matteo S, Alessandro C. Wearable electronics and smart textiles: A critical review. Sensors. 2014;**14**:11957-11992. DOI: 10.3390/s140711957

[29] Francisco SP, Eduardo M, Manuel FM, Eduardo PT. Growth of vertically aligned ZnO nanorods using textured ZnO films. Nanoscale Research Letters. 2011;**6**:524. DOI: 10.1186/1556-276X-6-524

[30] Pandiyarasan V, Suhasini S, Kalari HC, Omprakash M, Karthikeyan K, Tsunehiro T, Ikeda H. Morphology dependent thermal conductivity of ZnO nanostructures prepared via a green approach. Journal of Alloys and Compounds. 2017;**695**:888-894. DOI: 10.1016/j.jallcom.2016.10.196

[31] Awais K, Mazhar HP, Muhammad M, Max W. A review on developments in dyeing cotton fabrics with reactive dyes for reducing effluent pollution. Journal of Cleaner Production. 2015;**87**:50-57. DOI: 10.1016/j.jclepro.2014.09.017

[32] Manikandan E, Murugan V, Kavitha G, Babu P, Maaza M. Nanoflower rod wire-like structures of dual metal (Al and Cr) doped ZnO thin films: Structural, optical, and electronics properties. Materials Letters. 2014;**131**:225-228. DOI: 10.1016/j.matlet.2014.06.008

[33] Ying J and Feng C. Microwave-assisted preparation of inorganic nanostructures in liquid phase. ACS Chemical Reviews. 2014;**114**:6462-6555. DOI: 10.1021/cr400366s

[34] Kolthoff IM . Theory of coprecipitation. The formation and properties of crystalline precipitates. Journal of Physical Chemistry. 1932:**36**:860-881. DOI: 10.1021/j150333a008

[35] Ngom BD, Mpahane T, Manikandan E, Maaza M. ZnO nano-disc by lyophilization process: Size effects on their intrinsic luminescence. Journal of Alloys and Compounds. 2016;**656**:758-763. DOI: 10.1016/j.jallcom.2015.09.230

[36] Tina H, Majid M. A review on textile sonoprocessing: A special focus on sonosynthesis of nanomaterials on textile substrates. Ultrasonic Sonochemistry. 2015;**23**:1-10. DOI: 10.1016/j.ultsonch.2014.08.022

[37] Hangxun X, Brad WZ, Kenneth SS. Sonochemical synthesis of nanomaterials. RSC Chemical Society Reviews. 2013;**42**:2555-2567. DOI: 10.1039/C2CS35282F

[38] Kaviyarasu K, Manikandan E, Kennedy J, Maaza M. A comparative study on the morphological features of highly ordered MgO: AgO nanocube arrays prepared via a hydrothermal method. RSC Advances. 2015;**5**:82421-82428. DOI: 10.1039/C5RA15132E

[39] Oliver SH, Karen JE, Daniel TB, Laura TM. Deep eutectic-solvothermal synthesis of nanostructures ceria. Nature Communications 2017:14150. DOI: 10.1038/ncomms14150

[40] Kaviyarasu K, Fuku X, Mola GT, Manikandan E, Kennedy J, Maaza M. Photoluminescence of well-aligned ZnO doped CeO_2 nanoplatelets by a solvothermal route. Materials Letters. 2016;**183**:351-354. DOI: 10.1016/j.matlet.2016.07.143

[41] Drik E, Hideto S, Yuji K, Hiroki S, Akira Y, Tsuguo F. Solvothermal growth of ZnO. Progress in Crystal Growth and Characterization of Materials. 2006;**52**:280-335. DOI: 10.1016/j.pcrysgrow.2006.09.002

[42] Kaviyarasu K, Manikandan E, Kennedy J, Jeyachandran M, Maaza M. Rice husks as a sustainable source of high quality nanostructured silica for high performance Li-ion battery requital by sol-gel method- A review. Advanced Materials Letters. 2016;**7**(6):100-150. DOI:10.5185/amlett.2016.6059

[43] Pimente A, Rodrigues J, Duarte P, Nunes D, Costa FM, Monterio T, Martins R, Fortunato E. Effect of solvents on ZnO nanostructures synthesized by solvothermal method assisted by microwave radiation: A photocatalytic study. Journal of Materials Science. 2015;**50**:5777-5787. DOI: 10.1007/s10853-015-9125-7

[44] Mohana PS, Geetha A, Ramamurthi K, Pandiyarasan V, Ikeda H. Effect of pH and annealing temperature on the properties of tin oxide nanoparticles prepared by sol–gel method. Journal of Materials Science: Materials Electronics. 2018;**29**:658-666. DOI: 10.1007/s10854-017-7959-2

[45] Saravanan L, Manivel RM, Prabhu D, Pandiyarasan V, Ikeda H, Therese HA. Perpendicular magnetic anisotropy in mo/Co2FeAl0.5/Si0.5/MgO/mo multilayers with optimal Mo buffer layer thickness. Journal of Magnetism and Magnetic Materials. 2018;**454**:267-273. DOI: 10.1016/j.jmmm.2018.01.097

[46] Merina PD, Jeyanthi RL, Pandiyarasan V, Das J. Exploration of Wedelia chinensis leafassistedsilver nanoparticles for antioxidant, antibacterial and in vitro cytotoxic applications. J. Journal of Food and Drug Analysis. 2018;**26**:917-925. DOI: 10.1016/j.jfda.2017.07.014

[47] Kennedy J, Murmu PP, Manikandan E, Lee SY. Investigation of structural and photoluminescence properties of gas and metal ions doped zinc oxide single crystals. Journal of Alloys and Compounds. 2014;**616**:614-617. DOI: 10.1016/j.jallcom.2014.07.179

[48] Miller PH. The electrical conductivity of zinc oxide. APS Physical Review Letters. 1941;**60**:890. DOI: 10.1103/PhysRev.60.890

[49] Hahn EE. Some electrical properties of zinc oxide semiconductor. AIP Journal of Applied Physics. 1951;**22**:855. DOI: 10.1063/1.1700063

[50] Boor JD , Muller E. Data analysis for Seebeck coefficient measurements. Review of Scientific Instruments. 2013:**84**:065102. DOI: 10.1063/1.4807697

[51] Andres S, Maria A, Shahed R, Gabriel C, Miguel AT, Torres JC. Diez: A thermoelectric by any other name...Materials today. 2012;**15**:P415. DOI: 10.1016/S1369-7021(12)70178-5

[52] Zhi-Gang C, Guang H, Lei Y, Lina C, Jia Z. Nanostructured thermoelectric materials: Current research and future challenge. Progress in Natural Science: Materials International. 2012;**22**:535-549. DOI: 10.1016/j.pnsc.2012.11.011

[53] Khan F, Pandiyarasan V, Sakamoto S, Navaneethan M, Shimomura M, Murakami K, Hayakawa Y, Ikeda H. Seebeck Coefficient of Flexible Carbon Fabric for Wearable Thermoelectric Device. IEICE Transaction on Electronics. 2018;**101**(5):343-346. DOI: 10.1587/transele.E101.C.343

[54] Xue, Yin W, Jia ST, Ma JZ. UV-durable superhydrophobic textiles with UV-shielding properties by coating fibers with ZnO/SiO_2 core/ shell particles. IOP Nanotechnology. 2011;**41**:415603. DOI: :10.1088/0957-4484/22/41/415603

[55] Omer KA , Tao Z. Review on : Developing UV protection for cotton fabric. The Journal of the Textile Institute. 2015;**11**:1-13. DOI: 10.1080/00405000.2017.1311201

[56] Wang L, Zhang X, Li B, Sub P, Yang J, Xu H, Liu Y. Superhydrophobic and ultraviolet-blocking cotton textiles. ACS Applied Materials and Interfaces. 2011;**3**(4):1277-1281. DOI: 10.1021/am200083z

[57] Sandeep KL, Hafeez YH, Pandiyarasan V, Ganesh V, Anish K, Ikeda H, Neppolian B. Enhanced Photocatalytic degradation and hydrogen production activity of in situ grown TiO_2 coupled $NiTiO_3$ Nanocomposites. Journal of Applied Surface Science. 2018;**449**: 790-798. DOI: 10.1016/j.apsusc.2018.02.136

[58] Shanthi S, Poovaragan S, Arularasu MV, Nithya S, Sundaram R, Magdalane CM, Kaviyarasu K, Maaza M. Optical, magnetic and photocatalytic activity studies of Li, mg, Sr doped and undoped zinc oxide nanoparticles. Journal of Nanoscience and Nanotechnology. 2018;**18**:5441-5447. DOI: 10.1166/jnn. 2018.15442

[59] Rai P, Kwak WK, Yu YT. Solvothermal synthesis of ZnO nanostructures and their morphology-dependent gas-sensing properties. ACS Applied Materials Interfaces. 2013;**5**:3026-3032. DOI: 10.1021/ am302811h

[60] Ayudhya SKN, Tonto P, Okorn M, Pavarajam P, Praserthdam P. Solvothermal synthesis of ZnO with various aspect ratios using organic solvents. Crystal Growth. Designs. 2006;**6**:2446-2450. DOI: 10.1021/cg050345z

[61] Matinise N, Fuku XG, Kaviyarasu K, Mayedwa N, Maaza M. ZnO nanoparticles via *Moringa oleifera* green synthesis: Physical properties & mechanism of formation. Journal of Applied Surface Science. 2017;**406**:339-347. DOI: 10.1016/j.apsusc.2017.01.219

[62] Pandiyarasan V, Suhasini S, Archana J, Navaneethan M, Majumdar A, Hayakawa Y, Ikeda H. Fabrication of hierarchical ZnO nanostructures on cotton fabric for wearable device applications. Applied Surface Science. 2017;**418**:352-361. DOI: 10.1016/j.apsusc.2016.12.202

[63] Suhasini S, Greeshma G, Prashanth K, Kathirvel V, Saranya TS, Pandiyarasan V, Karolien DW, Ikeda H. Tailoring the functional properties of polyurethane foam with dispersions of carbon nanofiber for power generator applications. Journal of Applied Surface Science. 2018;**449**:507-513 DOI: https://doi.org/10.1016/j.apsusc.2018.01.088

[64] Pandiyrasan V, Suhasini S, Faizan K, Aranya G, Majumdar A, Hayakawa Y, Ikeda H. Incorporation of ZnO and their composite nanostructured material into a cotton fabric platform for wearable device applications. Journal of Carbohydrate Polymers. 2017;**157**:1801-1808. DOI: 10.1016/j.carbpol.2016.11.065

[65] Ming H, Feng Q, Zhiqun L. Towards high-performance polymer-based thermoelectric materials. Energy & Environmental Science. 2013;**6**:1352-1361. DOI: 10.1039/C3EE24193A

[66] Mario C, Clara MG, Andres C. Review on polymers for thermoelectric applications. Materials. 2014;**7**:6701-6732. DOI: 10.3390/ma7096701

[67] Mariappan P, Karthikeyan K, Jeyasubramanian K, Govindasamy M. Selective toxicity of ZnO Nanoparticles toward gram-positive bacteria and cancer cells by apoptosis through lipid peroxidation. Nanomedicine: Nanotechnology Biology, and Medicine. 2011;**7**:184-192. DOI: 10.1016/j.nano.2010.10.001

[68] Junjie L, Xinfeng T, Han L, Yonggao Y, Qingjie Z. Synthesis and thermoelectric properties of hydrochloric acid-doped polyaniline. Synthetic Metals. 2010;**160**:1153-1158. DOI: 10.1016/j.synthmet.2010.03.001

[69] Jiansheng W, Yimeng S, Wen-Bo P, Ling H, Wei X, Qichun Z. Self-assembled free-standing polypyrrole nanotube membrane as an efficient FTO- and pt-free counter electrode for dye-sensitized solar cells. ACS Applied Materials Interfaces. 2014:**196**:173-177. DOI: 10.1021/am404265q

[70] Olga B, Zia UK, Abdellah M, Slawomir B, Mats F, Magnus B, Xavier C. Optimization of the thermoelectric figure of merit in the conducting polymer poly(3,4-ethylenedioxythiophene). Nature Materials. 2011;**10**:429-433. DOI: 10.1038/nmat3012

Chapter 8

Conductive Yarn Embroidered Circuits for System on Textiles

Jung-Sim Roh

Additional information is available at the end of the chapter

http://dx.doi.org/10.5772/intechopen.76627

Abstract

With the recent convergence of electronics and textile technology, various kinds of smart wearables are being developed, such as heating clothes, health monitoring clothes, and motion sensing clothes. In this study, the novel conductive embroidery yarns for touch sensing and signal transmission for system on textile (SoT) are introduced. The conductive yarn for touch sensing can be used as a user interface of smart clothes by constructing an embroidery circuit. The conductive yarn for signal transmission can be embroidered on smart clothing and used as a transmission line to transmit power and signal. The conductive yarns and their embroidered circuits were characterized and SoT prototypes using the embroidered circuit of these conductive yarns were presented. These e-textiles based on touch sensing and signal transmission can be comfortably applied for SoT and maintain electrical performance without being damaged by tensile force generated by the movement of the wearer.

Keywords: smart wearables, system on textiles, e-textile, conductive yarns, embroidered circuits

1. Introduction

Wearing smart performance must be done very naturally and smoothly in order to be accepted by the public. Because textiles are breathable, lightweight, flexible, and robust, smart textiles are the most comfort wearables that can provide users with an intelligent environment by sensing, data processing and communication, and actuating. Thus, many studies on textile-based wearables have been conducted recently, and these studies cover topics such as textile-based full-body implementations, multifunctionality, customizability, and robust and seamless integration.

Jost et al. emphasized on devices made from textiles, suggesting the design concept of a smart power body suit where a piezoelectric patch of the knees, textile antennas between shoulder blades, and textile electrochemical energy storage at the waist back are integrated as part of a fabric with conductive yarns that act as leads to transmit energy or information throughout the garment. Body-worn sensors are unobtrusively embedded into a garment and distributed all over the body [1]. Lorussi et al. reported piezoresistive material-loaded, sensorized garments, such as gloves, leotards, and seat covers capable of reconstructing and monitoring body shape and postures [2].

PowerMatrix of Sefar AG is one of the leading multi-functional smart textiles. It is a hybrid fabric consisting of polyester monofilaments and metal monofilaments in warp and weft, which is possible to utilize for various purposes such as custom-made LED panels, temperature sensors, and heating fabrics [3, 4]. The e-fibers that combine various functionalities into the yarn structure simplify the fabrication process and interconnection-related issues and improve wearability [5, 6].

Customizability is another important issue in the field of e-textiles. The peregrine USB glove can be customized for over 30 user-programmable actions and allows the wearer to carry out those time-sensitive gaming commands with a switch of a finger by utilizing the touch sensitive pads embedded in the fingers and palm of the glove [7]. Buechley introduced a do-it-yourself (DIY) wearable electronics construction kit which includes a set of stitchable controllers, sensors, and actuators that enable novices to build their own system [8].

Issues with robust and seamless integration are critical to the feasibility of commercial manufacture and widespread consumer adoption. The EU-funded FP7 Place-it project focuses on the true integration of LED technology with flexible, stretchable, and textile substrates [9]. Likewise, the FP7 PASTA project focuses on electronic packaging and module interconnection to increase the robustness of integration by development of stretchable interposers to provide strain relief between components and fabrics [10].

In this context, embroidery techniques using conductive threads on substrate fabrics have become a very attractive approach for fabricating textile-based circuits for smart wearables because of their circuit design freedom and ease of manufacture compared to other processes such as weaving, knitting, and printing. Circuits can be easily patterned by employing a computer numerically controlled (CNC) embroidery using conductive threads. Embroidery also offers much more tailorability and customizability than other processes because they can be embedded in clothing at the end of the garment manufacturing process. Above all, it is a great advantage that the connection between the circuit embroideries and the small electronic components is possible to some extent during the embroidery process [11–14].

Two CNC embroidery methods, standard embroidery and tailed fiber placement (TFP), can be used to form the embroidered circuit [15]. In the standard embroidery method, the needle thread and the bottom thread form a double lock stitch forming a technically identical appearance on the upper side and the bottom side, respectively (**Figure 1a**). The CNC standard embroidery process allows for sophisticated work and enables the connection of flexible printed circuit board (FPCB) or small electronic elements and embroidered circuits during the embroidery process. And a one-stop production process for electronic textiles is also possible [16].

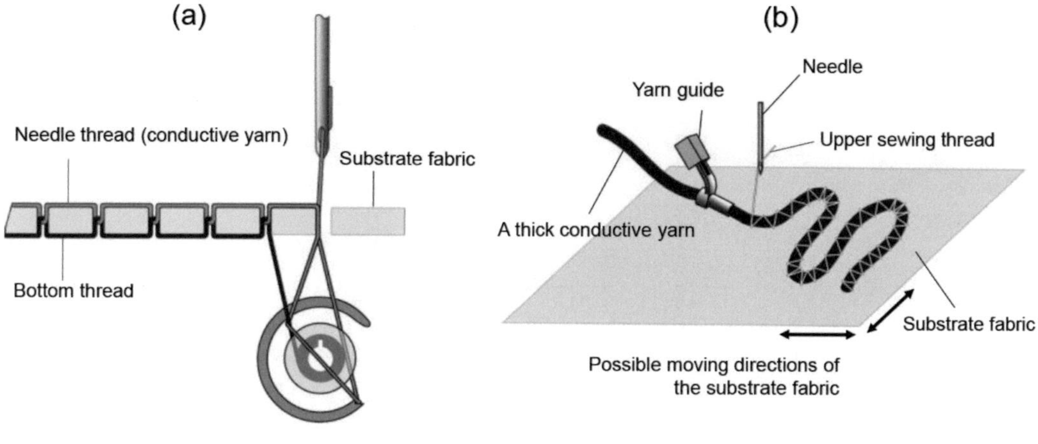

Figure 1. Embroidery methods: (a) standard embroidery forming double lock stitch and (b) the tailored fiber placement method.

On the other hand, the needle thread of the CNC standard embroidery machine is required to be thin and flexible and to have strength and yield strength to withstand stress during the embroidery process. On the other hand, TFP method is a three-thread system. The TFP method is used when the thread is very stiff, nonelastic and very thick like glass fiber or carbon fiber and cannot be worked on standard embroidery machine (**Figure 1b**). A thick conductive yarn is fixed on the substrate fabric with a set of top and bottom threads forming zigzag stitches. In the TPF method, the direct connection of the small electronic element and the embroidered circuit is impossible during the embroidery process.

Therefore, this section introduces metal composite embroidery yarns (MCEYs) that are very thin, flexible, very conductive, solderable and having suitable strength for the CNC standard embroidery process. In addition, several system on textiles (SoT) prototypes manufactured by the MCEY embroidery circuit method will be addressed.

2. Metal composite embroidery yarns

To make a very conductive, thin, flexible, robust, and solderable conductive embroidery yarn, we have developed metal composite embroidery yarns (MCEYs) consisting of superfine metal filaments and polyester filament yarns. During high-speed machine embroidery, the needle thread is subject to high stresses so that the needle thread requires high mechanical properties unlike the bottom yarn [17]. To ensure that the metal filaments contained in the MCEY are not stretched or broken during the embroidery process, the MCEY should have sufficient strength and yield strength as high as possible. In order to satisfy these conditions, two methods of type I and type II were developed (**Figure 2a**) [13, 18]. The metal filament is helically wrapped around the polyester yarn to minimize elongation against the longitudinal force. The type II method utilizes reinforcing polyester yarn in the plied yarn fabrication stage to increase the yield strength of MCEY over the type I method for external forces applied in the longitudinal

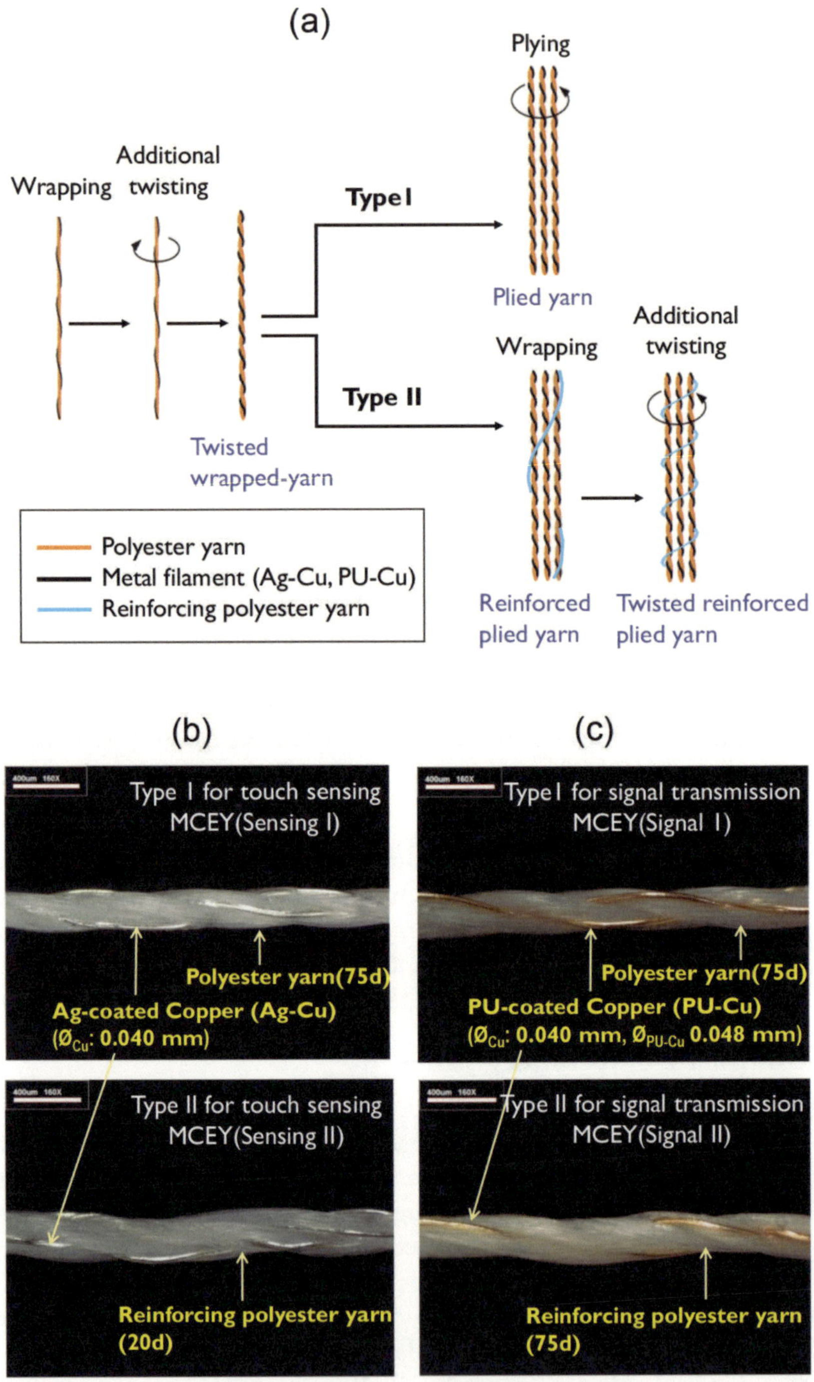

Figure 2. Structure of the MCEYs: (a) two types of metal composite yarn fabrication process [13], (b) side views of the MCEYs for touch sensing (type I and II) [18] and (c) side views of the MCEYs for signal transmission (type I and II) [13].

direction. On the other hand, two types of MCEYs were produced depending on the application. Ag-coated copper (Ag-Cu, $Ø_{Cu}$: 0.040 mm) and PU-coated copper (PU-Cu, $Ø_{Cu}$: 0.040 mm, $Ø_{PU\text{-}Cu}$ 0.048 mm) were used as metal filaments for MCEY for touch sensing and MCEY for signal and power transmission respectively. In this way, the four MCEYs introduced in **Table 1** were developed with the optimum twist conditions obtained from the experiment.

2.1. Morphologies of MCEYs

Figure 2 illustrates the structures and side views of MCEYs. The smooth surface results in less friction on the yarn and the spiral structure of the metal filament reduces elongational deformation of the metal filament that can occur during the machine embroidery process.

Yarn type	Composition (wt% of Metal: Polyester)	Resistance (Ω/cm)	Linear density (denier)/ Yarn thickness (μm)	Load at yield (N)	Max. load (N)	Strain (%)
Sensing(I)	56.4: 43.6	0.0487	547/199	3.92	13.3	13.1
Sensing(II)	54.5: 45.5	0.0492	570/231	4.45	13.2	12.4
Signal(I)	57.3: 42.7	0.0521	549/200	3.92	13.3	13.1
Signal(II)	50.3: 49.7	0.0510	634/221	4.48	17.4	15.2

Table 1. Characteristics of the metal composite embroidery yarns (MCEYs).

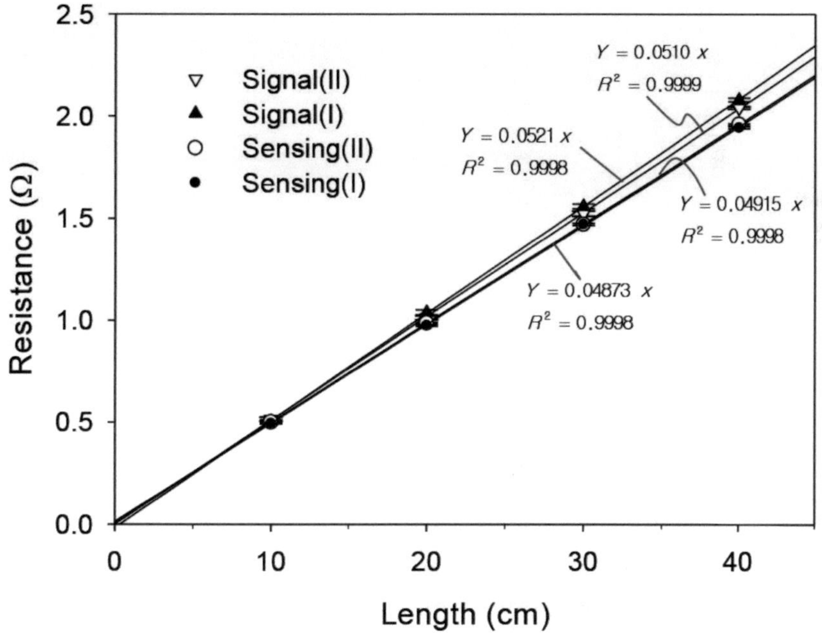

Figure 3. Electrical resistance linearity of four MCEYs as a function of the yarn length.

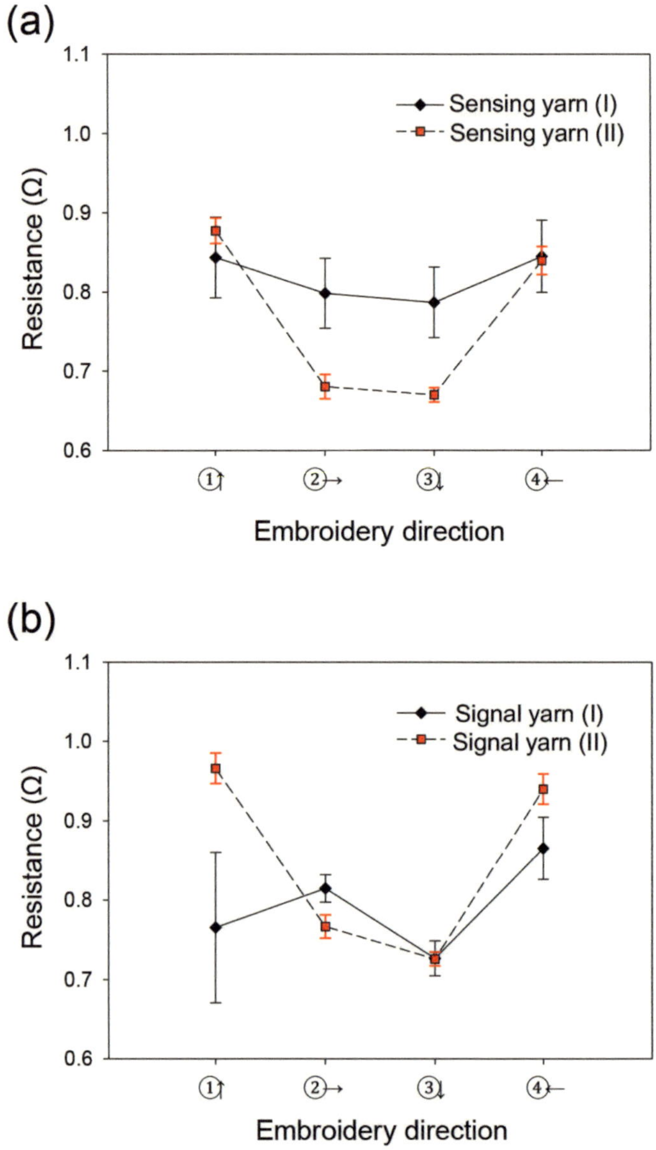

Figure 4. Electrical resistance of the MCEY embroidered circuit lines (10 cm circuit length, 2.5 mm running stitches): (a) touch sensing yarns (type I and II) embroidered circuits and (b) the signal yarns (type I and II) embroidered circuits.

Moreover, for type II structures, the reinforcing polyester yarn can increase the yield strength of MCEY and protect the metal filament of MCEY from friction during the embroidery process. Especially in the case of a sensing yarn, a metal filament spirals around the polyester yarn surface, so the twisted wrapped yarns made of metal filament and polyester yarn are electrically coupled in the final plied yarn, which provides a great advantage for touch sensing and electrical contact.

2.2. Tensile properties of MCEYs

As shown in **Table 1**, all MCEYs showed a very high tensile strength of 13 N or more. Especially for type II yarn, yield strength was increased by about 14% compared to type I yarn due to additional reinforcing polyester yarn.

2.3. Electrical properties of MCEYs

MCEYs were very thin and light but showed very low electrical resistance of about 0.05 Ω/cm (**Table 1**). The electrical resistances of MCEYs were measured 20 times for each length, and **Figure 3** shows the mean value, standard error, and linear regression equation for the electrical resistances of four MCEYs. As a result, electrical resistance increased very linearly as the length of yarn increased.

2.4. Electrical properties of MCEY embroideries

Figure 4 shows the effect of embroidery directions on the electrical resistance of the MCEY embroidery for both type I and type II. The values are the average of 20 samples. There were differences in resistance for each direction. D1 (front-to-back) and D4 (right-to-left) showed higher resistance than D2 (left-to-right) and D3 (back-to-front) in both type I and type II. Type II had a larger difference between resistance values in each direction than Type I. On the other hand, Type II had higher uniformity of electrical resistance than type I.

3. Sensing applications

3.1. Textile touch sensor manufactured by a one-stop production process

For wearables, textile interfaces are essential for the user to interact naturally with the smart function of the product [19, 20]. Electronic textiles, which measure the touched position or contact pressure, are the most widely used textile interface technologies [18, 21–24]. In most cases, a multilayer structured textile touch sensor that can sense the location of the user's touch can be used to manipulate the smart fashion system by mapping the command to the sensed location.

Figure 5 shows an example of a multilayer structured textile touch sensor that senses a touched position [16]. It shows a multilayer structure of a textile touch sensor composed of potentiometric resistive embroidery and an MCEY circuit pattern, a spacer fabric, and a conductive fabric for touch sensing. The spiral structure of the metal filament of MCEY provides the best conditions for electrical contact between the MCEY embroidered sensing pad and conductive fabric when pressed, and the MCEY satin stitch enables electrical connection with the resistive embroidery of the Ag-coated yarn. In addition, the standard CNC embroidery techniques enable a one-stop production of a multilayer structured textile touch sensor according to the four steps described in **Figure 5**. The conductive top layer and the MCEY circuit on the base layer can be electrically connected via a hole in the intermediate spacers and

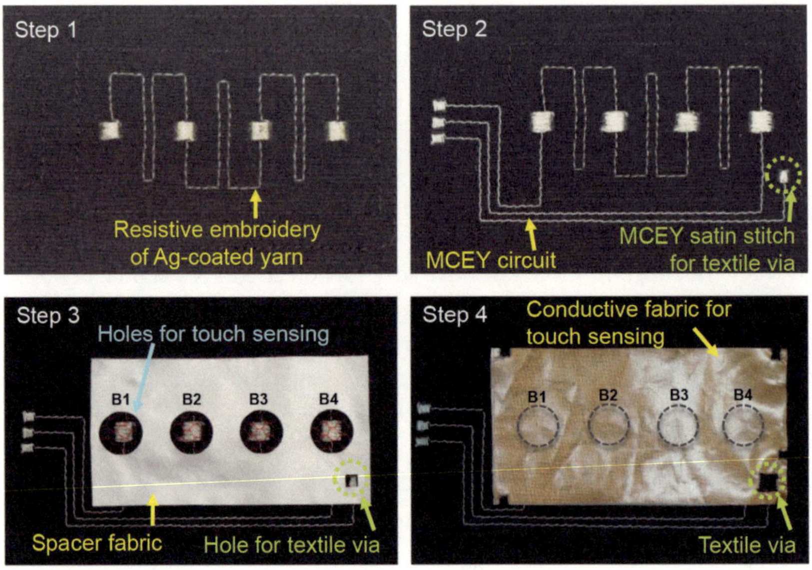

Figure 5. A one-stop production process on a standard CNC embroidery machine for fabricating the multilayer structured textile touch sensor that senses a touched position [16].

tightly fixed by a covering satin stitch of normal yarn. As a result, the connection portion with the external device can be formed in one layer. According to the contact resistance test for three textile via sizes (2 × 2, 3 × 3, and 4 × 4 mm), the overall contact resistance was lower than 0.15 Ω. The smaller MCEY satin stitches of the textile exhibit lower contact resistance, which is more desirable for use. Similarly, by finger touch sensing test, the MCEY sensing pads of 3 × 3 mm–5 × 5 mm showed a low contact resistance of less than 1 Ω because they provide sufficient contact area during finger touch sensing.

In conclusion, the proposed one-stop production process allows a CNC embroidery machine to form precise circuit patterns while fabric layers are piled up one by one; the inner layer or interlayer circuits are electrically connected by embroidery, and individual fabric layers are also fixed to the base layer by embroidery. The multilayer fabric-structured electronic textiles constructed as such can contribute to securing the market competitiveness of smart textile systems.

3.2. Multitouch and pressure-sensing textiles

Figure 6 is another example of a multi-layer-structured textile touch sensor, which is a piezoresistive, multitouch sensing surface that senses pressure as well as touched position. Conductive-polymer-infused non-woven fabric by Eeonyx was use for piezoresistive substrate. The resistance of the piezoresistive fabric can be detected by sandwiching it between two conductors. To sense the position and pressure of the finger touch, the MCEY can be simply arranged in parallel on both sides of the piezoresistive fabric at right angles to each other. A 5 mm spacing was chosen to estimate the finger position with sufficient resolution

http://dx.doi.org/10.5772/intechopen.76627

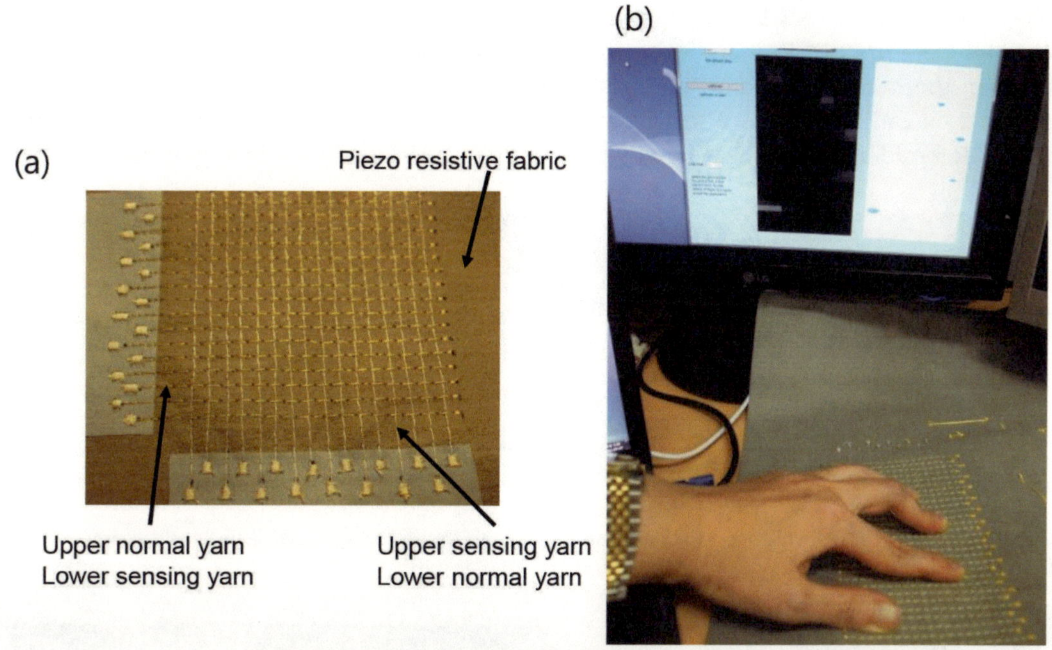

Figure 6. Multitouch and pressure sensing textiles using MCEY embroidery and piezoresistive fabric: (a) prototype of multitouch array sensing and pressure sensing textiles using standard MCEY embroidery and piezoresistive fabric and (b) display that expresses the touched positions and pressure intensities with five fingers.

to capture a vibrato-like gesture well. A standard sewing machine is set up with a conductive MCEY (sensing I or II) top thread and an insulating bottom thread, or vice versa. After one side is completed, the fabric is turned upside down and an orthogonal array of lines is embroidered in [23]. A sandwich structure of MCEY array fabric-Piezoresistive fabric-MCEY array fabric can be proposed for mass production of multitouch and pressure sensing textiles because the MCEY array fabric can be fabricated in large sizes and can be cut freely in various sizes and shapes to meet customer needs.

4. Signal and power transmission applications

4.1. Temperature sensing and heating textiles

MCEY embroidered circuits can be used for signal communication and power transmission in smart textiles systems. **Figure 7** shows the embroideries made with MCEY (signal II) to heat the outdoor jacket [24]. It can be customized to the most appropriate form for each application part of the jacket such as chest, belly, back, neck, ears, and so on. This pure textile component can be easily and unobtrusively embedded anywhere on a garment in any desired shape. It also offers comfort due to its low volume, light weight, flexibility, and breathability.

The thin and flexible MCEY can be embroidered at a distance of about 2 mm, allowing uniform temperature distribution during heating. In addition, since there is a linear correlation between the temperature and the resistance of the MCEY circuit, the MCEY embroidery circuit itself can be used as a temperature monitoring sensor [25]. The temperature of the heated embroidery can be measured by measuring its resistance value in a short time interval.

(a)

(b)

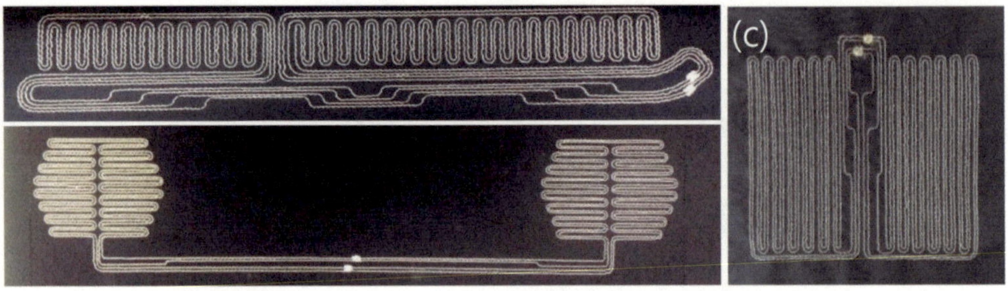

Figure 7. MCEY heating embroideries to be placed on neck (a), ears (b), and back, chest or belly (c) of the outdoor jacket [24].

(a)

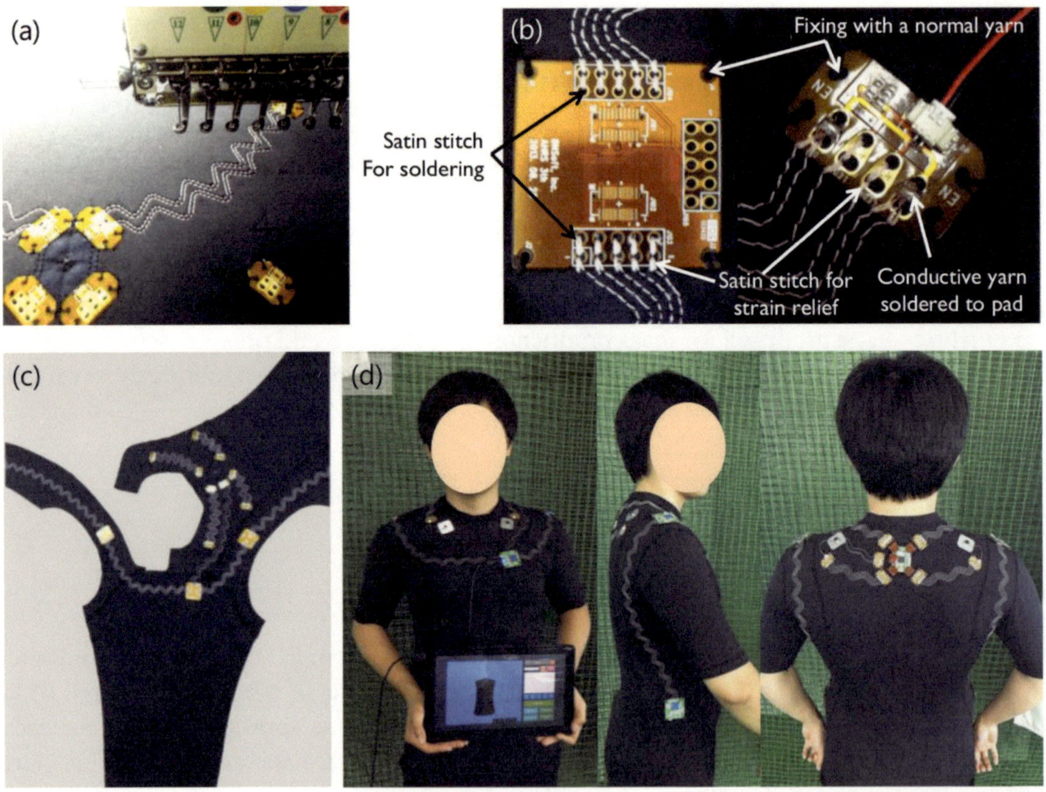

Figure 8. Posture monitoring smart clothing: (a) MCEY circuit patterning process using CNC embroidery method, (b). Embroidered contact with FPCBs on a standard embroidery machine [13], (c) circuit patterning embroidery fabric circuit board, and (d) smart clothing finished by sewing.

A power on–off switching system that refers to real-time-measured temperatures is able to maintain a comfortable set temperatures at all times regardless of changes in internal microclimate, external climatic conditions, and battery voltage levels.

4.2. Smart clothing for posture monitoring and corrective feedback

To develop "smart clothing" that collects postural data from the sensor and communicate with the host computer in real time, a wearable circuit patterning technique with low resistance and high elasticity is required, along with robust and reliable fabric circuit-sensor interconnection technology. **Figure 8** shows the fabrication process of smart clothing to monitor spine posture and to provide vibration feedback for bad posture. **Figure 8a** shows the process of forming stretchable circuit networks with MCEY (signal II) on a highly stretchable knit fabric (89% polyamide, 11% spandex). FPCBs designed to allow easy attachment and detachment of sensor modules are integrated together during circuit pattering process (**Figure 8b**) [13]. **Figure 8c** shows the fabric circuit board with the circuit patterning process completed on the CNC embroidery machine. Finally, the sensors were mounted on the FPCB to complete the smart clothing (**Figure 8d**).

The design and work order of the smart clothing were designed considering the embroidery process of the 3D wearable circuit networks and the feasibility of garment sewing. FPCB is fixed to the fabric by satin stitch of MCEY (signal II) in hole for fixing and hole for soldering (**Figure 8b**). The FPCB and MCEY embroidered circuit are electrically connected to each other by soldering, while the satin stitch for fixing prevents the strain generated during wearing from being transferred to the soldering connection site.

5. Conclusions

In conclusion, novel metal composite embroidery yarns (MCEYs) have been introduced for touch sensing, textile-based interconnection, signal communication and power transmission for smart wearables. Fabrication methods of the MCEYs and their electrical and physical characterization and circuit patterning and device integration during the CNC embroidery process have been described in detail.

Two methods have been proposed for MCEY fabrication. MCEYs with an electrical resistance of 0.05 Ω/cm showed very low electrical resistance compared with the thickness of the yarn. The electrical resistance of the MCEYs produced by these two methods increases linearly with the increase in length, and if a reinforcing polyester yarn was additionally used at the plying stage (type II method), the uniformity of the electrical resistance of the MCEY embroidered circuit line was greatly improved. As the number of the twisted wrapped yarn of a metal filament and polyester yarn forming the MCEY increases, the electrical resistance of the finally fabricated MCEY will become lower.

Since MCEY has a superfine metal filament covering the surface of the polyester in a spiral shape, it is strong against tensile force, minimizes change of electrical resistance according to elongation, and is very advantageous for touch sensing. Thus, robust and reliable MCEY embroidered circuits can be applied for sensing, interconnection, and signal and power transmission, and several prototypes have been introduced in this chapter. These examples demonstrate the feasibility and usability associated with the customizability, tailorability, easy and precise circuit formation, and device integration during circuit formation of the MCEY circuit patterning method.

Acknowledgements

This work was supported by the National Research Foundation of Korea (NRF) grant funded by the Korean Government (MSIP: Ministry of Science, ICT & Future Planning) (No. NRF-2015R1C1A1A02037064).

Author details

Jung-Sim Roh

Address all correspondence to: jungsimroh@smu.ac.kr

Department of Fashion and Textiles, Sangmyung University, Seoul, Korea

References

[1] Jost K, Diona G, Gogotsi Y. Textile energy storage in perspective. Journal of Materials Chemistry A. 2014;**2**:10776-10787. DOI: 10.1039/C4TA00203B

[2] Lorussi F, Scilingo EP, Tesconi M. Strain sensing fabric for hand posture and gesture monitoring. IEEE Transactions on Information Technology in Biomedicine. 2005;**9**(3):372-381. DOI: 10.1109/TITB.2005.854510

[3] Sefar. Smart Fabrics [Internet]. 2018. Available from: http://www.sefar.com. https://www.sefar.us/en/818/Smart-Fabrics.htm?Folder=1463166 [Accessed: January 25, 2015]

[4] Locher I, Kirstein T, Tröster G. Temperature profile estimation with smart textiles. In: Proceedings of the International Conference on Intelligent Textiles; September 2005; Smart Clothing, Well-being, and Design, Tampere, Finland; 2005. pp. 19-20

[5] Cherenack K, Zysset C, Kinkeldei T, Münzenrieder N, Tröster G. Woven electronic fibers with sensing and display functions for smart textiles. Advanced Materials. 2010; **22**(45):5178-5182. DOI: 10.1002/adma.201002159

[6] Dias T, Ratnayake A. Integration of micro-electronics with yarns for smart textiles. Electronic Textiles. Elsevier. 2015:109-116. DOI: 10.1016/B978-0-08-100201-8.00006-0

[7] The Pregrine Usb Glove. [Internet]. Available from: http://theperegrine.com/peregrine-glove/ [Accessed: January 25, 2015]

[8] Buechley L, Eisenberg M, Catchen J, Crockett A. The lilyPad Arduino: Using computational textiles to investigate engagement, aesthetics and diversity in computer science education. In: Conference on Human Factors in Computing Systems–Proceedings, 26th Annual CHI Conference on Human Factors in Computing Systems; 5-10 April 2008; Italy: Florence; 2008. pp. 423-432. DOI: 10.1145/1357054.1357123

[9] Place-it Project. [Internet]. Available from: http://www.place-it-project.eu. [Accessed: January 25, 2018]

[10] Pasta Project. [Internet]. Available from: http://www.pasta-project.eu. [Accessed: January 25, 2018]

[11] Linz T, Kallmayer C, Aschenbrenner R, Reichl H, editors. Fully untegrated EKG shirt based on embroidered electrical interconnections with conductive yarn and miniaturized flexible electronics. In: International Workshop on Wearable and Implantable Body Sensor Networks (BSN'06); 3-6 April 2006; Cambridge, MA, USA. Los Alamitos: IEEE; 2006. pp. 23-26. DOI: 10.1109/BSN.2006.26

[12] Linz T, Gourmelon L, Langereis G. Contactless. EMG sensors embroidered onto textile. In: 4th International Workshop on Wearable and Implantable Body Sensor Networks (BSN 2007); 2007; Berlin, Heidelberg: Springer; 2007. pp. 29-34

[13] Kang SW, Choi H, Park HI, Choi BG, Im H, Shin D, Jung YG, Lee JY, Park HW, Park S, Roh JS. The development of an IMU integrated clothes for postural monitoring using conductive yarn and interconnecting technology. Sensors. 2017;**17**(11):2560. DOI: 10.3390/s17112560

[14] Jeong MJ, Yun TI, Baek JJ, Kim YT. Wireless power transmission using a resonant coil consisting of conductive yarn for wearable devices. Textile Research Journal. 2016;**86**:1543-1548. DOI: 10.1177/0040517515586163

[15] Eichhoff J, Hehl A, Jockenhoevel S, Gries T. Textile Fabrication Technologies for Embedding Electronic Functions into Fibres, Yarns and Fabrics. Multidisciplinary Know-how for Smart-Textiles Developers. Cambridge, UK: Woodhead Publishing Limited; 2013. pp. 191-226

[16] Roh JS. All-fabric interconnection and one-stop production process for electronic textile sensors. Textile Research Journal. 2017;**87**(12):1445-1456. DOI: 10.1177/0040517516654108

[17] Sundaresan G, Hari P, Salhotra K. Strength reduction of sewing threads during high speed sewing in an industrial lockstitch machine: Part I-mechanism of thread strength reduction. International Journal of Clothing Science and Technology. 1997;**9**(5):334-345. DOI: 10.1108/09556229810205303

[18] Roh JS. Textile touch sensors for wearable and ubiquitous interfaces. Textile Research Journal. 2014;**84**(7):739-750. DOI: 10.1177/0040517513503733

[19] Gilliland S, Komor N, Starner T, Zeagler C, editors. The textile interface swatchbook: Creating graphical user interface-like widgets with conductive embroidery. In: 2010

International Symposium Wearable Computers (ISWC), 10-13 October 2010. Seoul: IEEE; 2010. pp. 1-8

[20] Dunne LE, Profita H, Zeagler C, Clawson J, Gilliland S, Do EY, Budd J. The social comfort of wearable technology and gestural interaction. In: Engineering in Medicine and Biology Society (EMBC), 2014 36th Annual International Conference of the IEEE; 26-30 August 2014; Chicago. IEEE; 2014. pp. 4159-4162. DOI: 10.1109/EMBC.2014.6944540

[21] Google. Project Jacquard [Internet]. 2015. Available from: https://www.google.com/atap/project-jacquard/ [Accessed: December 1, 2015]

[22] Min SD, Yun Y, Shin H. Simplified structural textile respiration sensor based on capacitive pressure sensing method. IEEE Sensors Journal. 2014;**14**(9):3245-3251. DOI: 10.1109/JSEN.2014.2327991

[23] Roh JS, Mann Y, Freed A, Wessel D. Robust and Reliable Fabric, Piezoresistive multitouch sensing surfaces for musical controllers. In: Proceedings of the International Conference on New Interfaces for Musical Expression (NIME); 30 May–1 June 2011; Oslo, Norway. pp. 393-398

[24] Kim S, Roh JS, Lee EY. Development and wearability evaluation of all-fabric integrated smart jacket for a temperature-regulating system based on user experience design. Fashion & Textile Research Journal. 2016;**18**(3):363-373. DOI: 10.5805/SFTI.2016.18.3.363

[25] Roh JS, Kim S. All-fabric intelligent temperature regulation system for smart clothing applications. Journal of Intelligent Material Systems and Structures. 2016;**27**(9):1165-1175. DOI: 10.1177/1045389X15585901

Chapter 9

A Wearable Heating System with a Controllable e-Textile-Based Thermal Panel

Senem Kurşun Bahadir and Umut Kivanc Sahin

Additional information is available at the end of the chapter

http://dx.doi.org/10.5772/intechopen.76192

Abstract

Flexible textile heating systems present great advantage due to their ability to bend and hence could ensure uniform heating for irregular geometries. In cooler outer environment, the user requires his/her body to be kept warm for monitoring vital body functions within realistic thermal body balance constraints. In this chapter, heated vest with controllable e-textile-based thermal panel has been studied. Several e-textile-based thermal panels with different conductive yarns were produced using hot air welding technology under different manufacturing parameters. E-textile-based thermal panels were tested for their heating behaviors at varying direct current (DC) power levels. Based on the experimental results, the optimum e-textile-based thermal panel design was chosen considering its flexibility and uniform heating behavior. Moreover, a control algorithm with electrical circuit and electrical connection network was designed and assembled in an electronic control module. Finally, the electronic module consisting of power control and management system was integrated to attachable e-textile-based thermal panel in order to form a wearable heating vest.

Keywords: heating textiles, e-textiles, thermal panel, temperature control, welding, wearable electronics

1. Introduction

The rapid interest on Internet of Things (IoT) technologies together with electronics has brought up new, interesting and challenging fields of crossover electronics researches to our work, homes and daily life with an abundance of new viewpoints.

Wearable electronics, newly growing area in the global market, are being fundamentally designed to be worn to function, which makes them different from mobile devices. In the mid-1990s, mobile phones and internet technologies penetrated into human life with a dramatic increase and the mobile devices became smaller, more portable and affordable. In all over the world, most people particularly aged 15–35 know how to operate computers and mobile phones and they are becoming the focus of attention of IT industry incredibly. Thus, the body-worn computing devices are believed to be a hit in the global market within a decade [1].

The integration of wearable technology is expected to set and enhance the design concept of apparel sector in the near future; however, the functioning alone is not enough in the global market to sell a kind of "wearable electronic" product. Without compromising any traditional characteristics such as wearability and washability, the product will be meaningless to be purchased by the consumers in the market. Besides, some wearable electronics require the user interface to be present and available all the time which makes them very obtrusive compared to other wearable devices such as wrist belt for healthcare. In addition, the appearance of clothes, their color, shape, style, etc. are very important to enhance the desirability of the product from the viewpoint of marketing strategies.

Ideally, an intelligent garment presents functioning unlike traditional garment, such as health monitoring, detection of fall, identification of emergency, etc. It collects the data and transmits the data either wirelessly or wired solutions to an external computing unit, where the data is processed and interpreted; thus results in a response or feedback to a wearer/external center.

In this chapter, a design of an intelligent garment with a controllable e-textile (electronic textile)-based thermal panel for heating purpose is presented. Firstly, textile-based heating approaches will be described. Then, the design of a wearable heating system will be given in detail. Design of e-textile-based thermal panels with their production process will be issued. The electrical connection network of the heating jacket with thermal panels and an electronic control module consisting of power control and management system will be addressed. The chapter goes on the integration of electronic control module to textile structure and describes the development of a whole wearable heating vest system. The chapter then addresses the real-time monitoring of the heating vest performance. Throughout this chapter, the development stage of a kind of intelligent garment particularly; "wearable heating vest" from design concept to prototyping will be understood.

2. Textile-based heating approach

2.1. Principle of electrical resistance heating in textiles

An electric current passing through any conductive material will generate heat. In an electric resistance heating system, the electric current passing through the resistor generates an amount of heat depending on its level; however, the current flow is limited by the resistance and applied voltage in a circuit based on a basic principle of Ohm's law (**Figure 1**) [2].

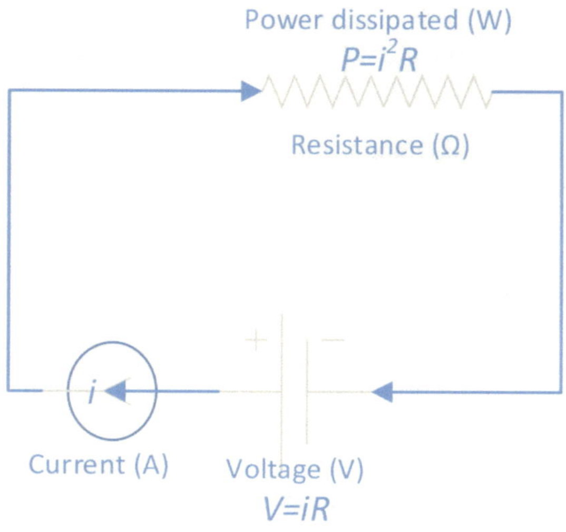

Figure 1. Ohm's law illustration.

Resistors with specific value can be manufactured in various shapes and sizes within a fabric. The area of the fabric in which the resistors are located has not any effect on the resistor's value. Independent from the area and the surface of the fabric, conductive elements can be placed in the form of parallel and series in order to form electrical networks inside/over the fabric structure. The resistance values can change depending on the conductive element type particularly from a few ohms to many kilo ohms. In fact, the linear resistance of conductive yarns/tracks defines the resistance values of electrical networks constructed in a fabric structure. The way of formation and insertion of conductive tracks are critical in a textile-based heating system since the manufacturing process can damage the conductive tracks hence may result in a decrease in conductivity level [3, 4]. Moreover, those conductive tracks used for wearable systems should be thin enough to be bend and to allow flexibility, and at the same time strong enough for not to be broken and provide an efficient heating [5].

Every square centimeter of fabric will dissipate heat in terms of several watts of power depending on the resistor conductivity level. In order to obtain higher power dissipation, the resistor can be scaled up in the fabric area, however in the case of high power dissipation, high level of power supply is mainly required.

For an efficient heat dissipation, to broaden the electrical network on a thin surface of the fabric is necessary. Indeed, the efficiency level at the dissipating heat due to the electrical network configuration on the fabric is one of the most important factors to have the efficient heating system [6]. For having flexible heating within a textile structure, the typical fabric constructions such as plain, twill weave patterns or single jersey, interlock etc. with typical fabric weight of 250 g/m^2 can be chosen. Composition of the fibers affects the washability and heating behavior of the system. For instance, the heating performance of wool fabric compared to cotton fabric is higher. For this reason, in a design of textile-based electrical heating system, if the cotton fabric is used as a base fabric for electrical network configuration then the system's heating capacity apparently will be less compared to those from woolen fabrics. Therefore, to choose appropriate fiber composition plays an important role on the heating performance of the system.

In battery-powered portable applications, to specify the required wattage value rather than its rated voltage gives great flexibility to design heating system. Especially, to determine wattage value with a specified resistance value of the conductor will present great freedom for having e-textile-based thermal panels such that to calculate the required level of distance for insertion of conductive yarns/tracks and hence, the size of textile-based heating panels will become easy.

Overall, the area requiring heat should be wrapped with a flexible surface for having efficient heating performance due to heat delivery by conduction and convection mechanisms [7–9]. For this reason, textile-based electrical heating systems present great advantage for maintaining human body at vital required heat level by satisfying more contact with flexible architecture, thus little heat loss occurs. Moreover, the researches on textile-based electrical heating systems recently focused on the development of a product that can monitor and regulate human body temperature according to climate change particularly for outdoor activities like skiing, snowboarding. Although heated garments have an obvious appeal for outdoor activities, the limitations of battery technology restrict the scale of electrical heating in those applications.

2.2. Textile-based heating systems

Household textile objects, such as seating, carpets, bedding and towels can be heated based on e-textile design approach, without intrusive heater cables. In our urban living spaces, textile-based heating systems like heated curtains and carpets might eliminate the use of radiators and vents. Apart from household applications, outdoor apparel market is clearly very promising and affluent, i.e., jackets and gloves. However, the market is currently very small and has niche applications for heated wear from motorcyclists and scuba divers to cold water fish hunters, pylon riggers and cold-store workers.

The aim of the textile-based heating systems is to provide necessary warmth to the user/environment in a cooler environment. Warmth condition is necessary to keep the human body system to function [10]. The constant temperature of a human body for maintaining body functions is 37°C; anything below resulted by prolonged exposure to cold temperatures can cause hypothermia and can be fatal for the survival of human body [11–13]. Therefore, heating systems play a major role to keep human body maintained at vital required heat levels. The heat can be applied to different areas on the human body. However, the flexibility of the heating system is very important since the users'/system structures are mostly irregular shaped. Heating with non-flexible systems may cause great heat losses due to fewer contact. For this reason, textile-based heating systems are an added advantage because of efficient heat delivery mechanism by wrapping the structure due to its flexibility.

Textile-based heating systems can be grouped into two categories: polymer-based and metal-based textile heaters. Metal-based textile heaters include metals as heating elements while polymer-based textile heaters include polymer materials to generate heat [14].

http://dx.doi.org/10.5772/intechopen.76192

As a metal-based textile heater, wires and sometimes metal sheets are used. Dorman® - Seat Heater Pad seen in **Figure 2a** and metal wires placed inside the car seat in **Figure 2b** are examples for textile-based car seat heaters.

Major applications of metal-based heating systems can be found in automotive, construction, sports and recreation sectors. Flexible, reliable and controllable heating systems have been manufactured using textiles. The common design approach is to insert metal wires onto woven fabrics [17]. In those systems, copper and nickel-chromium alloys are mainly used as heating elements. They are integrated into a layered textile system in order to be protected from damages when the seat is in use. Those systems include different controllable heating levels, from high to low which can be controlled by the user. Appropriate control to regulate the amount of heat and the duration of heating is very important in order to avoid excessive heat application to the user which could cause discomfort [18–20]. For this reason, on and off times are added to power control and management systems in today's automotive textile-based heating systems. In sports, recreation activities and in buildings, the similar design approach is taken into account. The main difference is only the size of application such that heating elements are placed on a much bigger scale, and so the power applied is also high [5]. Those applications do not require much flexibility, nor is weight an issue.

However, when the flexibility and drapability of a textile becomes important, polymer-based textile heating approaches take place. Indeed, how to create a comfortable wearable heating system is also a challenging issue and is a very important area of research.

With the introduction of polymer-based conductive yarns, polymer-based heating systems become great of interest for researchers and product developers. In a polymer-based heating system, when the direct current is applied, polymer-based conductive yarns carry the current and while carrying current they produce heat. In **Figure 3**, commercial flexible polymer-based products are seen. The presented commercial products by EXO2 uses FabRocTM technology as it reduces automatically the use of power when the temperature increases. The power consumption and heat generated are easily controlled and make the product intrinsically safe. In this system, carbon-loaded silicone yarn which is made by the combination of silicone and carbon polymers, and then extruded in a patented process to produce FabRoc® yarn is used [5]. This system can be applied to the various range of product applications where the

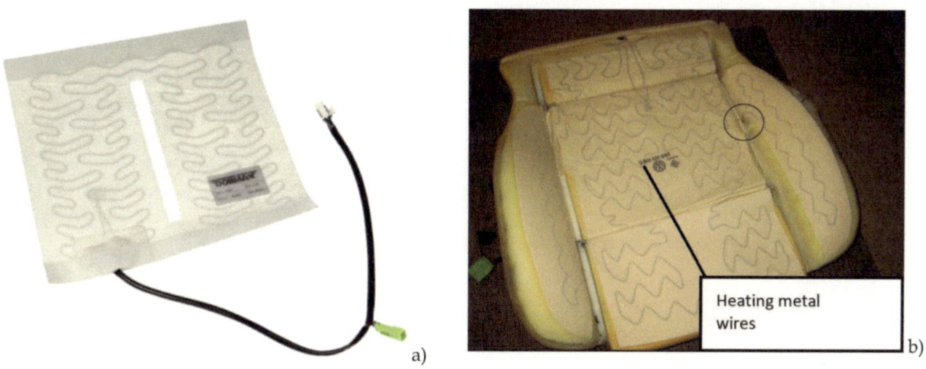

Figure 2. Textile-based car seat heaters (a) [15] (b) [16].

Figure 3. Flexible polymer-based (a) heated back support by EXO2 [21], (b) storm Walker jacket by EXO2 [22], and (c) DelphyneWomens 5 zone heated jacket [23].

flexibility is needed. Indeed, this system does not allow any yarn breakages or overheat as in use of wires and solid panels to accumulate.

The main applications of polymer-based textile heating systems are wearable heated fabric structures; heated vest, jacket, gloves, insoles, slippers, socks, therapeutic body warmers. In heated gloves (**Figure 4a**), extreme flexibility is required compared to jackets and vest since the arms and fingers exhibit excessive bending. For this reason, flexible conductive yarns, such as stainless steel yarns, silver plated polyamide yarns can be a good option as the heating element in those structures. The production technique can be knitting however the compatibility of the conductive yarn and its tolerance in terms of conductivity level for processing with knitting machine should be clearly investigated before starting production.

In a heated sock system (**Figure 4b**), polymer-based conductive yarns are knitted to the bottom of the foot through the heel and the toes to provide warmth.

In the slippers (**Figure 4c**), a heating system is generally made of woven structure composed of conductive fibers throughout the insole and placed between the foam and foot bed lining. In all systems, rechargeable lithium ion batteries are used as power supply to generate heat with a controller adjusting warmth level.

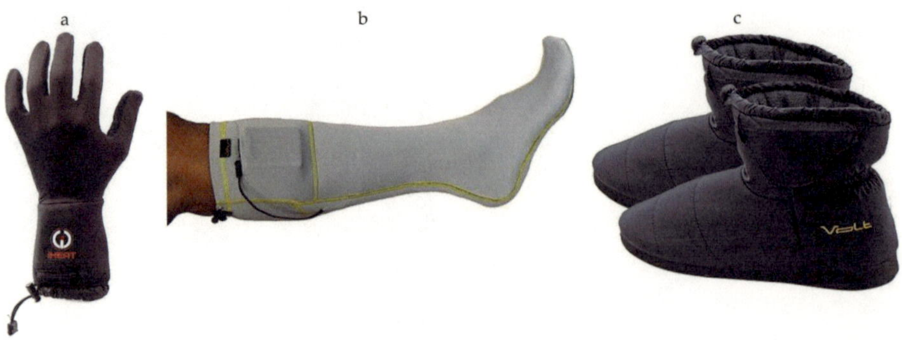

Figure 4. Flexible polymer-based heated glove, sock and slipper [24].

http://dx.doi.org/10.5772/intechopen.76192

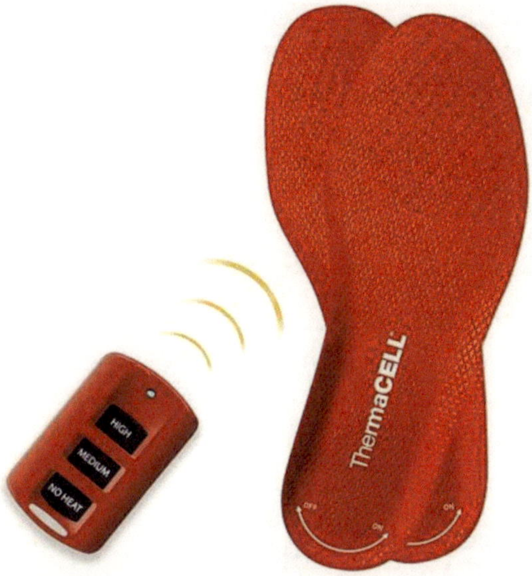

Figure 5. Flexible polymer-based heated insoles [24].

The heated insoles (**Figure 5**) provide a steady heat inside shoes and boots and they are operated by a wireless remote control. The insulated fabric structure does not allow the warmth to leak through the soles of footwear.

Flexible polymer-based therapeutic products provide heat to muscles, tendons and ligaments to increase blood circulation thereby relieving the sensation of pain [25] (**Figure 6**).

By positioning polymer-based heating structures behind the upholstery layers, protection against abrasion can be satisfied. Polymeric conductive yarns can be inserted on textiles structures using not only knitting techniques but also weaving, embroidery and stitching techniques. Then, the multilayered composite structures can be provided by coating, welding and lamination techniques according to specific usage area. For instance, in case waterproof property is needed, the textile structure (i.e. knitted, woven) including conductive yarns as heating element can be coated by different layers in order to provide waterproof property.

Wearable polymer-based heating systems are mostly advantageous because of their efficient heating and flexible structure for human body and they require less electrical power with

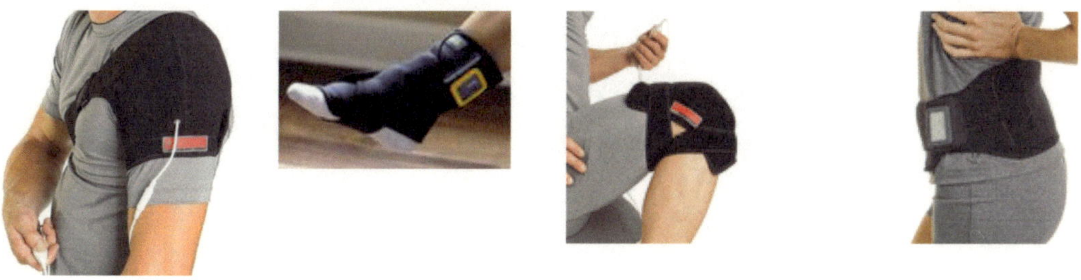

Figure 6. Flexible polymer-based therapeutic products [25].

small rechargeable batteries. They can be folded and compacted when they are not in use. They can be found in various range of products which can be worn in between layers of clothing and provide a controllable heating effect when required.

3. A wearable heating system with controllable e-textile-based thermal panel

This section describes the major considerations followed in the design and manufacture of the wearable heating vest. It starts with the design of e-textile-based thermal panel, followed by production process, electrical connection network and electronic control module. Finally, the assembly of all demanded components to reach the wearable heating vest is given in detail.

3.1. Design of e-textile-based thermal panel

During the design of the e-textile-based thermal panels, a number of factors are taken into consideration. First one among those was the selection of conductive yarn, examples of which are presented in **Figure 7**. In order to better investigate the effect of the resistance of the conductive yarn on resulting surface temperature, a total of seven stainless steel conductive yarns having resistance values of 4, 5, 9, 18, 30, 39 and 51 Ω/m were selected and used.

Secondly, the textile fabric was considered. For easier processability and their physical and mechanical superiority during processing and use, two PU-coated thermoplastic fabrics, namely Polyester and Polyamide fabrics, were used. Finally, three different welding tapes were selected for protection and waterproofness of the circuits made of conductive yarns, namely ST-604 with two PU layers, ST-306 with two PU and one Nylon layers, and ST-318 with two PU and one Polyester layers (see **Figure 8**). All three welding tapes selected are strong, resilient, and easy to work with. Since they do not contain plasticizers or volatile organic compounds, they are suitable for use in the production of e-textile transmission lines via a hot air welding technique. Welding was also previously shown to be an effective technique in the production of e-textile transmission lines [26].

Figure 7. Conductive yarns.

http://dx.doi.org/10.5772/intechopen.76192

Figure 8. Welding tapes.

Depending on the fabric and welding tape to be used, processing temperatures and feeding rates were carefully selected in order to achieve proper adhesion of the tape while avoiding any harm on the fabric, the welding tape or the conductive yarn. The levels of processing temperatures (450–550°C) and feeding rates (1.5–12 ft./min) are carefully selected by preliminary trials.

Following the experimental trials, demanded amounts of conductive yarns are placed over the selected fabric, either polyamide or polyester, and covered by the selected welding tape using H&H AI-001 hot air welding machine, as presented in **Figure 9**. Thanks to the unique

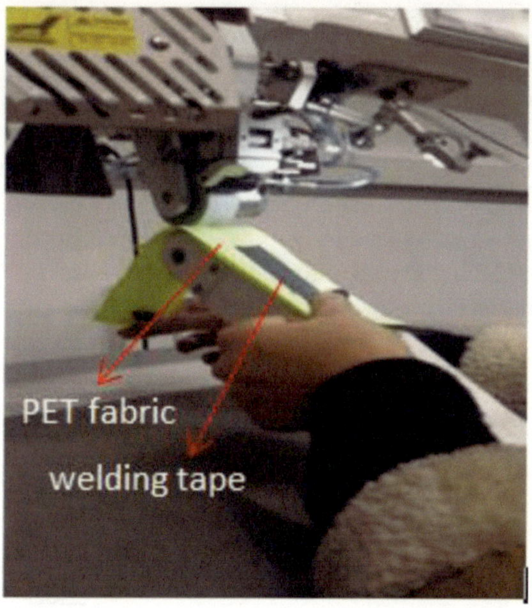

Figure 9. Manufacturing e-textile-based thermal panel using hot air welding machine.

excellent flexibility, lamination capability and soft hand properties of the welding tape, slippage of conductive yarns, distortion of overall fabric structure and thus formation of short circuits are avoided. Moreover, being manufactured from multilayered thermoplastic films, water leakages are prevented and edge fraying is eliminated by over bonded seams on conductive yarns.

Lithium batteries of three different potentials of 7.4, 9 and 12 V are used for application of potential difference to the samples. Details of the batteries are given in **Table 1**.

In order to avoid any interference during testing of the samples for their thermal performance via Fluke® Thermal Camera Ti200, a testing set up is prepared in a dark room compatible with AATCC TM 128 [27]. The dark room avoids any light or heat reflection from the walls, the floor or the ceiling, and does not have any other heat or light source other than those demanded during testing. An opaque 45° inclined background block is set into a dark testing cubicle. The thermal camera is locked to the tripod with its apparatus and they are placed together in front of the dark testing cubicle. At this point, the thermal camera was perpendicular enough to the inclined background block, and the distance between the thermal camera and the specimens is adjusted to 34–38 cm so that the errors were negligible, as instructed by the manufacturer of the thermal camera. Demanded amount of potential difference, 7.4, 9 or 12 V, is applied using the related battery, and thus a current flowing through the conductive yarns is created. SmartView® Software is used to analyze the specimens after they are tested. Emissivity value of opaque background block is estimated to have a value of 0.95. **Figure 10** shows the experiment set up to test electrical heating in which the e-textile structure is connected to a battery (power supply) by the electrical crocodiles.

An example of the thermal image of the e-textile sample together with sample's original image is given in **Figure 11**.

For the selection of better e-textile-based thermal panel, some criteria were applied. When the performance of the final product is considered, the leading factors to be taken into account were selected as a deviation of tested surface temperature from optimal surface temperature, battery life and battery weight. An optimal surface temperature of 60°C is considered, and any sample having tested surface temperature below 55°C or over 65°C are excluded. Considering battery life and battery weight; 12 V applied Polyester – ST306–550°C – 6 ft./min with 5 Ω resistance seems to be the optimal sample. It has 17 h of battery life and can heat the garment up to 57–58°C. The weight of the battery that is used for this sample to supply 12 V potential difference was 295 g.

Potential (V)	Type	Electric power (Ah)	Weight (g)	Price (USD)
7.4	Lithium	2.5	120	14.5
9	Lithium	10	254	16.5
12	Lithium	6.8	295	15.0

Table 1. Characteristics of lithium batteries.

Figure 10. Testing set up to measure surface temperature.

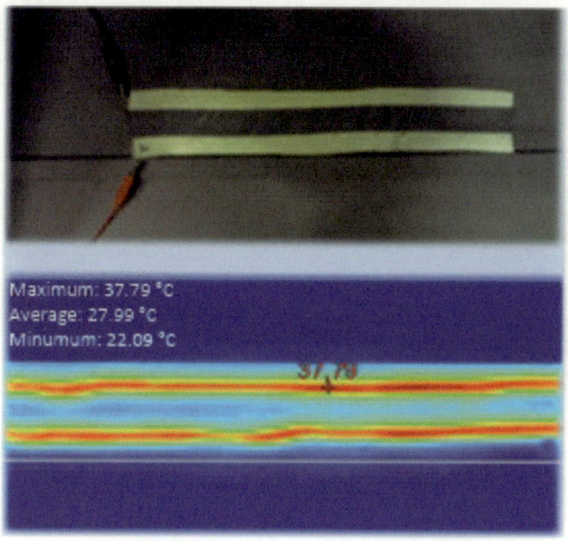

Figure 11. Original e-textile sample and its thermal image acquired by the thermal camera.

3.2. Production process

For manufacturing e-textile-based thermal panel to be placed inside the wearable heating vest, stainless steel yarn of 5 Ω/m resistance is placed over approximately 35 × 45 cm of PVA interlining. PVA interlining is easily dissolved when in contact with water at over 50°C, and was selected on purpose in order to observe the waterproof property of the thermal panel. The placement of the conductive yarn is as given in **Figure 12**.

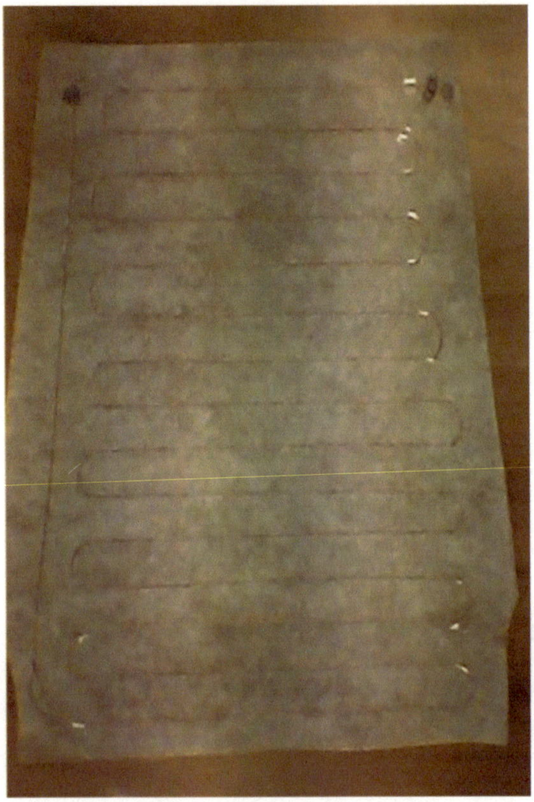

Figure 12. Placement of conductive yarn over interlining.

Figure 13. E-textile-based thermal heating panel to be placed inside the heating vest.

http://dx.doi.org/10.5772/intechopen.76192

Maximum care is taken during placement of the conductive yarn in order to avoid the formation of the short circuit. Transmission lines made of conductive yarns are then covered with ST306 welding tape as shown in **Figure 9**. For reaching 550°C of temperature during adhesion of ST306 welding tapes over the conductive yarns, hot press is used. Hence, the structure including conductive yarns covered by welding tapes became flexible, soft and waterproof. Having high bending strength aside with flexibility, the distortion and slippage of the conductive yarns are prevented. Thus, over bonded e-textile structure eliminates short circuit formation and hot/cold spot formation when folded. Finally, the interlining comprising conductive yarns is placed between two folds of 40 × 50 cm of PU-coated polyester fabrics and stitched on all four sides to reach the final e-textile-based thermal heating panel. Moreover, a pocket for inserting the electronic control module is attached to the left bottom side of the panel, at the bottom of which two snap fasteners are placed on which alternating ends of the conductive yarn are soldered. A total of eight brit buttons are stitched all around the panel for easier attachment and displacement of the panel when desired. The final e-textile-based thermal panel to be placed inside the vest is presented in **Figure 13**.

3.3. Electrical network and electronic control module

For supplying the demanded potential difference an electronic control module is designed and manufactured. Three 3.7 V batteries are connected in series, and used as power supplies. The technical drawing for control module circuit is as presented in **Figure 14**. The circuit brings protection against short circuits and electrical overload. To prevent electrical overloads, circuit breakers are used to stop the flow of current to the overloaded transmission

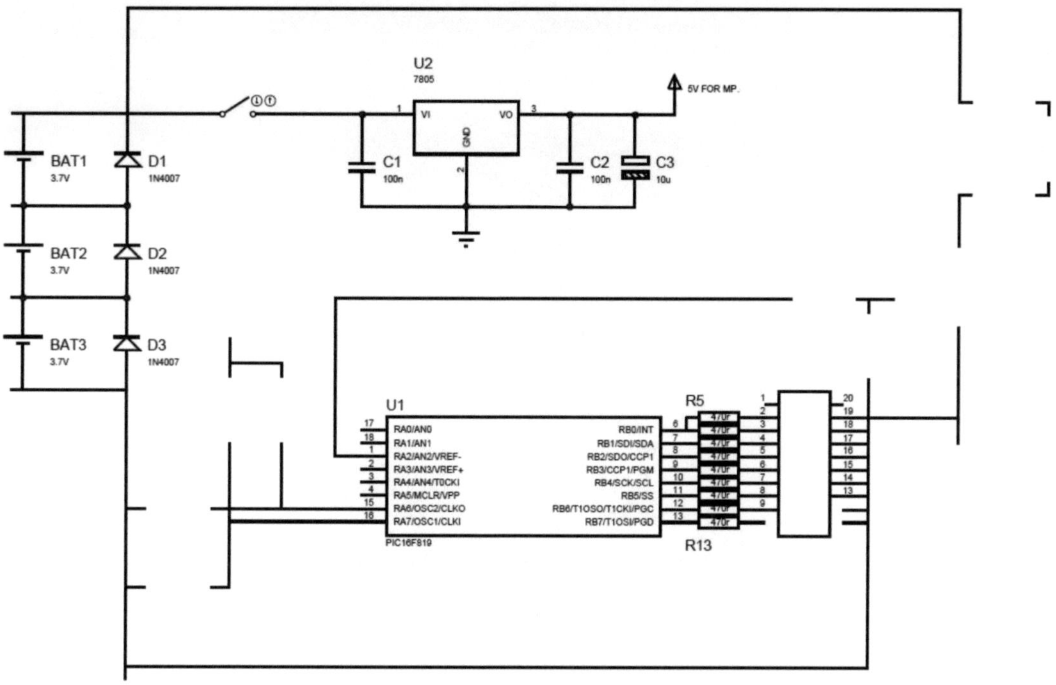

Figure 14. Technical drawing for control module.

lines. Hence, the control module with its circuit provides electrical safety. Moreover, thanks to its special design, the circuit is able to feed the system with constant energy output by automatically leveling the energy input from the batteries during use.

Figure 15. Electronic control module.

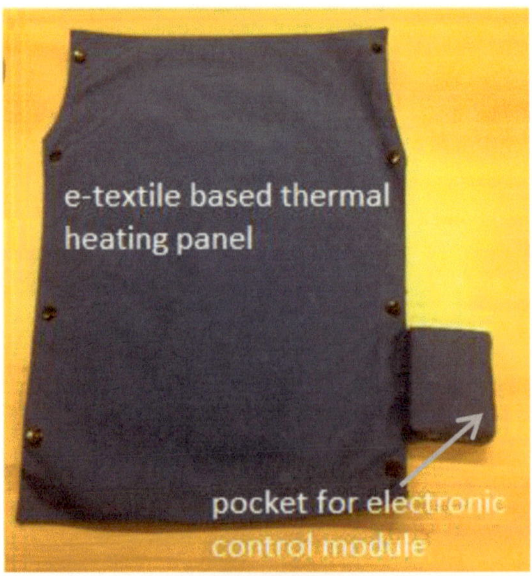

Figure 16. Insertion of electronic control module inside the pocket on the panel.

The manufactured circuit card is placed in an HH-0055 model 76 × 112 × 26 mm standard box. The control module, as shown in **Figure 15**, has 10 heating levels equipped with 10 LED displays which show the total energy output adjusted by the user at the desired comfort level. It has a total weight of less than 200 g.

Through the two snap fasteners at the bottom of the box, the control module is easily fastened after being inserted inside the pocket at the bottom left of the e-textile-based heating panel as shown in **Figure 16**. It has an on/off switch which is easily reachable by the user of the heated vest without being taken out of the pocket in which the module is inserted.

When connected to DC via its charger as shown in **Figure 17**, the control module can be fully recharged in 11 h, and keeps on supplying the heated vest with demanded energy at 10th level (up to 22 W) for as long as 4.5 h when used outdoors at an ambient temperature

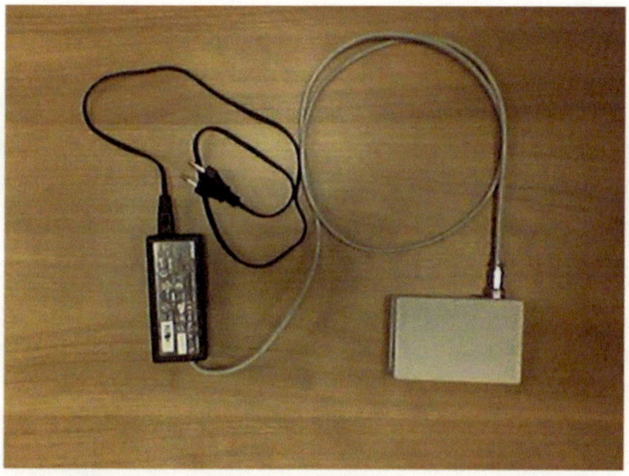

Figure 17. Connection of electronic control module and its charger.

Figure 18. Wearable heating vest (a) view from inside and (b) view from outside.

of 10°C. Thanks to the special design of the pocket in which it is inserted and fastened, the control module can be recharged without removal from the pocket.

3.4. A wearable heating vest

The integration of e-textile-based thermal heating panel to the vest is presented in **Figure 18a**. Eight buttons are stitched inside the vest on the corresponding locations where they will be fastened to the brits on the panel. Thanks to the eight brit buttons stitched all around the panel, the panel can be attached and displaced easily, and does not slip or cause discomfort of the wearer while putting on. Because of the proper selection of the place of the inner pocket in which the electronic control module is inserted, there is no distortion on the heated vest as shown in **Figure 18b**.

4. Real-time monitoring of the heating vest performance

For observation and monitoring of the heating performance of the heated vest in real-time, the vest was worn by a volunteer, and thermal camera images from the front and back of the torso were taken after wearing for 30 s, and from the back torso when the heating vest is on for 10 min and immediately after putting off at an ambient temperature of 21°C. As the ambient temperature was 21°C, the heating vest was set for a heating level of 5. As presented in **Figures 19** and **20**, the front temperature on the surface of the vest reached 23.5°C, and back temperature on the surface of the vest reached 25.5°C just 30 s after wearing the heating vest, respectively. It is apparent that the heating vest quickly starts heating up, and the heat is evenly distributed even after 30 s.

As presented in **Figures 21** and **22**, the back temperature on the surface of the vest reached 30.2°C, 10 min after wearing, and the surface temperature on the top clothing immediately after putting off was observed as high as 38.4°C, respectively.

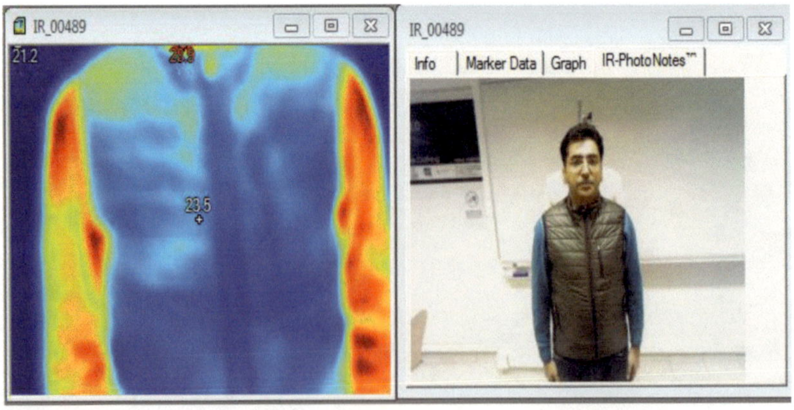

Figure 19. Thermal and actual images on first 30 s after wearing the heating vest (Front surface temperature on the vest: 23.5°C, ambient temperature: 21°C).

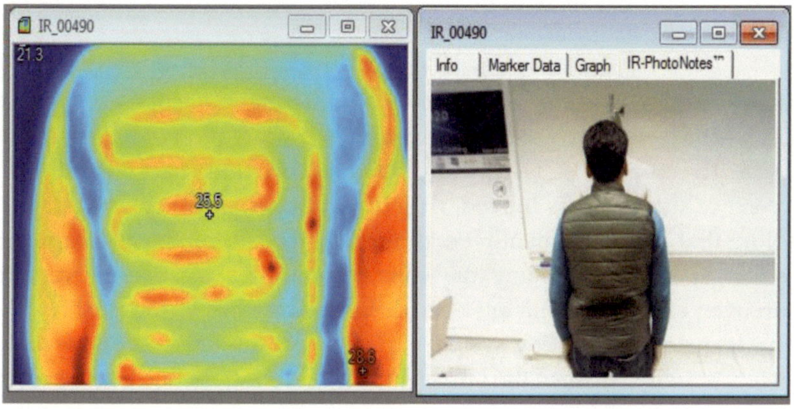

Figure 20. Thermal and actual images on first 30 s after wearing the heating vest (Back surface temperature on the vest: 25.5°C, ambient temperature: 21°C).

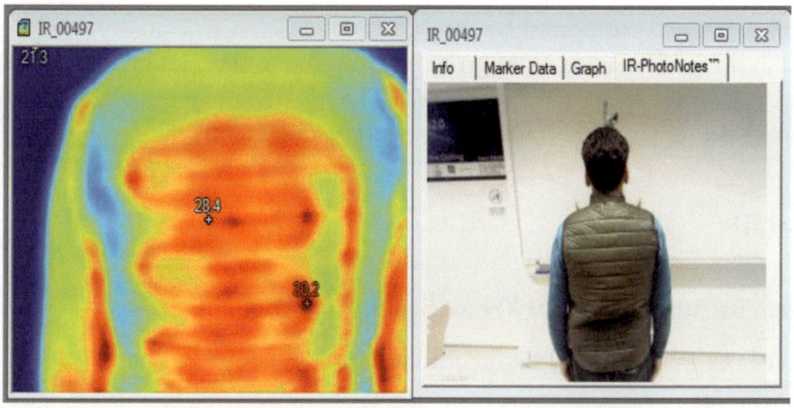

Figure 21. 10 min after wearing the heating vest (Back surface temperature on the vest: 30.2°C, ambient temperature: 21°C).

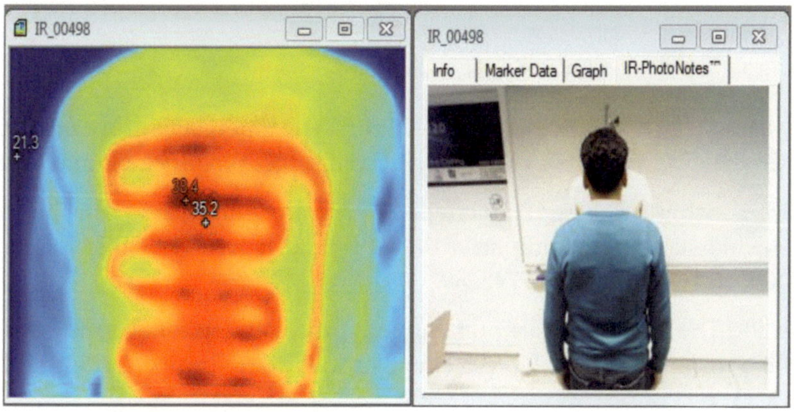

Figure 22. 10 min after wearing the heating vest (Back surface temperature on the top clothing of wearer: 38.4°C, ambient temperature: 21°C).

5. Future trends

The textile-based heating systems are one of the influential developments in wearable electronic textiles. Heating with textile structure requires conservation of battery life with respect to power output. To compromise wearability between cost, bulk, heat output and the lifespan of the battery is not an easy task. Indeed, the current battery technology restricts the scale of electrical heating that can conveniently be applied. In the future, the development of flexible fuel cells or supercapacitor technology may soon become a realistic alternative to rechargeable batteries. Moreover, another promising effector technology lies in the washability concept of those wearable textile-based heating systems. In addition, attractive efforts would be on solutions for interconnectivity between the power supply and the polymer-based heating elements.

Acknowledgements

This research work was supported by the Ministry of Science, Industry and Technology of TURKEY, with the project titled "Design of e-textile-based thermal heating panels via welding technology, Project No.0034.TGSD.2015-2". The researchers would like to thank CETEKS Elektronik Tekstil San. Tic. Ltd.Şti for final prototype development.

Author details

Senem Kurşun Bahadir* and Umut Kivanc Sahin

*Address all correspondence to: kursuns@itu.edu.tr

Faculty of Textile Technologies and Design, Department of Textile Engineering, Istanbul Technical University, Istanbul, Turkey

References

[1] Malmivaara M. In: McCann J, Bryson D, editors. The Emergence of Wearable Computing, Smart Clothes and Wearable Technology. Cambridge: Woodhead Publishing; 2009

[2] Bahadir SK, Sahin UK, Kiraz A. Modeling of surface temperature distributions on powered e-textile structures using an artificial neural network. Textile Research Journal. 2017. DOI: 10.1177/0040517517743689

[3] Bahadir SK. The effect of textile pretreatment processes on signal transferring capability of textile transmission lines. Fibres & Textiles in Eastern Europe. 2015;**23**(2(110)):55-62

[4] Bahadir SK, Jevsnik S, Fakin D, Sahin UK. Color and electrical resistance evaluation of cotton fabrics composed of stainless steel yarns treated with direct and reactive dyes. Textile Research Journal. 2016;**86**(13):1356-1371

[5] Mbise E, Dias T, Hurley W. Design and manufacture of heated textiles. In: Dias T, editor. Electronic Textiles: Smart Fabrics and Wearable Technology. Cambridge: Woodhead Publishing; 2015

[6] Swallow S, Thompson AP. In: Mattila HR, editor. Intelligent Textiles and Clothing. Cambridge: Woodhead Publishing; 2006

[7] Xu D-H, Cheng J-X, Zhou X-H. A model of heat and moisture transfer through parallel-pore textiles. Journal of Fiber Bioengineering and Informatics. 2011;**3**:250-255

[8] Hsieh Y-L. Liquid transport in fabric structures. Textile Research Journal. 1995;**65**:299-307

[9] Sousa LHCD, Monteiro AS, Perri VR, Motta Lima OC, Pereira NC, Mendes ES. Generalization of the drying curves in convective and conductive/convective textilefabric drying. In: 14th International Drying Symposium, São Paulo; 2004. pp. 710-717

[10] Sampath M, Aruputharaj A, Senthilkumar M, Nalankilli G. Analysis of thermal comfort characteristics of moisture management finished knitted fabrics made from different yarns. Journal of Industrial Textiles. 2012;**42**:19-33

[11] Au KF. Advances in Knitting Technology. Cambridge: Woodhead Publishing Series in Textiles; 2011

[12] Kar F, Fan J, Yu W, Wan X. Effects of thermal and moisture transport properties of T-shirts on wearer's comfort sensations. Fibers and Polymers. 2007;**8**:537-542

[13] Kissa E. Wetting and wicking. Textile Research Journal. 1996;**66**:660-668

[14] Altmann D, Haupt E, Knuppel M. Heated seat. Google Patents; 1990

[15] Available from: https://www.carid.com/dorman/seat-heater-pad.html?make=Chevy&model=Avalanche [Accessed on February 3 2018]

[16] Available from: http://forums.tdiclub.com/showthread.php?t=423547 [Accessed on February 3 2018]

[17] Weiss M. Flat heating element. Google Patents; 2013

[18] Buie D, Buie J. Temperature controlled seat cover assembly. Google Patents; 1997

[19] Fangueiro R, Filgueiras A, Soutinho F, Meidi X. Wicking behavior and drying capability of functional knitted fabrics. Textile Research Journal. 2010;**80**:1522-1530

[20] Sweeney M, Branson D. Sensorial comfort. Textile Research Journal. 1990;**60**:371-377

[21] Available from: http://www.exo2theheatinside.com/markets/medical-therapeutic-rehab/heatwave-back-support.html [Accessed on February 3 2018]

[22] Available from: http://www.exo2theheatinside.com/markets/hunting-fishing-outdoors/stormwalker-jacket.html [Accessed on February 3 2018]

[23] Available from: https://dragonheatwear.com/collections/womens/products/delphyne-womens-heated-puffer-jacket [Accessed on February 3 2018]

[24] Available from: http://www.staywarmed.com [Accessed on February 3 2018]

[25] Available from: https://www.thewarmingstore.com/heating-pads.html [Accessed on February 3 2018]

[26] Bahadir SK, Kalaoglu F, Jevsnik S. The use of hot air welding technologies for manufacturing E-textile transmission lines. Fibers and Polymers. 2015;**16**(6):1384-1394

[27] AATCC TM128-2017. Wrinkle Recovery of Fabrics: Appearance Method. Research Triangle Park, NC, USA: Association of Textile, Apparel & Materials Professionals Technical Manual

Section 3

Technology TI

Chapter 10

The Comparison of Wearable Fitness Devices

Kanitthika Kaewkannate and Soochan Kim

Additional information is available at the end of the chapter

http://dx.doi.org/10.5772/intechopen.76967

Abstract

The wearable devices or wearable trackers help to motivate you during daily exercise or workouts. It gives you information about your daily routine or fitness by using wearable technology in combination with your smart phone to track your daily activities and fitness without the manual calculations or records that can be intrusive. Generally, companies display advertising for these kinds of products and depict them as good, user-friendly, and accurate. However, there are no subjective research results to prove the veracity of their words. Four popular wrist band-style wearable devices currently in the market were selected at the devices which are most popular (Withings Pulse, Misfit Shine, Jawbone Up24, and Fitbit Flex). The accuracy of tracking was one of the key components for fitness tracking, with some devices performing better than others. Accuracy in the tracking of daily activities such as walking, running, and sleeping is important. This research showed subjective and objective experiment results, which were used to compare the accuracy of four wearable devices in conjunction with user-friendliness. Satisfaction levels, the accuracy of tracking, and the opinion of each subject while using wearable device to track their daily activity were compared. The results determined that the cost-effectiveness was the Withings Pulse, followed by the Fitbit Flex, Jawbone Up24, and Misfit Shine.

Keywords: wearable devices, Withings, Fitbit, misfit, jawbone, best tracker

1. Introduction

There are many online reviews of wearable trackers, typically presenting different perspectives on the rankings. However, objective and factual information cannot match the subjective findings which convey further details about the devices, the participants in the experiments, or the particular reviewers involved. Further, there is no quantitative comparison table to show the results of subjects reviewed. For example, from the site of "Top Ten reviews" [1, 2], the best wearable device reviewed was the Fitbit Flex, followed by the Withings Pulse, Jawbone, and Misfit. From this site, the table score was shown and compared but provided no details about where the information originated. Another example for a wearable tracker review site is "Best fitness tracker 2015." From Wearable Tech for your connected self [3], the top four trackers were Jawbone, Misfit, Fitbit, and Germinly, respectively. Online reviews rarely present objective and measurable comparison data, and the review content is usually only the opinions of bloggers. While the information is subjective in such cases, it is useful for customers who are considering purchasing a tracker, since the content can guide the customer to find the right product to meet their needs. However, customers would benefit greatly if the information offered included subjective comparison results which could let customers know which trackers would be the best fit for their requirements [4–6].

This research study provides a comparison among the leading wearable fitness-tracking devices available today, covering the accuracy, user-friendliness, and customer satisfaction. All the selected devices were of the wristband type and ranked within the top ten of the best 2015 products from reviews [1–3]. Four of these products were selected randomly: the Fitbit Flex (Fitbit Inc., San Francisco, California, USA) [7], the Withings Pulse (Consumer Electronics, Issy-les-Moulineaux, France) [8], the Misfit Shine (Msfit Inc., Apple Inc., Apple, Mitten Rd., Burlingame, California, USA) [9], and the Jawbone Up24 (Consumer Electronics, San Francisco, California, USA) [10–13]. The results will be shown as subjective and objective research results for the trackers with the best accuracy and user-friendliness by physical information from real users.

This paper reviews the following: (1) the overall specs of four devices were compared, (2) the user satisfaction for all four devices and compares the results of the satisfaction scores, (3) the opinion from user in experiments were explored, (4) the reviews of wearable devices of blogger or reviewer from internet site were compared with linking to the users' opinion in experiments, and (5) the accuracy and repeatability of activity tracking for each model were also recorded and compared.

2. Material and methods

2.1. Wearable devices used in the experiment

The four wearable devices in the experiments were done randomly for wrist band devices available in Korea, chosen from ten devices in the top ten review rank [1–3] (see **Figure 1**). The four devices are Jawbone Up24, Misfit Shine, Fitbit Flex, and Withings Pulse. **Table 1** provides the comparable features of the four wearable devices.

2.2. The user interface application (UI app) of each devices

Most wearable devices differ in their user interface. The UI design for wearable devices should be simple, clear, and quick to navigate for users' comfort [3]. This is not an easy design feature since wearable wrist devices have to be small. As a result, devices which link to a smartphone though a UI app have become more popular among users. The smartphone apps which work

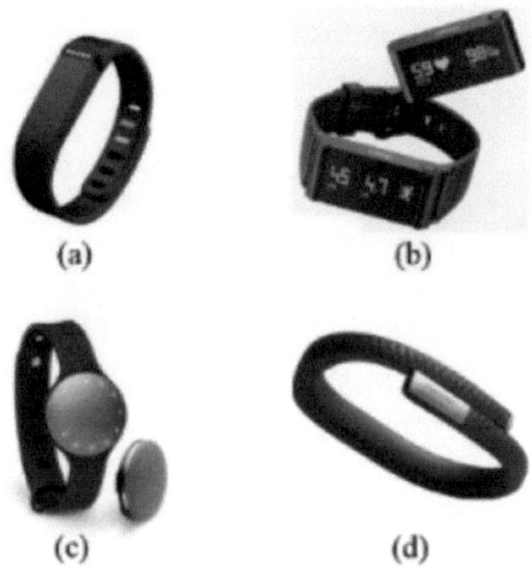

Figure 1. The wearable devices in the experiments: (a) Fitbit flex, (b) Withings pulse, (c) misfit Shine, and (d) jawbone up.

with wearable devices must therefore be easy to download. The app also serves to process the collected data, store the data, and perform network activities [6].

2.3. Experiment methods

2.3.1. Subjects

Seven healthy subjects participated in the experiments, comprised of six healthy men (adults aged 27–50 years, mean age of 31 years old, mean height 171.5 cm, and mean weight 68.18 kg) and one healthy woman (adult aged 30 years, height 160 cm, and weight 42.1 kg).

Each participant wore each of the four devices for 1 week in turn, taking notes throughout of the results of their usage, their satisfaction levels, and their point of view regarding the benefits and shortcomings of each product. Upon completion of the testing cycle, the four devices were all tested in order to confirm the accuracy of the recorded data. The experiment details are outlined in Section 2.3b.

2.3.2. Experimental methods

2.3.2.1. Subjective satisfaction of wearable device users

During the course of this study, each participant wore each of the four wearable devices for 1 week periods, completing the evaluation form to assess their satisfaction levels upon completion of the week. The total test duration was 1 month. The evaluation form had two sections.

Section 1. The Likert scale used to evaluate the devices.

The participants provided scores using the five-point Likert scale for each category on each device, assessing the design, functions, and features after the 1-week test period when the device was worn every day. A score of five indicated a very positive assessment, while one represented the poorest performance. The evaluation form comprised two parts:

Features	Specifications	Jawbone Up24	Fitbit Flex	Withings Pulse	Misfit
Company detail	Company name	Consumer Electronics	Fitbit Inc.	Consumer Electronics	Misfit Inc.
	Country	San Francisco, California, USA	San Francisco, California, USA	Issy les Moulineaux, France	Apple Inc., Apple, Mitten Rd., Burlingame, California, USA
	Information Link	http://jawbone.com	http://www.fitbit.com	www.withings.com	www.withings.com
Product released	Released(USA)	13-Nov-13	6-May-13	27-Jun-13	16-Sep-14
	Announced(USA)	13-Nov-13	7-Jan-13	6-Jan-13	16-Sep-14
	Present availability	Available	Available	Available	Available
Type	Smart watch	Watch style	Wearable/clip-on	Wearable/clip-on	Wearable/clip-on
General	Price in market	$150	$100	$100	$95
	Dimension (W×D×H)	6.1 × 6.1 inch	Small, 5.5 × 0.6 inch	1.7 × 0.87 × 0.33 inch	1.08 × 0.13 × 1.08 inch
			Large, 6.3 × 8.2 inch		
	Weight	Small, 19 g	Small, 16.4 g	8 g	9.4 g
		large, 23 g	large, 18.9 g		
Battery	Type	$LiMnO_2$ 225mAh	Lithium polymer battery	Lithium-ion polymer	CR2032 coin cell
	Battery life	4–6 months	4–6 months	6 months	3 months
	Rechargeable battery	Yes	Yes	Yes	No
	Changeable battery	No	No	No	Yes
	Usable per one time charged (advertised)	Up to 10 days	Up to 14 days	Up to 14 days	Up to 180 days
	Full charging time	3 hours	4 hours	2 hours	No
Tracking Metric	Motion	Yes	Yes	Yes	Yes
	Step counting	Yes	Yes	Yes	Yes
	Distance	No	Yes	Yes	Yes
	Calories	Yes	Yes	Yes	No
	Sleep	Yes	Yes	Yes	Yes
	Heart rate	No	No	Yes	No
	Fitness analytics	Yes	Yes	Yes	Yes
	Wind	No	No	No	No
	3D mapping	No	No	No	No
	Speed	No	No	No	No
	SpO2	No	No	No	Yes
	Goal tracking	No	Yes	No	No

Features	Specifications	Jawbone Up24	Fitbit Flex	Withings Pulse	Misfit
Resistance function	Water resistance	Not too high	Yes	No	Yes (up to 30 m)
Synchronization	Sync type	Wireless (Bluetooth)	Wireless (Bluetooth)	Bluetooth	Wireless (Bluetooth)
	Sensor network	Bluetooth	Bluetooth	Bluetooth	Bluetooth
Screen and display	Screen type	Dual LED	Five LEDs	OLED (black lit)	12 LED and blink
	Touchscreen	Capacitive finger	Capacitive finger	Capacitive finger	Capacitive touch
	Screen size (inch)	No (LED bar)	No (LED bar)	1.69	No (12 LED Blink)
Sensor type	Three-axis accelerometer	Yes	Yes	Yes	Yes
	Three gyro sensors	No	No	No	No
	Magnetometer	No	No	No	No
	Pressure sensor	No	No	No	No
	GPS	No	No	No	No
	Altimeter	No	No	No	No
Alarm function		Yes	Yes	Yes	Yes
Data sharing		Yes	Yes	Yes	Yes
Material	Wearable body type	Rubber	Rubber	Rubber	Anodized aircraft-grade aluminum
Smartphone	Smartphone operation system	iOS 5.1 or greater, Android 4.0 (Ice Cream Sandwich) or later	Window Xp/Vista/7/8 Mac OS X 10.5 or above iOS/Android	Android 2.3.3 over or iOS	Pair to iOS only
UI interface	History tracking (days)	270 days	30 days	10 days	30 days
	Computer data storage (Web app)	No	Yes	Yes	Yes
Social network data Sharing	Data sharing	Only friends who you already have known	Yes	Yes	Yes

Table 1. The comparison table of features and function of four wearable devices.

Part 1. Satisfaction assessment of properties and features.

This section invited participants to provide a satisfaction rating score for the properties and features of each of the four devices. Factors to consider included the hardware, or general design, the user interface and UI app, the synchronicity, the battery, and the user-friendliness.

Part 2. Satisfaction scores for the device metric function.

This section invited participants to provide a satisfaction rating score for the metric function on each of the devices. This encompasses measures such as step count, distance, calories, sleep, and analysis of nutrition.

Section 2. Participants' personal opinions about the devices.

This section allowed participants to record their opinions on the positive and negative aspects of each device. The personal comments of the participants can then be presented subsequently.

2.3.2.2. Testing the devices for accuracy and repeatability

The functionality offered by each of the four devices is similar; the differences lie within the user interfaces, applications, and the algorithms used for calculations. The most important criteria from the perspective of the user are accuracy and repeatability, since these aspects will guide the users to reliability achieve their targets. However, the accuracy and repeatability of any of these devices will also depend to a certain extent of personal factors such as the weight, height, gender, and age of the user. The physical data will therefore be required along with the subjective perceptions of the users in order to determine the accuracy and repeatability of the four devices.

To conduct the test, the devices were placed on the participants' wrists as shown in **Figure 2**. Following the recording of test data during the experiment, the real data were then measured in terms of distance so as to compare with the recorded data from the devices in order to determine the accuracy.

The percentage of accuracy and repeatability for the four devices are presented in this paper. The repeatability was calculated using Cronbach's Alpha, SPSS program (SPSS V.2012, IBM Corporation, USA). Subsequently, we scaled scoring among the four devices from the best to the lowest, as explained in **Table 2**.

Experiment 1. Distance traveled and step counting of indoor walking.

Subjects wore the four devices (**Figure 2**) and then walked straight across the indoor experiment court. Total distance was 48 meters for ten trials per person. The data for step counting and distance represented for each device were collected.

Experiment 2. Distance traveled and step counting of treadmill running (jogging).

Subjects wore the four devices (**Figure 2**) and then ran or jogged on a treadmill at 8 km/h [13, 14] for 1 minute, repeated for five trials. The real data record from the treadmill was collected to compare with the real distance calculation from the treadmill's LCD.

Experiment 3. Step counting when walking up and down stairs.

Subjects wore four devices as shown in **Figure 2**; then walked up four flights of stairs, repeated the experiments for five times; and then walked down the stairs, repeated for five times.

Upon completion of the data gathering stage, scores were assigned to each device for accuracy and repeatability using a scale rating of one to four, where four indicates the best performance among the four tested products, as shown in **Table 3**.

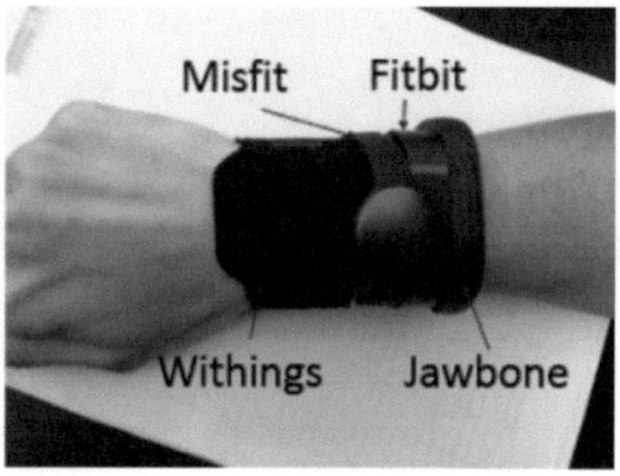

Figure 2. The subjects wore all four devices to measure the accuracy and repeatability of results.

Scale (point)	Meaning
4	The highest accuracy or repeatability among the four devices
3	The second highest accuracy or repeatability among the four devices
2	The second lowest accuracy or repeatability among the four devices
1	The lowest accuracy or repeatability among the four devices

Table 2. Scale of accuracy and repeatability when compared among four devices for each experiment.

3. Results and discussion

3.1. User satisfaction

After each of the four 1-week test periods, the participants completed the evaluation form providing their Likert scores concerning the device attributes and qualities until UI application. The satisfaction scale applied is described in detail in **Table 3**.

Figure 3a shows the mean score for the five conditions of features, including device design, battery use, smartphone synchronization, UI applications, and ease of use. On the other hand, **Figure 3b** shows the mean and standard deviation score of the satisfaction when using the four main functions of each device, including step counting, sleep tracking, distance tracking,

Scale (point)	Meaning
5	Very useful and very satisfied
4	Moderately useful and moderately satisfied
3	Slightly useful and slightly satisfied
2	Less useful and less satisfied
1	Not useful and not satisfied

Table 3. The scale of evaluation and corresponding meanings.

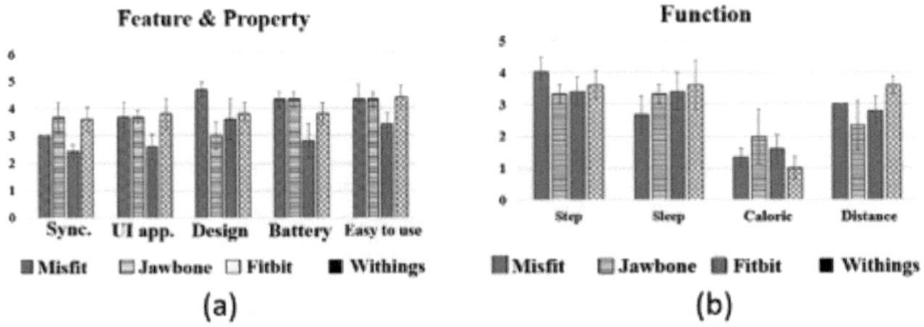

Figure 3. (a) Bar graph comparison of mean and standard deviation of the satisfaction score by subjects when using the devices. (b) Bar graph comparison of mean and standard deviation of the satisfaction score by subjects when using the devices.

and caloric analysis. The case of heart rate analysis does not exist in the evaluation score because only the Withings Pulse possessed this function.

From the subjective results, the Withings Pulse device had the highest satisfaction score, followed by the Misfit Shine, Jawbone Up24, and Fitbit Flex.

The results in **Table 4** show the opinions, which imply that subjects gave similar answers to two persons about the features and functions of each device.

Opinion about features	Jawbone Up24	Fitbit flex	Withings pulse	Misfit Shine
Design	Light and good for any sport	Device design is good and sleek; it is good for any sports	Design is not attractive, but the fabric band can help to hold it as wrist band	Design is very attractive, beautiful, and fashionable
Display	Easy to tap the screen to active	Easy to tap on screen to active	Display is big and shows the activity tracking without any smart phone sync	Display is as clock; it can also be used as a watch, but in the sunshine it is hard to see the LED display
Water resistant	It is water resistant, but according to the manual, it is less water proof	It can be used to take shower without worrying	According to manual, it is not water resistant	It is designed for sports as swimming; water resistant is too high
UI app	1. Tips of app and how to use always shown on home screen	1. UI app is colorful and has fun display, easy to use	1. Display is easy to use and colorful	1. Display is beautiful and easy to understand
	2. Enjoyable fitness tracker	2. Nutrient analysis is very detailed	2. Dashboard log is easy to check all activity	2. It has goal tracker to lead you to know your daily activity
	3. Dashboard shows the overall daily activity	3. Dashboard shows the overall daily activity	3. The heart rate function is good to check your health status	3. App can be shared to your friends; it shows how your friends seek the goal

http://dx.doi.org/10.5772/intechopen.76967

Opinion about features	Jawbone Up24	Fitbit flex	Withings pulse	Misfit Shine
Metric function	Sleep tracking is its main function because the sleep tracking report is very detailed, but it is difficult to use	Food and nutrient calculation is its main function; it is very easy to use	The Pulse O2 measurement is its main function; it can help you detect your heart status	The goal tracking is its main function; you can check how your status to seek the goal
Battery	It can be charged only on the USB cable	It has the battery indicator to check the battery status, but it has high battery consumption	It has battery indicator to check battery status; battery can be used in too many days	It is comfortable; no need to charge the battery
Synchronization	Slow synchronization	Slow synchronization but always lost connection	Fast synchronization, data can send via Bluetooth and WiFi	Fast synchronization but easy to lose connection
Others (disadvantage/ cons)	1. The devices required smartphone to display	1. It required smartphone	1. Design is not modern	1. It required smartphone
	2. No display on itself	2. Slow synchronization	2. If the battery of witlings is low, the device cannot connect. The data transfer which is shown on the smartphone is inaccurate	2. It has slow synchronization, not always updated real time
	3. Sleep tracking results are difficult to use and nonautomatic	3. The device is confusing sometimes; it needs to be reset	3. The sleep tracking is not automatic	3. Sometimes it gave inaccurate display
	4. Cannot share the data through social network	4. Tracking problem when walking up- or downstairs	4. Not water proof	4. Tracking problem when walking up- or downstairs (inaccurate)
	5. Most expensive among four devices	5. High battery consumption	5. Automatic loss of syncing	5. The display does not always respond to finger tapping
	6. It is not fully waterproof	6. Data is not updated sometimes	6. Screen is difficult to see in sunlight	6. No nutrient analysis
	7. Slow Synchronization	7. Calories count is not easy to use and only European foods are in the database	7. No Nutrient analysis	7. Always disconnected from mobile phone

Table 4. Comparison table of user feedback (summarized from seven subjects for each device).

From **Table 4**, it is apparent that all four devices were both satisfactory and unsatisfactory to the subjects. As mentioned in Topic 3.2, the summary of the opinions came from the similar meaning answers from two or more subjects. The most clearly apparent problem across all devices was the automatic loss of synchronization, which presents a problem in updating data, and leads to incorrect reports. However, all of the participants were able to use all of the devices easily with minimal instruction, or less, so the user-friendliness was good in all cases. The summary of the different claims from the reviewers on various commercial websites reviewing the devices is shown in **Tables 5–8**. Considering the five top-ranked sites from a Google search, it is clear that the leading reviews are well known due to the large numbers of people visiting those top-ranked sites to investigate wearable devices. The comments and descriptions of the reviewers can inform customers who might wish to purchase one of the devices for themselves. The drawback is that although the reviews can appear helpful, it is difficult to know whether the views have been influenced by the companies themselves or if they are in fact genuine independent perspectives. It is possible for an opinion to come only from one reviewer who used a product.

This section is explored because of the pros and cons of using reviewer claims, whether or not they may be similar or different than the customer's and seven subjects' opinions in this study. **Tables 5–8** shows the summarized data of advantages (pros) and disadvantages (cons) for each of the four devices from reviewers on the websites.

Jawbone UP24 subjects' opinions and the reviewers imply that it has a good design and fits comfortably. The UI app is colorful and easy to understand. The sleep tracker is very smart and also has good alarm functions. However, disadvantages of the device (cons) include the design lacking a screen; it is not fully water proof, and the battery charger is complex.

Withings Pulse has good primary features, such as the heart rate function. The display itself is big and can show the results tracking. The data log uses Bluetooth syncing or wireless for automatic updating. The Withings design does not, however, provide an attractive case, with direct sunlight making it hard to read the display. Furthermore, there is no automatic sleep tracking function.

Fitbit Flex has a slender and attractive design, is wholly waterproof, and is equipped with good social features. However, the Fitbit Flex has weak points in that it has no screen; the food log and calories tracking are not easy to use, and the tapping screen is sometimes confusing.

Misfit Shine looks both fashionable and elegant. It is ideal for watersport enthusiasts since it is fully waterproof. It offers a goal tracking feature which motivates users to achieve their targets. The battery is exchanged rather than recharged. However, the device works only with iOS, and while there are plans to introduce Android compatibility, this has not yet taken place. A smartphone is necessary to check the tracking status, and there are sometimes problems which arise when poor syncing from device to smartphone results in inaccuracies.

3.2. Accuracy and repeatability of the four devices

From **Table 9**, it can be seen that the Withings Pulse achieved the best results for the accuracy and repeatability of measurements for indoor walking, at 99.9% for accuracy and 86% for repeatability. **Figure 4** shows the results for all of the four devices. The lowest scores for accuracy and repeatability were measured for Misfit.

Reviewer name	Site name	Reviewed date	Reference site	Advantages (pros)	Disadvantage (cons)
Weebly	Jawbone Up24 review	8-Nov-14	[11]	1. Wireless syncing	1. Social sharing: can only add friends you already know
				2. Can be used in the shower	2. No website interface, only phone app
				3. Deep sleep and light sleep data	3. Hair can stuck in cap button
				4. "Smart Wake" alarms for naps	4. Overcount arm movement as steps
				5. Usability design	5. No screen
				6. Holds battery charge for up to 7 days	—
Raphael Mumford	Smart activity tracker, Jawbone Up24 review	Not mentioned	[12]	1. Well-designed band fits your wrist comfortably	1. No measurement of stairs climbed
				2. Long lifetime battery (up to a week for per full-energy charging)	2. Just for iOS devices
				3. Wireless syncing via Bluetooth without any hassle	—
				4. Quick charging	—
				5. Smart alarm	—
				6. Expandable storage capability	—
				7. Price is cheap	—
Matt Swider	Jawbone Up24 review	24-Mar-14	[13]	1. Wireless syncing added	1. No display for on-demand stats
				2. Stylish and lightweight	2. Doesn't have a web app
				3. Very soft rubber for comfort	3. Works with just ten Android phones
				4. iOS and Android compatible	4. 2.5 mm stereo jack for charging
Michael Sawh	Jawbone Up24 review	26-Mar-14	[13]	1. Bluetooth smart support for real-time syncing	1. No built-in screen
				2. Slim, stylish design	2. Shorter battery than Jawbone UP
				3. Great silent alarm feature	3. App is sluggish at times
				—	4. Not waterproof

Reviewer name	Site name	Reviewed date	Reference site	Advantages (pros)	Disadvantage (cons)
Matthew Miller	Jawbone UP24	6-Dec-14	[14]	1. Well-designed band that fits comfortably long battery life	1. No altimeter to measure stairs climbed
				2. Flawless syncing via Bluetooth	2. Limited just to iOS for now
				3. Integrated Microsoft Office software	3. Hangs up on jackets and long sleeve shirts
				4. Charges up quickly	—
				5. Great sounding front facing stereo speakers	—
				6. Expandable storage capability	—

Table 5. Summary of pros and cons from reviewer opinions for the jawbone UP24.

Reviewer name	Site name	Reviewed date	Reference site	Advantages (pros)	Disadvantages (cons)
Weebly	Withings Pulse smart activity tracker review	Not mentioned	[15]	1. Captures heart rate information	1. Not shower-safe
				2. Captures flights of stairs climbed and elevation climbed	2. Easy to misplace (leave in pockets, etc.)
				3. Check running stats (duration and distance traveled) in real time	—
				4. Automatic wireless syncing	—
				5. Captures sleep (duration, quality, light versus deep sleep, interruptions)	—
				6. Screen with constant feedback	—
				7. Discreet and multiple ways to wear	—
				8. Battery charge lasts up to 14 days	—
				9. App also pulls in data wirelessly	—
				10. Internet site available for Withings devices	—

Reviewer name	Site name	Reviewed date	Reference site	Advantages (pros)	Disadvantages (cons)
Raphael Mumford	Withings Pulse wireless activity tracker review	Not mentioned	[16]	1. Easy to clip on something thanks to a silicone and metal clip	1. Not automatically detect the time and switch to sleep mode
				2. Large OLED screen touch screen and easy history browsing	2. Lack of silent alarm
				3. Free iOS and Android apps	3. The screen is not easy to read under the sunlight
				4. Accurate heart rate monitoring	4. Syncing to PC is impossible
				5. Low power consumption. The battery has a long lifetime of 2 weeks	–
				6. Charging easily via a standard micro-USB to USB power cable on a computer or on a smartphone	–
				7. Free account on Withings.com to store your health and fitness data	–
				8. Worn as a wristband to track your activity and sleep	–
				9. Log your foods and weight, and get the perfect balance between activity and nutrition	–
DC Rainmaker	Withings Pulse In-Depth Review	21-Nov-13	[17]	1. Can record resting heart rate quickly and easily	1. The unit is a bit chubbier than some others
				2. Display is clear and easy to understand	2. Does not track heart rate all 24 hours, only on demand
				3. Good battery life	3. Does not automatically go from sleeping mode to nonsleeping, must switch over manually
				4. Good ability to connect to third-party platforms/sites	–

Reviewer name	Site name	Reviewed date	Reference site	Advantages (pros)	Disadvantages (cons)
Mikey Campbell	Review: Withings Pulse with built-in heart rate monitor	4-Nov-13	[18]	1. Variety of sensors	1. Lack of meaningful data presentation
				2. Impressive data accuracy	2. Display lag, touchscreen issues
				3. Flexible carry options	3. Wear ability limited to belt clip
Julie Strielmeier	Withings Pulse Activity tracker review	23-Aug-13	[19]	1. Size of the pulse	1. Syncing problem
				2. Can see all the important info right on the device itself instead of like some devices	2. Sleep data is not always accurate, and the detailed data could use some beefing up to show more info
				3. Wireless syncing is a real plus too	3. It does not work with a stand along computer
				4. The built-in heart rate sensor is super easy to use	—

Table 6. Summarized data of pros and cons from reviewer opinions for the Withings pulse.

Reviewer name	Site name	Reviewed date	Reference site	Advantages (pros)	Disadvantages (not so cons)
Weebly	Fitbit Flex review	10-Aug-14	[20]	1. Comfy wristband form factor	1. Does not track flights of stairs (like the FitBit One)
				2. shower-safe water resistance	2. Always visible if worn with short sleeves
				3. Very adjustable wristband	3. No screen on device to show you detailed information on goal progress
				4. Progress lights tell you how close you are to reaching your daily goal	4. Very hard to attach to the wrist and can pop off (while canoeing, for me)
				5. Wireless syncing	5. Have to tap band repeatedly to enter/exit sleep mode or to stop the silent alarm
				6. Great integration with existing fitness apps like MyFitnessPal	6. Chopping veggies can trigger sleep mode
				7. Strong social features including adding friends with a FitBit device or other FitBit users, a competition	—

Reviewer name	Site name	Reviewed date	Reference site	Advantages (pros)	Disadvantages (not so cons)
Raphael Mumford	Fitbit Flex wireless activity Review	Not mentioned	[21]	1. Slim and stylish design for the perfect fit	1. Lack of stair tracking
				2. Track everything relating to your activities and sleeps, except for stairs quantity	2. A tiny LCD display is not included
				3. Waking up silently thanks to gentle vibration	—
				4. Five built-in LED indicator lights for better monitoring of your progress	—
				5. Water resistance is included	—
				6. The Fitbit App for iPhone and Android devices to track your real-time stats, set goals, log food and other workout information, and then represent your sleep trends	—
				7. Connecting and competing with other athletes for a better motivation	—
				8. A long lifetime battery of 5–7 days per charge	—
				9. Two options in size and two options in color you can choose from	—
Articles by Suzie	Fitbit Flex review	Mar 15	[22]	1. It is easy to wear all the time	1. Only charge the tracker with the USB cable
				2. Water resistant	2. It takes a lot of work in the beginning to establish your food menu
				3. Upload status automatically through the Bluetooth or dongle	3. Sometimes have trouble tapping the tracker into sleep mode
				4. Notification alert to let me know when my battery is running low.	—
				5. Learning curve to get the most from it; the Dashboard is a colorful and has fun display of my activity	—
Bethany Gordon	Fitbit Flex	Only year mentioned (2015)	[23]	1. Excellent interface	1. This device does not have a screen
				2. Excellent app	2. Only view your data from your computer or your phone.

Table 7. Summarized data of pros and cons from reviewer opinions for the Fitbit flex.

Reviewer name	Site name	Reviewed date	Reference site	Advantages (pros)	Disadvantage (cons)
Weebly	Misfit shine activity tracker review	8-Nov-14	[24]	1. Waterproof	1. Sleep data is basic
				2. Wireless data transfer (when placed near the device)	2. Shine attachment can come unsecured (can pop out of sports band)
				3. Can track swimming and cycling	3. Time-telling feature suggests this could replace a watch, yet it lacks all other watch features including alerts
				4. Elegant aluminum design	4. Limited info on "screen": does not have a full digit-based display
				5. On-device feedback to let you know how close you are to reaching a goal	5. Shine's tapping-based interface can be frustrating to use
				6. No recharging. Just replace the watch battery when it runs out (~4–6 months)	—
				7. Partnership with Pebble watch allows you to use the Pebble as a Misfit Shine	—
				8. Social features including a leaderboard, profile, and newsfeed	—
Bethany Gordon	Misfit Shine	Only year mentioned (2015)	[25]	1. The interchangeable design	1. Tapping the screen is the only way to see your progress
				2. Comfortable band making the Misfit Shine extremely easy to use	2. The Shine does not always respond to tapping
				3. Convenient to wear	3. It has to sit on your arm a certain way to display time and daily progress
				4. Water resistant	—
Kristen Buck	Misfit Shine	Only year mentioned (2015)	[26]	1. Misfit Shine is about the size of a quarter and undeniably attractive.	1. The Misfit Shine only works for iOS
				2. Water resistant	2. Does not have an altimeter to count how many flights of stairs you climb
				3. Great activity monitor for swimmers and surfers	3. Not compatible with Android devices
				4. Can wear it in different ways to track different activities more accurately	—

Reviewer name	Site name	Reviewed date	Reference site	Advantages (pros)	Disadvantage (cons)
Jill Duffy	Misfit Shine	10-Dec-13	[27]	1. Best looking activity tracker	1. Limited data analysis
				2. Includes clip and wristband mounts	2. No integration with other services
				3. Functions as a watch	3. No Web app
				4. Fully waterproof for swimmers	4. No syncing between iOS and Android apps
				—	5. Dashboard lacks weight-tracking and calorie counting
Mikey Campbell	Review: Shine activity monitor	12-Nov-13	[28]	1. Great design	1. Clunky tagging method
				2. Easy to understand graphical readout	2. Light in features
				3. Long battery life	3. LEDs unusable in bright sunlight

Table 8. Summary of pros and cons from reviewer opinions for the misfit Shine.

Experiments and results	Devices	Accuracy (%)	Repeatability
Indoor walking straight	Jawbone	97.7	0.55
	Withings	99.9	0.86
	Misfit	92.4	0.69
	Fitbit	99.6	0.72
Walking up/downstairs	Jawbone	97	0.89
	Withings	97.2	0.83
	Misfit	97.8	0.79
	Fitbit	96.4	0.81
Walking on treadmill	Jawbone	97	0.89
	Withings	97.2	0.83
	Misfit	97.8	0.79
	Fitbit	96.4	0.81

Table 9. Comparison of accuracy and repeatability for the devices.

The total scores for each device are shown in **Table 9** and **Figure 4**. The Withings Pulse has the highest score for both repeatability and accuracy. The lowest accuracy and repeatability were recorded by Misfit. With regard to opinions from seven subjects and also the table of reviewers, we concluded that both the Fitbit and the Misfit have difficulties in detecting when the

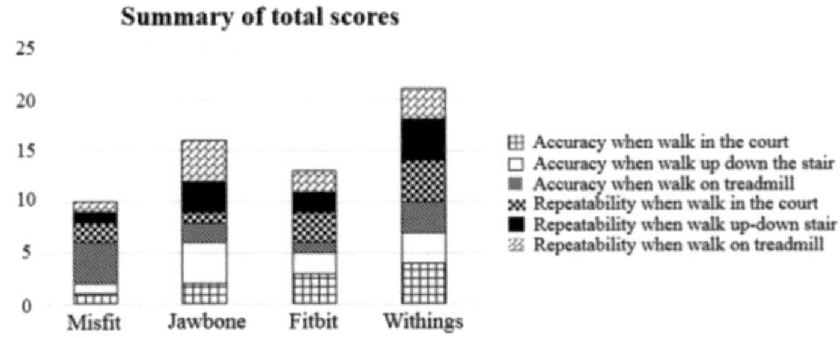

Figure 4. The summary of the accuracy and repeatability scores among the four devices.

wearer is ascending or descending steps. Along with the subject's experiments in **Table 5** and **Figure 4**, the scores from experiments of walking up and down stairs are both lowest in cases of accuracy and repeatability. This resulted in the overall scores for the Misfit and the Fitbit being the lowest among the four wearable devices in terms of repeatability and accuracy.

The total scores for each device are shown in **Figure 4**. The Withings Pulse has the highest score for both repeatability and accuracy. The lowest accuracy and repeatability were recorded by Misfit.

4. Discussion and conclusion

We attempted to evaluate the best wearable device from the four devices selected. This study focused on both objective and subjective methods to get the physical comparison results. The results are independent of manufacturers' claims. The main two sections of experiment testing verified the quality of the devices, both objectively and subjectively.

Section 1. Eight categories were classified in the form for evaluation of satisfaction levels: synchronization, hardware design, UI app, sleep tracking, step count, nutrition analysis, calories, battery, and user-friendliness. The most satisfying device based on the participants' rankings was the Withings, followed the Misfit, Jawbone, and Fitbit.

Further to the information gathered in Section 1, a summary was compiled from the viewpoints of the seven participants and online reviewers. The summary revealed that each device had its own strengths and weaknesses. The evaluation form and satisfaction scores thus allowed the subjective records of genuine users to be presented for each of the four devices. The opinions of the participants and those of the online reviewers were shown to be similar, leading to the following conclusions:

Jawbone UP24 is well designed and fits comfortably for the subjects. The UI app is colorful and easy to understand. The sleep tracker is very smart and also has good alarm functions. However, disadvantages (cons) include the design not having a screen, it is not water proof, and the battery charger is very complex.

Withings Pulse has good features such as the heart rate function, which can detect pulse rate. It is just one of the functions that the Withings has. The display on the Withings device is large enough to show the results tracking. The data log can be updated automatically via

Bluetooth syncing or wireless. The design, however, is not particularly attractive, since it is hard to read the display under exposure to sunlight and the sleep tracking feature does not work automatically.

Fitbit Flex has a slender and attractive design, is wholly waterproof, and offers a number of social features. However, the Fitbit Flex has weak points. It has no screen, the food log and calories tracking are not easy to use, the food log is hard to learn for beginners, and the tap screen is sometimes confusing.

Misfit Shine has an attractive, elegant, and fashionable design. It is fully water proof and especially good for water sports. The goal tracking function leads the user to achieve the goal, and the battery does not need to recharge, only exchange. In contrast, the Misfit Shine can only be used with iOS. Although compatibility with Android is a planned future feature, it has not yet been implemented. A smartphone is necessary to display the tracking status since the device does not have its own display. When loss of syncing with the smartphone occurs, this can result in data inaccuracies.

Section 2. The experiments compared the accuracy and repeatability of the devices awarded among four wearable devices. The score four points for the best accuracy and repeatability. The score three, two, and one point for the second, third, and fourth device, respectively. The most accurate and repeatable device was the Withings, second Jawbone, third Fitbit, and fourth Misfit.

In contrast, Misfit had the highest score for design and hardware. Thus, physical design is also appreciated by users in addition to other devices.

Therefore, the Withings device provided the greatest satisfaction and was the most user-friendly from the perspective of the users. It was also the highest ranked for accuracy and repeatability in step count and distance tracking. Tracking accuracy is vital in fitness tracking, but in personal tracking, there are differences which stem from age, gender, weight, and height. The tracking of daily activities including sleeping, waking, or running will also be important, and the results of the tests indicate that the greatest accuracy is achieved by the Withings device in all these categories for accuracy and repeatability.

As the results showed, the cause that made the Misfit Shine and Fitbit Flex have the lowest score of accuracy and repeatability was stair tracking. These two devices could not track activity when the subjects walked on stairs or climbed. For this reason, users were disappointed in these devices.

In this study, it was unclear by inspection of the deep or shallow sleep of the subjects whether the wearable devices could measure accurately. Thus, true value comparison was not shown in this research. However, future studies may advance this knowledge.

The results in this study relative to step counting and others are subjective enough to assist in the buying process for potential wearable device purchasers. One more fact from the seven subjects' opinions is that four of the seven subjects would not buy this kind of wearable device. It interferes with their arm while in use and is uncomfortable with syncing the data log for daily tracking. One more important reason is that data gave uncorrected reports because of automatic loss of syncing.

Nowadays, wearable technology has the greatest potential impact in the fields of health and fitness. However, it can also be influential for gaming and other forms of entertainment.

Wearable technology can create a vividly realistic and immersive environment in real time. This concept is not new. Modern prototypes are moving away from bulky technology, such as large goggles and backpacks, toward smaller, lightweight, and more mobile systems.

On the author's viewpoint, the most comparable of the wearable device is that it cannot display itself but needs the smartphone to involve the metric data and reports. However, the storage of mobile phone to store and display the results is bigger, but it is inconvenient to use both at the same time. And yet, presently many fitness-tracking applications are available through the online store for free without any special or specific device. It is very convenience to whom that focuses in their healthy or fitness tracker. Even though the report of it is not guarantee 100% of accuracy, it is the easiest way to track their activity without any payment. Thus, the companies who produced the fitness tracker or wearable devices to the market in this highly competitive market will continuously develop new eye-catching products and reduce errors using the voices and opinions of users from this study to reach a wider market in the future. The relationship between technology and esthetics must go together, such as unobtrusive design, very sleek and modern and light weight, waterproof function, and many choices to recharge batteries. Basic activities such as walking or climbing stairs require accuracy and repeatability, while it is also necessary to accurately measure physical parameters such as pulse, heart rate, body temperature, and breathing rates. These features should be added in cases where they are not yet available. At present, the market for wearable devices is growing rapidly, and this will drive the further development of the technology to deliver the features and attributes demanded by users. This study has therefore addressed the fact that consumers need access to accurate and reliable information with regard to the latest gadgets available on the market and the performance of those devices.

Acknowledgements

This work was supported by the National Research Foundation of Korea(NRF) grant funded by the Korea government(MSIP) (No. 2017M3A9C6029312).

Author details

Kanitthika Kaewkannate and Soochan Kim*

*Address all correspondence to: sckim@hknu.ac.kr

Department of Electrical and Electronic Engineering, Hankyong National University, Anseong-si Gyeonggi-do, South Korea

References

[1] Best Fitness Trackers. 2015: Jawbone, Misfit, Fitbit, Germin and More. Available on: http://www.wareable.com/fitness-trackers/the-best-fitness-tracker

[2] 10TopTen reviews "Fitness Tracker Review and Comparisons". Available on: http://fitness-trackers-review.toptenreviews.com/

[3] Best Fitness Tracker 2015 from Wearable Tech for your connected self. Available on: http://www.wareable.com/fitness-trackers/the-best-fitness-tracker

[4] Meredith A, Holland A, Kevin G. Accuracy of smartphone application and wearable devices for tracking physical activity data. JAMA. 2015:625-626. DOI: 10.1001/jama.2014.17841

[5] Rienzo M, Rizzo F, Parati G, Brambilla G, Ferratini M, Castiglioni P. MagIC system: A new textile-based wearable device for biological signal monitoring. Applicability in Daily Life and Clinical Setting. 2016. DOI: 10.1109/IEMBS.2005.1616161

[6] Kaewkannate K, Soochan K. A comparison of wearable fitness devices. BMC Public Health. 9, 2016;**16**:433. DOI: 10.1186/s12889-016-3059-0

[7] Fitbit Flex. Available on: https://www.fitbit.com/kr/flex

[8] Withings Pulse Ox". Available on: http://www.withings.com/eu/withings-pulse.html

[9] Misfit. Available on: http://misfit.com/?locale=en

[10] Jawbone. Available on: https://jawbone.com/

[11] Jawbone Up24 Review. Available on: http://www.bestfitnesstrackerreviews.com/jawbone-up24-review.html

[12] Smart activity tracker Jawbone Up Review. Available on: http://www.smartactivitytracker.net/jawbone-up24

[13] JawboneUp24 Review. Available on: http://www.techradar.com/reviews/gadgets/jawbone-up24-review-1230596/review

[14] Jawbone UP24 Review: Make Each Day Better Wirelessly. Available on: http://www.zdnet.com/pictures/jawbone-up24-review-make-each-day-better-wirelessly/

[15] Withings Pulse Smart Activity Tracker Review. Available on: http://www.bestfitnesstrackerreviews.com/withings-smart-activity-tracker-review.html

[16] Withings Pulse wirless Activity tracker review. Available on: http://www.smartactivitytracker.net/withings- pulse

[17] Withings Pulse In-Depth Review. Available on: http://www.dcrainmaker.com/2013/11/withings-depth-review.html

[18] Review: Withings Pulse with built-in heart rate monitor". Available on: http://appleinsider.com/articles/13/11/05/review-withings-pulse-with-built-in-heart-rate-monitor

[19] Withings Pulse Activity tracker review. Available on: http://the-gadgeteer.com/2013/08/27/withings-pulse-activity-tracker-review/

[20] Fitbit Flex Review. Available on: http://www.bestfitnesstrackerreviews.com/fitbit-flex-review.html

[21] Fitbit Flex Wireless Activity Review. Available on: http://www.smartactivitytracker.net/fitbit-flex

[22] Fitbit Flex Review. Available on: http://health.thefuntimesguide.com/2014/09/fitbit-flex.php

[23] Fitbit Flex. Available on: http://fitness-trackers-review.toptenreviews.com/fitbit-flex-review.html

[24] Misfit Shine Activity Tracker Review. Available on: http://www.bestfitnesstrackerreviews.com/misfit-shine-review.html

[25] Misfit Shine. Available on: http://fitness-trackers-review.toptenreviews.com/shine-activity-tracker-review.html

[26] Misfit Shine. Available on: http://pedometers-review.toptenreviews.com/misfit-shine-review.html

[27] Misfit Shine. Available on: http://www.pcmag.com/article2/0,2817,2423341,00.asp

[28] Review: Shine Activity Monitor". Available on: http://appleinsider.com/articles/13/11/13/review-shine-activity-monitor

Chapter 11

Bio-Inspired Wearable Antennas

Paulo Fernandes da Silva Júnior,
Alexandre Jean René Serres,
Raimundo Carlos Silvério Freire,
Georgina Karla de Freitas Serres,
Edmar Candeia Gurjão,
Joabson Nogueira de Carvalho and
Ewaldo Eder Carvalho Santana

Additional information is available at the end of the chapter

http://dx.doi.org/10.5772/intechopen.75912

Abstract

Due to the recent miniaturization of wireless devices, the use of wearable antennas is steadily increasing. A wearable antenna is intended to be a part of the clothing used for communication purposes. In this way, a lower visual cost may be achieved. Recently, biologically inspired design, a kind of design by cross-domain analogy is a promising paradigm for innovation as well as low visual cost. The shapes of the plants are structures optimized by nature with the primary goal of light energy capture, transforming it into chemical energy. In this case, they have similar behavior to that of parabolic reflectors; this enables microwave engineers design innovative antennas using bio-inspired concepts. One of the advantages of using bio-inspired plant shapes is the design of antennas with great perimeters in compact structures. Thus, we have small antennas operating in low frequencies. This chapter presents the recent development in bio-inspired wearable antennas, easily integrated to the clothes and accessories used by the body, built in denim, low-cost flexible dielectric, and polyamide flexible dielectric, that is flexible with high resistance to twists and temperatures, for wireless body area network (WBAN) applications, operating in cellular mobile (2G, 3G, and 4G) and wireless local area network (2.4 and 5 GHz) protocols.

Keywords: wearable flexible antenna, bio-inspired plant shape, wireless body area network

1. Introduction

Wireless body area network (WBAN) are sensor networks with elements located in the human body whose mobility and chemical composition impose challenges to wireless communication in such networks. Antennas are crucial elements for such communication and beyond the requirements of gain, directivity and others, for WBAN network wearable antennas as they are named have other important design requirements such as minimum destructive coupling between antenna and body, low visual impact, low-cost, flexible, and compact structure, ease of integration to the clothes and accessories used next to the body [1]. Its applications can be extended for continuous health and sports monitoring, safety, and security of people [2]. Research into development of wearable antennas has used several materials and shapes operating in different resonance frequencies [1–7].

Evolution has permitted animals and natural structures to adjust their behavior and formats to obtain optimum performance in various aspects. Engineering has observed and applied such aspects to solve problems. Examples are genetic algorithms and ant colony optimization. Natural plant leaves present similar characteristics to fractals, as, for example, reduction of dimensions with perimeter increase and also that they have complex and efficient light harvesting-reaction center, that is, an array of antennas capable of operating in the visible light range (400–700 nm) with characteristics analogous to satellite dishes.

The leaves characteristics are well suited to antenna design, including the ones for wearable antennas and this bio-inspiration (leaf-shaped antennas) open a vast research field for more compact and efficient antennas. In this chapter, we use a polar expression developed to represent the geometry of plants and other living beings known as Gielis formula [8] to design wearable antennas operating in cellular mobile systems and WBAN. Antennas were developed for various plant shapes and different substrates, and their superior performance compared to classical antennas developed to the same application is clear.

Projects of the bio-inspired wearable antennas were realized with patch and monopole antennas, built in denim and polyamide, with shapes generated by Gielis formula, for WBAN application operating in WLAN (2.4–2.4835 and 5.4–5.85 GHz), 2G (1.8–2.1 GHz), 3G (1.885–2.17 GHz), and 4G (2.5–2.69 GHz) bands.

2. Bio-inspired antenna design

2.1. Gielis formula

In plants, the process of photosynthesis uses visible light (electromagnetic waves) to produce chemical energy. The leaves have two sections for processing the light, the complex capture centers and the reaction centers. The complex capture centers are formed by an arrangement to perform the highest light energy capture [9]. Thus the leaves act as receiving electromagnetic wave antennas with similar structure to parabolic reflectors, formats of the leaves evolved to give the optimum performance in terms of energy capture; thus for antennas design plants shapes represent a great potential research field. Based on this characteristic, leaf shapes have been used as inspiration in the geometry of antennas via formulation projected to automatically reproduce such shapes as fractal geometry [10], the Fibonacci series, the Gold number [10], and the Gielis formula [8].

The formulation proposed by Johan Gielis in 2003 allows mathematically describing a wide variety of natural and abstract forms, such as leaf and flower shapes. Based on the concept of superellipses given by.

$$\left|\frac{x}{a}\right|^{n} + \left|\frac{x}{b}\right|^{n} = 1 \tag{1}$$

and modifying Eq. (1) considering idea that many natural forms can be interpreted as modified circles, Gielis obtained what is called a superformula (Eq. (2)) by using polar coordinates, replacing $x = r.cos(\theta)$ and $y = r.sin(\theta)$ in addition to inserting the argument $m/4$ to create specific rotational symmetry in some structures, and the possibility of using different values of exponent n for each term (n_1, n_2, n_3).

$$f = f(\theta)\frac{1}{\sqrt[n_1]{\left(\left|\frac{1}{a}\cos\left(\theta\frac{m}{4}\right)\right|\right)^{n_2}\left(\left|\frac{1}{b}\sin\left(\theta\frac{m}{4}\right)\right|\right)^{n_3}}} \tag{2}$$

From this expression, it is possible to generate and modify several shapes by the manipulation of the six parameters (a, b, m, n_1, n_2, n_3).

This expression can also be combined with other functions, (θ), generating other forms. In order to illustrate the possibilities of the Gielis formula some example of shapes can be seen with the respective parameters in **Figure 1**.

In [11] the Gielis formula is used to generate various metamaterial unit cells with resonant frequencies in range of 6–8 GHz. In [5], a wearable textile printed monopole antenna is bio-inspired in *Gingko biloba* leaf shape, generated by Gielis formula with bandwidth of 2.7 GHz, operating in 2G, 3G, and 4G bands.

Patch antennas using the Gielis formula were presented in [12], operating in cellular mobile and WLAN band, built in denim and fiber-glass (FR4). A parametrical analysis of bio-inspired leaf shape of jasmine flower, developed by Gielis formula, in printed monopole antennas with application in ultra-wide band, and X-band frequencies was presented in [13]. A transparent

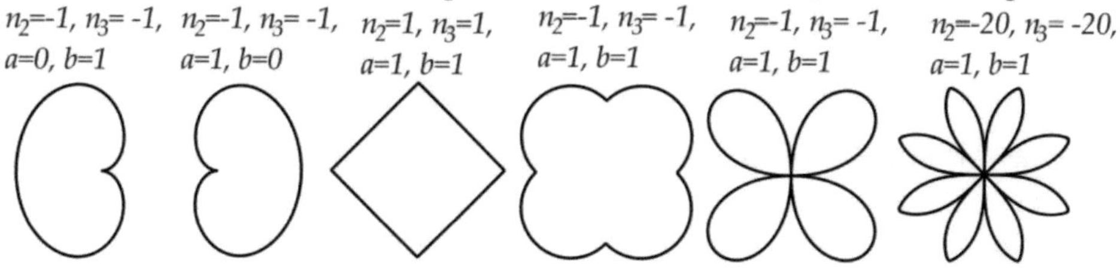

Figure 1. Shapes generated by Gielis formula with the respective parameters.

patch antenna operating in WLAN 5G, with bio-inspired plant shape of *Inga edulis mart* was performed in [14].

2.2. Procedure to design antenna via Gielis formula

Parameters of an antenna, such as resonance frequency, bandwidth, gain and radiation pattern are directly affected by the shape and materials used for its construction. Beyond such characteristics, wearable antennas should preferably have flexible structures for conductive material and dielectric substrate and must be as flat as possible [15]. In addition, characteristics of permittivity and thickness of dielectric substrate are crucial to determine the bandwidth and the antenna efficiency.

The design methodology of bio-inspired antennas in plants generated by the expression of Gielis was adapted from [16] for the design of wearable antennas, and consists of the following 11 steps:

1. Definition of the application;
2. Identification of operating frequencies;
3. Choice of antenna characteristics suitable for the application, broadband or narrowband, type of polarization, among others;
4. Selection of the conductor and dielectric material;
5. Characterization of the properties of the materials, using the technical data informed by the manufacturers or using some available characterization method;
6. Design of an antenna with Euclidean geometry (square, rectangular, or circular) in order to obtain total perimeter of the structure;
7. Selection of the bio-inspired shape between elliptic leaves that have total perimeter closest to the antenna with Euclidean geometry;
8. Generation of the image by the Gielis expression with the use of computer aided (CAD) techniques in a format exportable to the full-wave simulation software;
9. Simulation and optimization of the antenna characteristics, with adjustments to obtain the desired resonance frequency;
10. Construction of the bio-inspired antenna;
11. Validation using measurement and comparison with simulated results.

In order to choose the shapes of the elliptical sheets, in step 7, it may observed that for the patch antenna design, the width of the sheet will follow the same principle of the design of microstrip transmission lines [16, 17]. Thus, the use of substrate with lower thickness (h) and higher relative electrical permittivity (ε_r) will imply the design of smaller width sheets.

For the design of planar monopole leaves having width greater than the length allows the development of antennas with greater bandwidth. The bio-inspired antennas presented in this chapter use symmetrical sheets structures, with the purpose of providing diagrams of broadside radiation, that is, with maximum gain in the axial direction to the axis of the antenna. Depending on the application, non-symmetrical structures can be used.

In the choice of structure with composite sheets we must consider the perimeter identified in the design of the antenna with Euclidean geometry. The objective is to obtain a bio-inspired structure with total perimeter closest to the antenna with Euclidean geometry.

Microstrip patch antennas and monopole antennas are a well-known concept. In the following sections the formulation of bio-inspired shapes using the Gielis formula and the electrical characterization of wearable materials employed to design the antennas are presented.

2.3. Electrical characterization of wearable substrate

Researches in wearable antennas cover medical and non-medical applications, operating in several frequency ranges, and built on various substrates [2, 18–21]. For the development of wearable antennas and other electromagnetic wearable devices, it is crucial to know the electrical characteristics of the flexible substrate used. The objective is to identify the relative permittivity and the loss tangent of the materials. According to [22], the main methods employed to characterize dielectric materials are: coaxial probe, transmission line, free space, resonant cavity, and parallel plates.

The measurement of the flexible materials performed with coaxial probe method is the more popular technique to measure complex dielectric permittivity of many materials. This method is non-destructive, broadband and measurements at high-temperature can be performed with a commercial instrumentation easily available.

In this chapter, design of wearable bio-inspired antennas used two flexible substrates, namely the denim and the polyamide.

Research presented antennas and other devices in textile substrate with applications in different frequency bands. In [23], the performance of textile antenna under two-dimensional crumpling conditions for 2.45 and 5.8 GHz is shown. In [6], a shielded stripline made in textile materials is designed as wearable flexible transmission line for broadband operation until 8 GHz. Denim is porous material with planar structure whose properties are determined by its fiber arrangement, density, volume, and size. Jeans is a fabric of denim that is very thin with a planar dielectric structure.

Polyamide laminate is flexible material with thermal and mechanical resistance characteristics, which has possibility of used in development of the devices for monitoring in situations of high-temperature risks such as monitoring the level of blood oxygenation and cardiac beats in firefighters and workers of metal machining centers. A wearable rectangular patch antenna for medical body area network (MBAN) (2.36–2.4 GHz) built in polyamide was presented in [24].

The dielectric substrates were characterized in the Laboratory of Measurements of the Federal Institute of Paraíba (IFPB), Campus of João Pessoa using a Vector Network Analyzer (VNA) of Agilent model S5071C (300 kHz–20 GHz) and the Dielectric Probe Kit 85070 of Agilent. **Figure 2** shows the electrical characterization measurement setup of the flexible dielectrics used to design wearable antennas, polyamide and denim.

Figure 3 shows dielectric characterization, permittivity and loss tangent, of polyamide and denim substrates. The denim substrate was characterized with $\varepsilon_r = 2.03$, loss tangent of 0.2, and

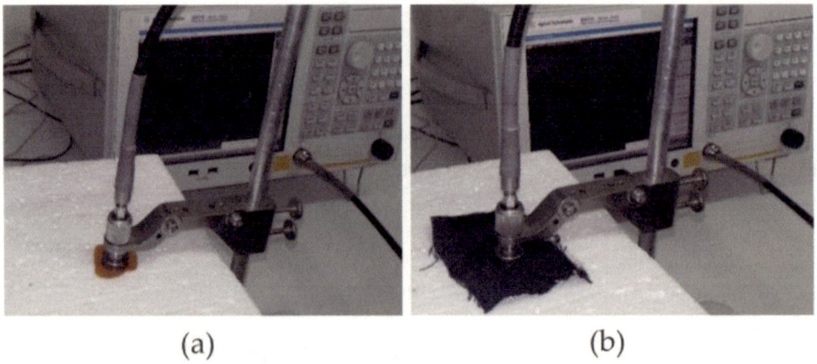

Figure 2. Electrical characterization measurement setup: (a) polyamide; and (b) denim.

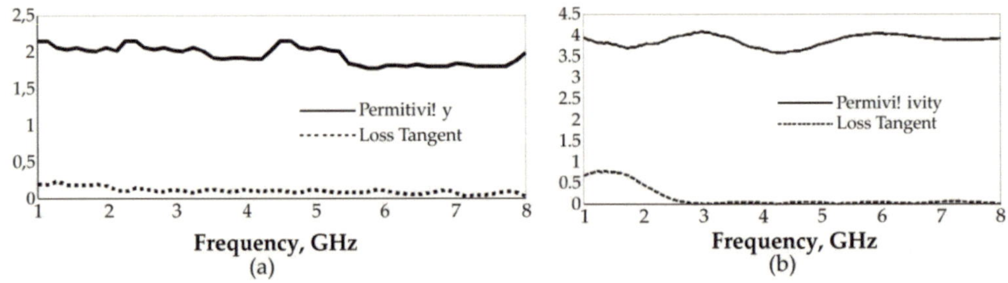

Figure 3. Electrical characterization: (a) denim; and (b) polyamide.

thickness of 1 mm and the polyamide substrate was characterized with $\varepsilon_r = 4$, loss tangent of 0.04, and thickness of 0.05 mm.

3. Bio-inspired wearable antennas

This section presents wearable bio-inspired antennas built in denim and polyamide operating in cellular mobile communication (2G, 3G, and 4G), and WLAN in 2.4 and 5 GHz applications.

3.1. Wearable bio-inspired antennas built in denim

3.1.1. Wearable monopole antenna bio-inspired in jasmine flower shape

The jasmine flower presented in **Figure 4(a)** was the bio-inspiration for antennas operating in mobile cellular system (2G, 3G, and 4G). **Figure 4(b)** presents the simulated antenna and **Figure 4(c)** the simulated ground plane. **Figure 4(d), (e)** presents the implemented antenna front and back respectively. The shape of the jasmine flower was generated by the Gielis formula with value: $m = 8$, $n_1 = -40$, n_2 and $n_3 = 20$, a and $b = 1$.

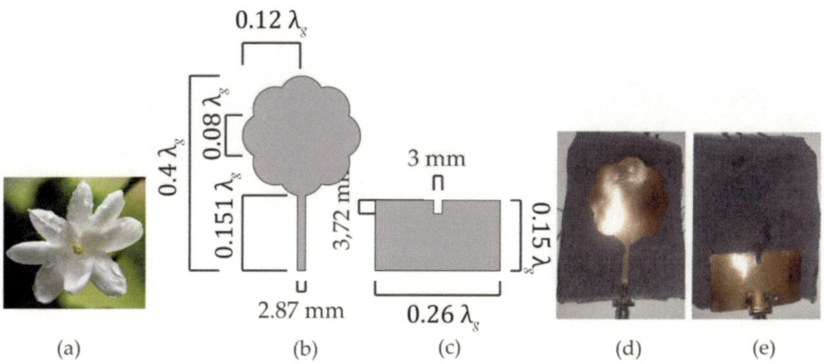

Figure 4. Developed wearable textile monopole antenna bio-inspired in jasmine flower shape: (a) jasmine flower [25]; (b) simulated patch element; (c) simulate ground plane; prototype, front vision (d); prototype, back vision (e).

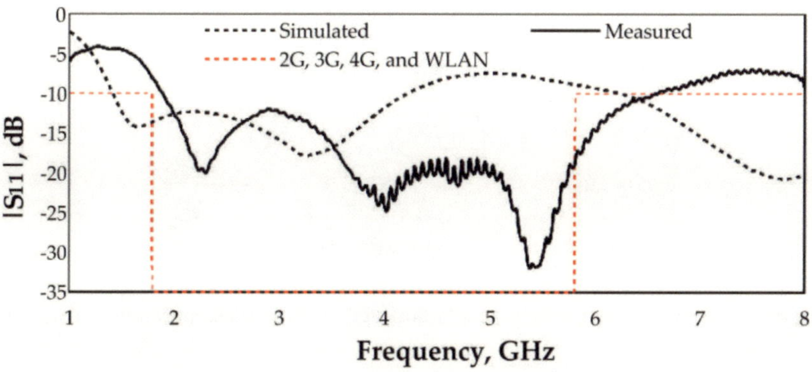

Figure 5. |S11| curves for the simulated and measured (prototype) jasmine flower wearable monopole bio-inspired antenna.

Figure 5 presents curves for the simulated and measured values of $|S_{11}|$ for the jasmine flower bio-inspired antenna parameters of wearable monopole antenna bio-inspired presented in **Figure 4**. In **Figure 5**, the frequency mask for 2G, 3G, 4G, and WLAN 2.4 and 5 GHz band are indicated. The mismatch between the simulated and measured results can be explained by the manufacture process of the antenna. However, a wider band at −10 dB can be observed with the built antenna covering all of frequency bands.

Considering that the projected antenna have to be wearable, it was necessary to perform measurements near the body, as also considering that in certain situations the antenna can be bent. **Figure 6** presents some of the positions for the near body situations, namely on the hand, in the pocket, and on the chest of the prototype proposed. **Figure 7** presents the $|S_{11}|$ curves in each configuration. For comparison, in **Figure 7**, the curve measured was obtained with the antenna in the free space, that is, with the minimum reflections in its neighborhood. It can be noticed when the antenna is close to the body its first resonance frequency modifies, and the highest variation compared to the Measured value (1.9 GHz) occurs with the antenna on the hand

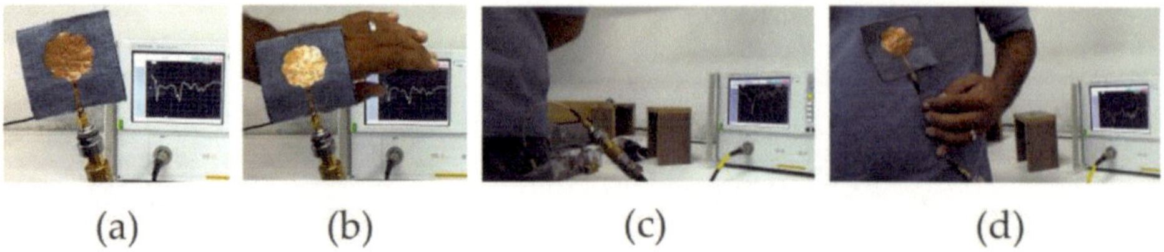

Figure 6. Positions of measurement for the jasmine flower antenna: (a) free space(measured), (b) on the hand, (c) in the pocket, and (d) on the chest.

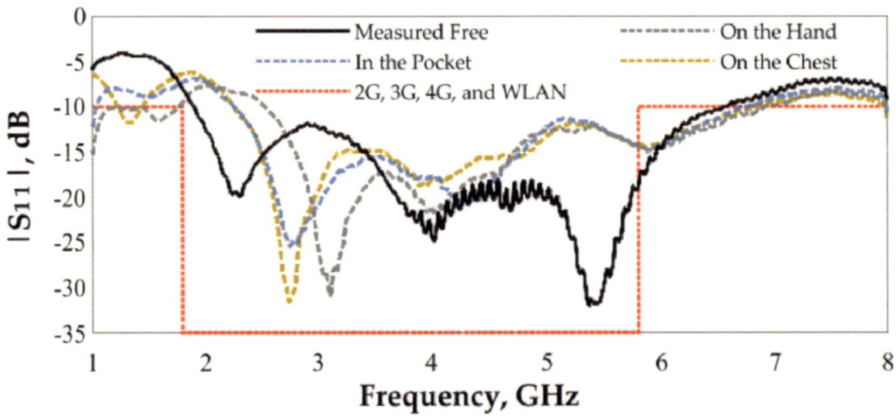

Figure 7. |S11| curves for the simulated and measured (prototype) of the wearable antenna with body interference.

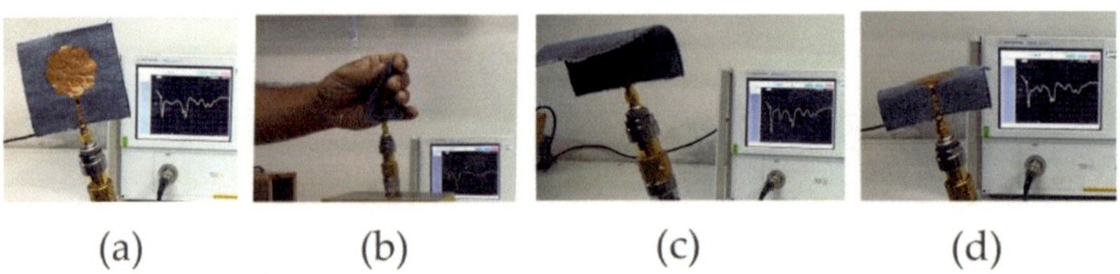

Figure 8. Position of measurement for bending the antenna: (a) free space, (b) creased, (c) forward bended, and (d) backward bended for the jasmine flower wearable antenna.

(2.52 GHz). The permittivity of the various parts of the body can be different according to the percent water and mainly modifying the resonance frequencies in the UHF band.

Another set of measurement considers that the antenna can be bent or even creased as occurs in clothes. **Figure 8(a)** presents the free space, **Figure 8(b)** creased, **Figure 8c)** forward bending, and **Figure 8(d)** backward bending positions. **Figure 9** presents the measured $|S_{11}|$ values for each configuration. Again, the free space (Measured curve) is the reference and it can be noticed that the greatest difference in the first resonance frequency occurs in the forward

http://dx.doi.org/10.5772/intechopen.75912

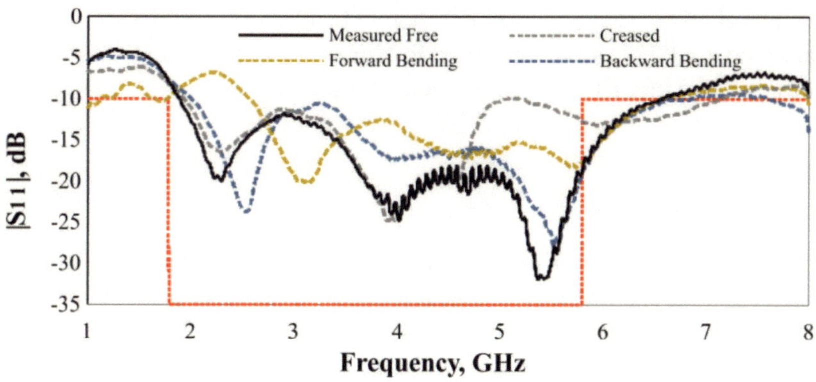

Figure 9. |S11| curves for bending positions of the jasmine flower wearable antenna.

bending configuration and a greater bandwidth (5.21 GHz) is observed in the creased configuration.

The different bending of the antenna generates a discontinuity of the current density on the radiating element, thus modifying its impedance. In spite of that, all the frequency bands are covered except for the forward bending in low frequencies.

In **Table 1** are presented values for the first resonance frequency (f_1), last resonance frequency (f_2) in −10 dB, and bandwidth for the proposed wearable antenna for each configuration presented above.

Observing values in **Table 1** we can notice that the Jasmine wearable antenna simulated, measured free and forward bending presents difference in first resonance frequency (f_1), resonance of 35.26%, and smaller bandwidth of 14.77%. The most significant points are the five similar results, with bandwidth ranging from 4.4 to 4.87 GHz. Only two results are outside these limits, with a bandwidth below of 4 GHz and greater than 5 GHz. It can be observed a difference about 1 GHz between the backward and forward bending that is important information for this kind of wearable antenna.

Antenna	f_1 (GHz)	f_2 (GHz)	Bandwidth (GHz)
Simulated	1.42	4.20	2.78
Measured	1.90	6.57	4.67
On the hand	2.52	6.92	4.40
In the pocket	2.25	6.75	4.50
On the chest	2.25	6.85	4.60
Creased	1.94	7.15	5.21
Forward bending	2.57	6.55	3.98
Backward bending	2.01	6.88	4.87

Table 1. Measured and simulated values of resonant frequencies and bandwidth of the wearable bio-inspired antenna.

The 2D and 3D radiation patterns, gain, half power beamwidth (HPBW), and current density simulated at 3 and 1.6 GHz can be observed in **Figure 10(a), (b)**. The results are according to the Federal Communications Commission (FCC) for a UWB antenna with a gain of 4.2 dBi at 3 GHz, omnidirectional pattern, and a HPBW greater than 75°.

3.1.2. Wearable antenna bio-inspired in Bidens pilosa *plant shape*

In **Figure 11(a)** is presented the *Bidens pilosa* plant that consists of a shape with three elliptical leaves, this format was used as the bio-inspiration for a narrowband wearable antenna that covers the WLAN range at 2.40 GHz (2.40–2.4835 GHz). **Figure 11(b), (c)** present the single leaf and the simulated antenna respectively, and **Figure 11(d)** the prototype antenna.

In the simulation and the prototype the under leaf was inclined at 20°, and the down leaves were inclined at 40° in relation to the geometry of *Bidens pilosa,* in order to provide fine-tuning of the resonance frequency. From the Gielis formula, the leaves were generated with the parameters, n_1 = 2, m = 400, n_2 and n_3 = 1200, a and b = 1. The final structure obtained total perimeter of 143.3 mm.

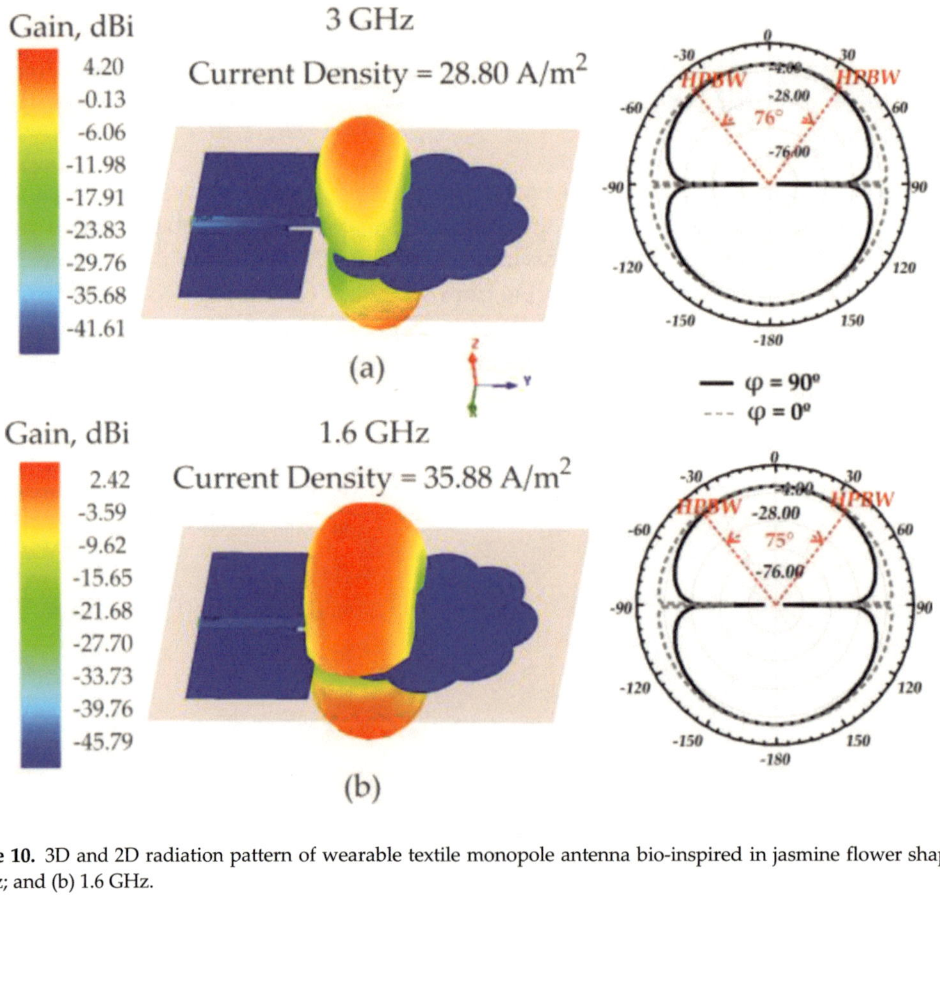

Figure 10. 3D and 2D radiation pattern of wearable textile monopole antenna bio-inspired in jasmine flower shape: (a) 3 GHz; and (b) 1.6 GHz.

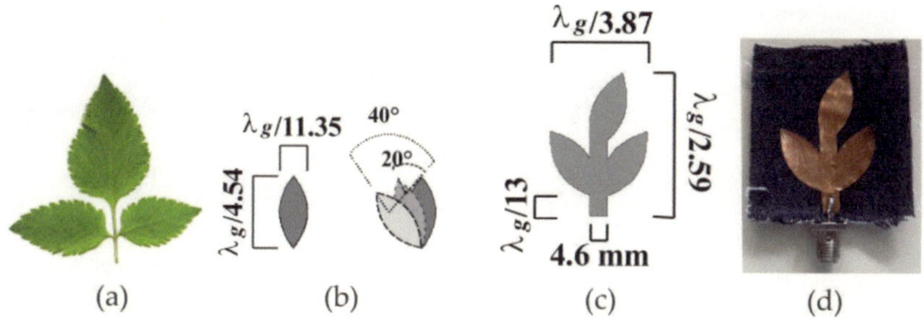

Figure 11. Project of wearable antenna bio-inspired in *Bidens pilosa* plant shape: (a) *Bidens pilosa* leaf [26]; (b) single leaf with inclination; (c) simulated antenna; (d) prototype.

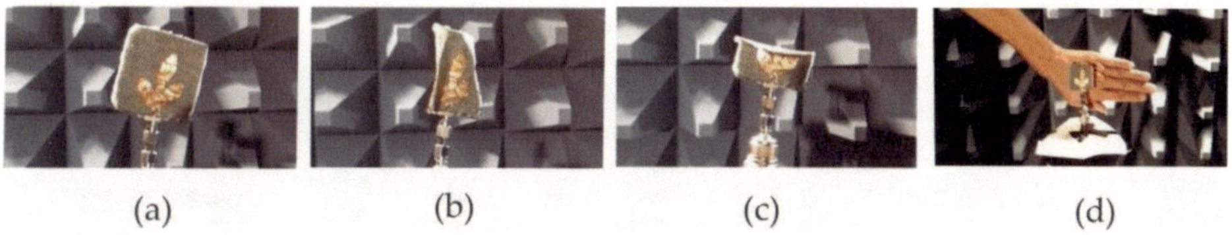

Figure 12. Measurement of prototype of the wearable textile antenna bio-inspired in *Bidens pilosa* plant shape: (a) free space; (b) creased; (c) bended; (d) on the hand.

Figure 12 illustrates the measurements positions of the wearable antenna bio-inspired in *Bidens pilosa* plant shape in anechoic chamber.

Simulated and Measured $|S_{11}|$ curves for the wearable antenna proposed are presented in **Figure 13(a)**. As it can be observed, the simulated and measured values present similar behavior. However the results present a difference of resonance frequency and bandwidth of 1.23% and 74.07%, respectively. The mismatch between the simulated and measured results can be explained by the manufacture process of the antenna and the dielectric permittivity variation of denim. Denim is a product used in the manufacture of clothing, that is, was not prepared by the factory to be used as a dielectric substrate in the construction of antennas. Difference in tenths of the dielectric permittivity may contribute to variations in MHz in the resonant frequency.

Figure 13(b) presents the measured $|S_{11}|$ curves to each configuration illustrated in **Figure 12**. The greatest difference for resonance frequency and bandwidth (14.28%) appears with the wearable antenna on the hand, but still fully covering the WLAN band. The different bending of the antenna generates a discontinuity of the current density on the radiating element thus modifying its impedance and return loss. The greater difference is observed with the antenna close to the hand with the modification of the ground plane characteristic.

Values of bandwidth (BW), resonance frequency (f_0), first (f_1) and last (f_2) resonance at −10 dB are presented in **Table 2**. The mismatch between the measured and simulated results can be explained by the rudimentary manufacture process. However, all measured results are coherent and demonstrate a certain immunity of the antenna when creased, bent, or close to the

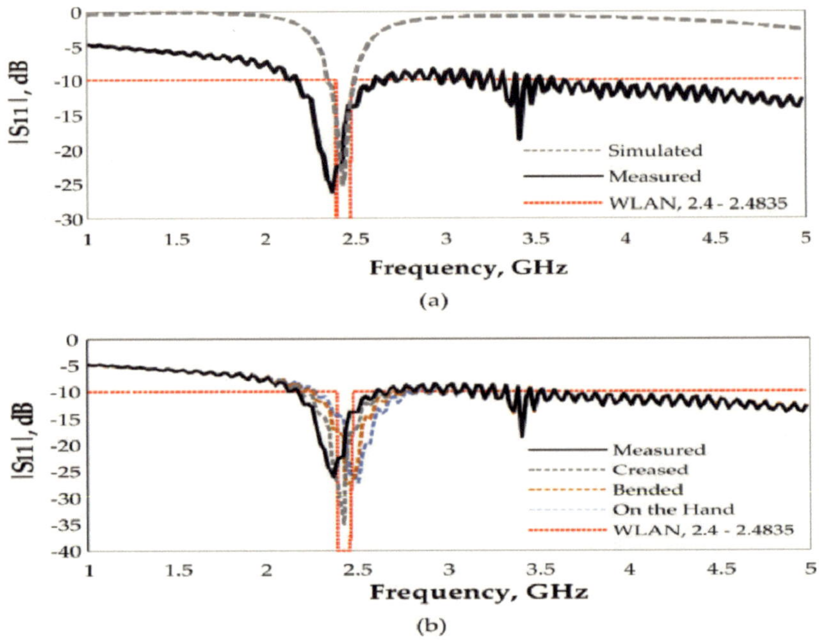

Figure 13. Comparison of |S11| parameters of wearable textile antenna bio-inspired in *Bidens pilosa* plant shape: (a) simulated and measured; (b) creased, bended and on the hand.

Antenna	f_0 (GHz)	f_1 (GHz)	f_2 (GHz)	*BW* (GHz)
Simulated	2.43	2.35	2.49	0.14
Measured	2.40	2.19	2.73	0.54
Measured creased	2.43	2.22	2.82	0.60
Measured bended	2.44	2.28	2.82	0.54
Measured on the hand	2.50	2.28	2.91	0.63

Table 2. Values for resonance frequency and bandwidth for the wearable textile antenna bio-inspired in *Bidens pilosa* plant shape.

body with 60, 0, and 90 MHz of difference compared to the free space measurement, respectively.

Figure 14 shows the 2D and 3D radiation patterns of the wearable antenna bio-inspired proposed. The maximum current density observed was 48.06 A/m^2 at 2.42 GHz, with gain of 6.73 dBi, HPBW of 92°, and relative front-to-back (F/B) of 25 dB. It can be noted than the radiation pattern measured and simulated presented similar behavior indicating good relationship between results. A higher concentration of current density in a smaller physical area is a characteristic of the bio-inspired antenna.

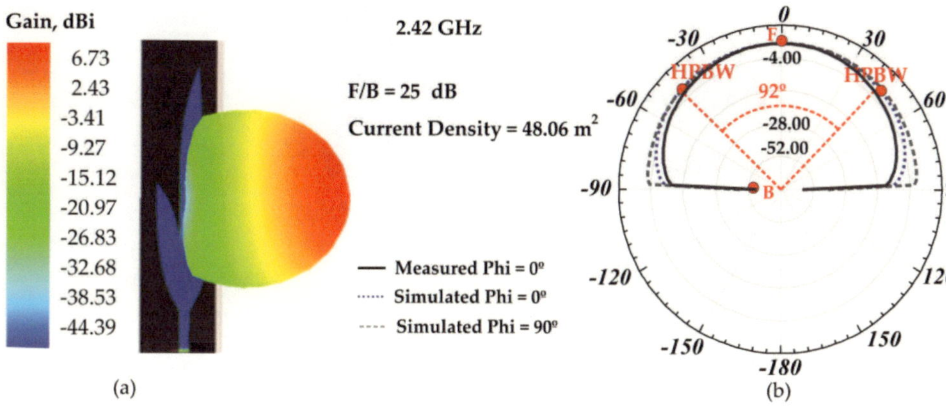

Figure 14. Radiation pattern of wearable textile antenna bio-inspired in *Bidens pilosa* plant shape [27]: (a) 3D; (b) 2D.

3.2. Wearable bio-inspired antennas built in polyamide

In this section, Gielis formula is used in the design of a wearable flexible antenna array bio-inspired in *Inga maritimus* plant shape, and it is designed both one leaf for 5.4 and 5.8 GHz, and two patch antenna arrays with two and four leaves covering the WLAN 5 GHz band.

3.2.1. Wearable antenna arrays bio-inspired in Inga maritimus *plant shape*

Figure 15 shows the leaves of *Inga maritimus*, the single elliptical leaf for 5.4 and 5.8 GHz, the dimensions of the antenna arrays with two and four leaves and the prototypes built in polyamide. The single leaf was generated by Gielis formula, with parameters: $m = 2$, $n_1 = 20$; n_2 and $n_3 = 12$, a and $b = 1$. To prevent mutual coupling between the radiating elements, a distance of one effective wavelength between the leaves was used. Position of leaves in the patch antenna array with four elements has been inverted, thus the radiation pattern stays in the center of the structure.

Figure 16 shows the curves for simulated and measured $|S_{11}|$ curves for wearable antenna bio-inspired in *Inga maritimus* plant shape with one leaf, with the indication in dashed lines of wireless local area network (WLAN) band in 5 GHz (5.15–5.85 GHz). As observed, the single leaf shape antennas cover part of the WLAN band. The greater difference in resonance frequency (3.85%) was observed for the single leaf antenna at 5.4 GHz, and bandwidth 57.8% was observed for the single leaf antenna at 5.8 GHz.

Figure 17(a) shows the simulated and measured $|S_{11}|$ curves of the wearable bio-inspired antenna arrays with two elements and four elements in **Figure 17(b)**. The greater difference in bandwidth (52.39%) was observed for the wearable antenna with two elements, with difference in first resonance less than 1.01%, **Table 3**.

The differences between the measured and simulated results of antennas and antenna arrays in polyamide can be the result of the variation in the dielectric properties of the material, and of the manufacture process. As the transmission lines are very thin, and the bio-inspired antennas

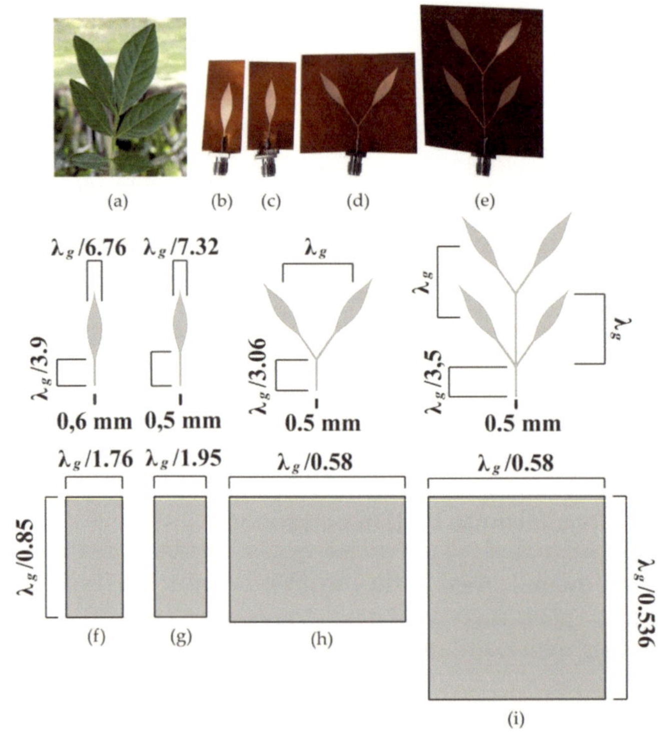

Figure 15. Project of wearable antenna arrays bio-inspired in Inga maritimus plant shape: (a) *Inga maritimus* leaves [28]; (b) prototype for 5.4 GHz; (c) prototype for 5.8 GHz; (d) prototype antenna array with two leaves; (e) prototype antenna array with four leaves.

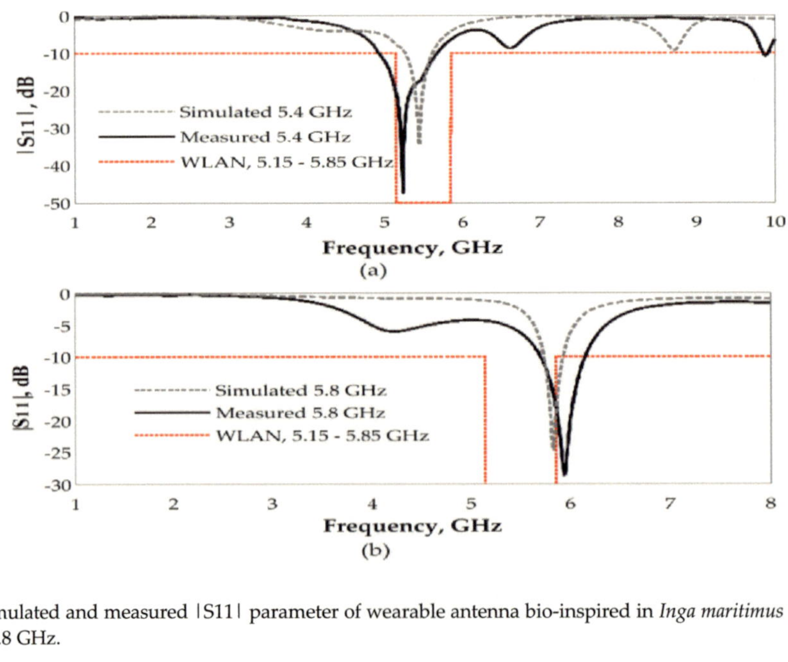

Figure 16. Simulated and measured |S11| parameter of wearable antenna bio-inspired in *Inga maritimus* plant shape: (a) 5.4 GHz; (b) 5.8 GHz.

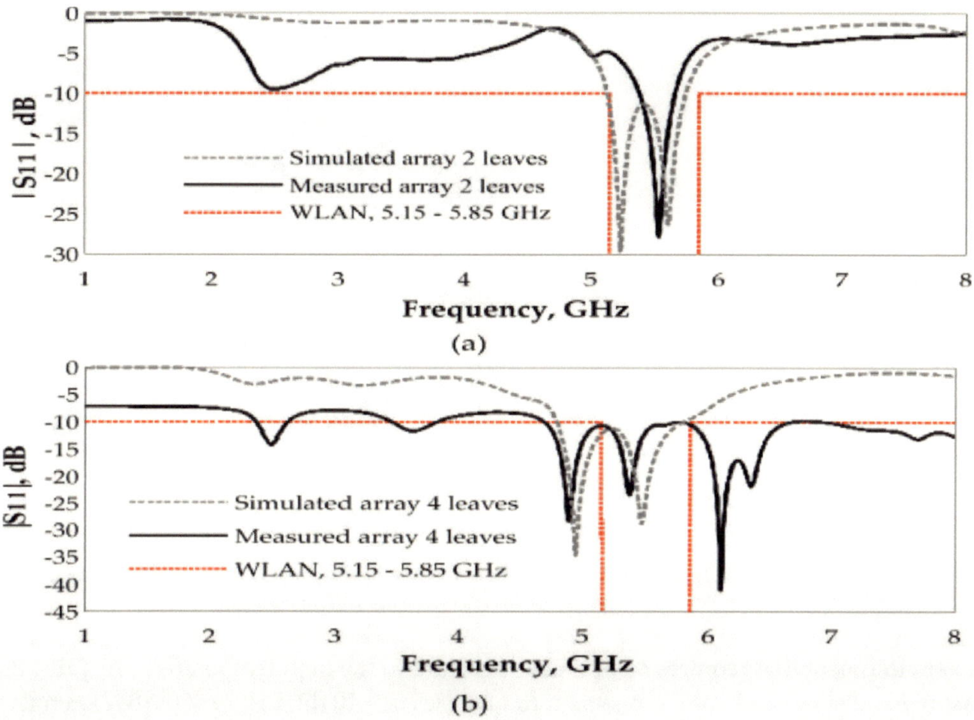

Figure 17. Comparison of simulated and measured |S11| parameters of wearable antenna arrays bio-inspired in *Inga maritimus* plant shape: (a) antenna array with two leaves; (b) antenna array with four leaves.

Antenna	f_0 (GHz)	f_1 (GHz)	f_2 (GHz)	BW (GHz)	RL (dB)
Simulated 5.4 GHz	5.45	5.27	5.62	0.35	34.63
Measured 5.4 GHz	5.24	4.93	5.68	0.75	47.25
Simulated 5.8 GHz	5.83	5.74	5.93	0.19	24.62
Measured 5.8 GHz	5.94	5.7	6.15	0.45	28.54
Simulated Array 2	5.24/5.61	5.12	5.75	0.63	26.39
Measured Array 2	5.54	5.37	5.67	0.3	27.94
Simulated Array 4	4.94/5.48	4.78	5.82	1.04	34.6/28.8
Measured Array 4	4.89/5.38	4.53	6.52	1.99	23.51/41.91

Table 3. Frequency responses of wearable antenna arrays bio-inspired in *Inga maritimus* plant shape.

in the Inga *maratimus* plant shape are compact and have sharp cuts, small variations in the built structure can cause significant variations in resonant frequencies and bandwidth, which can be numerically observed in **Table 3**. Even with the differences observed, the results indicate that the antennas presented antenna array characteristics, with increased gain (by increasing metal), and bandwidth, covering the frequency band indicated for WLAN technology at 5 GHz, indicating than there are possibility of using this type of antennas and substrate

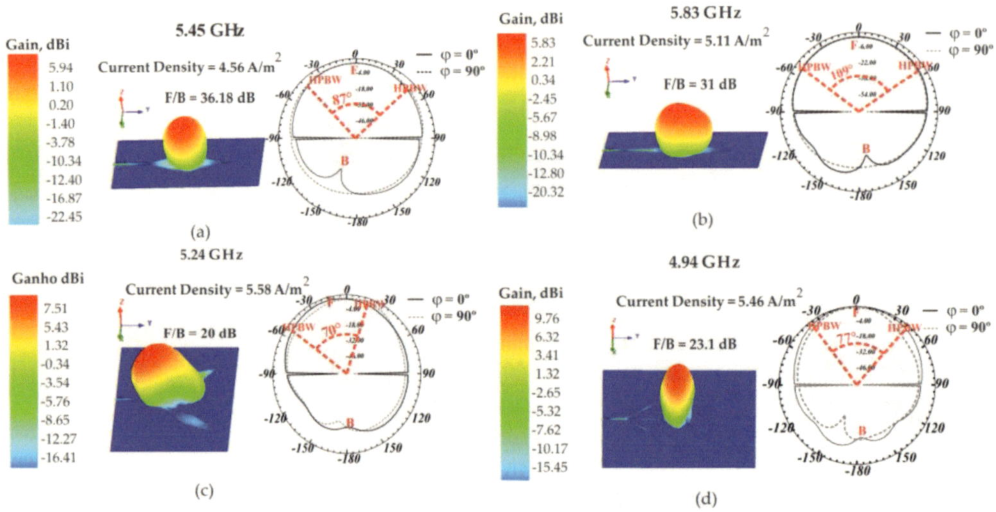

Figure 18. 3D and 2D radiation pattern of wearable antenna arrays bio-inspired in *Inga maritimus* plant shape: (a) 5.4 GHz; (b) 5.8 GHz; (c) antenna array with two leaves; (d) antenna array with four leaves.

in an environment that requires components resistant to the high temperature. In **Table 3**, the measured and simulated values, resonant frequencies at −10 dB, bandwidth (BW), return loss (RL), can be observed of each structure presented.

Figure 18 shows the 2D and 3D radiation patterns at respective resonance frequencies, with indication of gain, front-to-back (F/B) relation, half power beamwidth (HPBW), and current density of simulated wearable antennas bio-inspired in *Inga maritimus* plant shape. As observed, the use of bio-inspired leaf shape on the antenna array permits an increase in both bandwidth and gain. In comparison with wearable bio-inspired antenna with single leaf shape, wearable antenna arrays bio-inspired in *Inga maritimus* plant shape with four leaves presented gain 2.25 dBi greater than antenna array with two leaves, and 3.96 dBi greater than antenna with single leaf shape. From the radiation patterns of the bio-inspired antenna array with four elements, a lower HPBW (77°) can be observed, which indicates a higher concentration of radiated energy, added to the highest gain (9.76 dBi), and a higher front-to-back ratio (23 dB), this indicates that the arrangement can be used for communication of greater distances or with less signal intensity.

4. Conclusion

Wearable wireless systems impose new challenges to antenna design since the utilization in clothes, the proximity of the human body, and the possibility of format variations as bending implies in parameters like good esthetical appearance, low cost, integration to the clothes and accessories used next to the body, among others related to the wear like use of the antenna. In this chapter, we presented some trends for design innovative wearable bio-inspired antennas using plant leaves as inspiration parameterized by the Gielis formula to design antennas, and also we characterize different wearable flexible and low-cost dielectric materials (denim and polyamide). Designed antennas were analyzed by simulation and by measurement in implemented prototypes. The proposed bio-inspiration results in more compact antennas by the reduction of the antennas radiating element. However, compared to Euclidean shapes, two side effects were observed a reduction of the gain and an increase of the current density. On the other hand, the bio-inspired antennas present a higher concentration of the surface current and the decrease of gain can be prevented using leaf arrays with esthetic appeal. The gain can be improved by using thicker substrates and the current density can be regulated using plant shapes with flat geometries or the least sharp possible. These characteristics open a large research field for wearable embedded antennas.

Author details

Paulo Fernandes da Silva Júnior[1]*, Alexandre Jean René Serres[1], Raimundo Carlos Silvério Freire[1], Georgina Karla de Freitas Serres[1], Edmar Candeia Gurjão[1], Joabson Nogueira de Carvalho[2] and Ewaldo Eder Carvalho Santana[3]

*Address all correspondence to: paulo.junior@ee.ufcg.edu.br

1 Electrical Engineering – COPELE, Federal University of Campina Grande – UFCG, Campina Grande, Brazil

2 Electrical Engineering – PPGEE, Federal Institute of Paraiba – IFPB, João Pessoa, Brazil

3 State University of Maranhão, São Luis, Brazil

References

[1] Gupta NP, Maheshwari R, Kumar M. Advanced in ultra wideband antennas for wearable applications. International Journal of Scientific & Engineering Research. 2013;**4**:341-348

[2] Shafique K, Khawaja BA, Tarar MA, Khan BM, Mustaqim M, Raza A. A wearable ultra-wideband antenna for wireless body area networks. Microwave and Optical Technology Letters. 2016;**58**:1710-1715. DOI: 10.1002/mop.29888

[3] Baroni A, Nepa P, Rogier H. Wearable self-tuning antenna for emergy rescue operations. IET Microwaves, Antennas and Propagation. 2016;**10**:173-183. DOI: 10.1049/iet-map.2015.0315

[4] Cavalcante GA, Minervino DR, D'Assunção AG, D'Assunção AG. A compact multiband reject inverted double-E microstrip filter on textile substrate. Microwave and Optical Technology Letters. 2015;**57**:2543-2548. DOI: 10.1002/mop.29370

[5] Silva Júnior PF, Silva PHF, Serres AJR, Freire RCS. Bio-inspired design of directional leaf-shaped printed monopole antennas for 4G 700 MHz band. Microwave and Optical Technology Letters. 2016;**58**:1529-1533. DOI: 10.1002/mop.29853

[6] Xu A, Feumeaux C. Wearable textile shielded strapline for broadband operation. IEEE Microwave and Wireless Components Letters. 2014;**24**:566-568. DOI: 10.1109/LMWC.2014.2321060

[7] Silva Junior PF, Freire RCS, Serres AJR, Silva PHF, Silva JC. Wearable textile bioinsipired antenna for 2G, 3G and 4G systems. Microwave and Optical Technology Letters. 2016;**58**: 2818-2823. DOI: 10.1002/mop.30150

[8] Gielis J. A generic geometric transformation that unifies a wide range of natural and abstract shapes. American Journal of Botany. 2003;**90**:333-338. DOI: 10.3732/ajb.90.3.333

[9] Pakrasi H. Natural Antennas: Structure & Efficiency [Internet]. 2012. Available from: http://parc.wustl.edu/research/themes/biohybrid [Accessed: October 02, 2015]

[10] Falconer K. Fractal Geometry: Mathematical Foundations and Applications. 2nd ed. London: Wiley; 2003. p. 337

[11] Zarghooni B, Dadgarpour A, Pourahmadazar J, Denidni TA. Supershaped metamaterial unit-cells using the Gielis formula. In: IEEE International Symposium on Antennas and Propagation & USNC/URSI National Radio Science Meeting, Vancouver. 19–24 July 2015; pp. 458-459. DOI: 10.1109/APS.2015.7304615

[12] Serres AJR, Serres GKF, Silva Júnior PF, Freire RCS, Cruz JN, Albuquerque TC, Oliveira MA, Silva PHF. Bio-inspired microstrip antenna. In: Chattopladhyay B, editor. Trends in Research on Microstrip Antennas. Rijeka, Crotia: InTech; 2017. p. 88, 109. DOI: 10.5772/intechopen.69766

[13] Silva Junior PF, Freire RC S, Serres AJR, Silva PHF, Silva JC. Bio-inspired antenna for UWB systems. In: 2016 1st International Symposium on Instrumentation Systems, Circuits and Transducers (INSCIT), Belo Horizonte; 29 Aug–3 Sept 2016. pp. 153-157. DOI: 10.1109/INSCIT.2016.7598210

[14] Silva Júnior PF, Freire RCS, Serres AJR, Catunda SY, Silva PHF. Bio-inspired transparent antenna for WLAN application in 5 GHz. Microwave and Optical Technology Letters. 2017;**59**:2879-2884

[15] Khaleel H. Innovation in Wearable and Flexible Antennas. Boston: WIT Press; 2015. p. 215

[16] Balanis CA. Antenna Theory. Analysis and Design. 3nd ed. New York: Wiley; 2005. p. 941

[17] Stutzman WL, Thiele GA. Antenna Theory and Design. New York: Danvers; 2013. p. 820

[18] Mendes C, Peixeiro C. A dual-mode single-band wearable Microstrip antenna for body area networks. IEEE Antennas and Wireless Propagation Letters. 2017;**16**:3055-3058. DOI: 10.1109/LAWP.2017.2760142

[19] Ashyap AYI, Abidin ZZ, Darlan SH, Majid HA, Shah SM, Kamarudin MR, Alomainy A. Compact and low-profile textile EBG-based antenna for wearable medical applications. IEEE Antennas and Wireless Propagation Letters. 2017;**16**:2550-2553. DOI: 10.1109/LAWP.2017.27323

[20] Saeed SM, Balanis CA, Birtcher CR, Durgun AC, Shaman HN. Wearable flexible reconfigurable antenna integrated with artificial magnetic conductor. IEEE Antennas and Wireless Propagation Letters. 2017;**16**:2396-2399. DOI: 10.1109/LAWP.2017.2720558

[21] Lee H, Tak J, Choi J. Wearable antenna integrated into military berets for indoor/outdoor positioning system. IEEE Antennas and Wireless Propagation Letters. 2017;**16**:1919-1922. DOI: 10.1109/LAWP.2017.2688400

[22] Kassem H, Vigneras FLG. Characterization techniques for materials properties measurement. In: Minin I, editor. Microwave Wave Technologies from Photonic Bandgap Devices to Atenna and Applications. Rijeka, Crotia: InTech; 2010. pp. 289-314. DOI: 10.5772/9055

[23] Bai Q, Langley R. Crumpled textile antennas. Electronics Letters. 2009;**45**:436-438. DOI: 10.1049/el.2009.0341

[24] Rano D, Hashmi M. Design and analysis of wearable patch antenna array for MBAN applications. In: Twenty Second National Conference on Communication (NCC), Guwahati; 4–6 March 2016. pp. 1-6. DOI: 10.1109/NCC.2016.7561201

[25] Jasmine tattoo meaning [Internet]. 2015. Available from: http://vse-o-tattoo.ru/wp-content/uploads/2015/06/chinese-tea-jasmine-flower-jpg.jpg [Accessed: 30 November 2017]

[26] Gonçalves FF, Macedo CC, Pasin LAAP. Caracterização Morfológica da Bidens Pilosa (In Portuguese). Revista Científica da FEPI. 2015;**6**:1-3

[27] Serres AJR, Serres GKF, Silva Júnior PF, Freire RCS, Cruz JN, Albuquerque TC, Oliveria MA, Silva PHF. Bio-inspired micristrip antenna. In: Trends in Research on Microstrip Antennas. Rijeka, Crotia: Intech; 2017. pp. 87-109. DOI: 10.5772/intechopen.69766

[28] Garcia FCP, Fernandes JM. Inga. In: Listas de Espécies da Flora do Brasil (In Portuguese), Rio de Janeiro [Internet]. 2012. Avalilable from: http://floradobrasil.jbrj.gov.br/ [Accessed: 10 December 2016]

Chapter 12

Middleware-Driven Intelligent Glove for Industrial Applications

Farouq Muhammad Aliyu and Basem Almadani

Additional information is available at the end of the chapter

http://dx.doi.org/10.5772/intechopen.76382

Abstract

It is estimated that by the year 2020, 700 million wearable technology devices will be sold worldwide. One of the reasons is the industries' need to increase their productivity. Some of the tools welcomed by industries are handheld devices such as tablets, PDAs and mobile phones. However, handheld devices are not ideal for industrial applications because they often subject users to fatigue during their long working hours. A viable solution to this problem is wearable devices. The advantage of wearable devices is that they become part of the user. Hence, they subject the user to less fatigue, thereby increasing their productivity. This chapter presents the development of an intelligent glove, which is designed to control actuators in an industrial environment. This system utilizes RTI connext data distributed service middleware to facilitate communication over WiFi. Our experiments show very promising results with maximum power consumption of 310 mW and latency as low as 23 ms. These results make the proposed system a perfect fit for most industrial applications.

Keywords: RTI, DDS, RTI connext DDS, ubiquitous computing, Internet of Things, embedded systems

1. Introduction

In manufacturing industries, automated industrial systems that are intelligent and cheap are necessary in order to increase efficiency and productivity [1, 2]. Traditionally, such intelligent systems are wired. However, they are now wireless, thanks to the increase in reliability and flexibility of wireless communication systems [3]. Most wireless control systems are controlled through one type of handheld device or another. This is due to their computational capabilities and energy efficiency. But recent developments in embedded systems have given rise to smaller yet more powerful embedded processors. As such, researchers are now turning their attention toward wearable devices.

In wearable sensor network (WeSN), sensor nodes are designed to gather physiological and kinetic data from the wearers' bodies [4]. These devices extend the users' intelligence and their ability to interact with the environment. They also take advantage of the intimacy between the user, the computer and the environment [5]. Wearable sensors can be helpful in industrial applications that require mobile users [6]. They provide more natural avenues of interacting with the system [7, 8]. In some scenarios handheld devices are inapplicable or too expensive [9, 10]. Furthermore, embedded devices are design specific. Therefore, they are optimized for a given function, unlike their handheld counterparts. This gives them upper hand when it comes to accuracy and reliability [11].

Anliker et al. [12] discussed the properties of a good WeSN. The authors stated that a good WeSN should have: (1)The ability to act in a proactive manner rather than depend on user's commands; (2) the ability to gather complex information regarding the user's interaction with the world; (3) high connectivity and mobility; (4) the ability to process wide variety of data at high speeds with low power consumption; (5) the ability to combine I/O devices and sensors from different parts of the body into one heterogeneous system; and (6) inconspicuous appearance such that it does not change the users' look in an unacceptable way or hinder the user from carrying out his/her normal day-to-day activities.

Sadly, there has been little research in WeSN with focus on industrial applications. This is due to the fact that distributed embedded systems at an industrial scale are not easy to maintain and control. In this chapter, we present the development of an intelligent hand glove for an industrial-based networked control system (NCS). NCS is a distributed control system (DCS) where sensors, actuators and controllers communicate over a network [13]. In order to ensure real-time performance, real time innovations (RTI) connext middleware is used. The middleware is used to manage the heterogeneous distributed system at real time. Also, the project is designed to investigate whether intelligent wearable sensors can replace the present handheld devices used in industrial NCS applications. We do believe that many users are in support and keen to use wearable devices than handheld devices. This is because they are lighter, cheaper, consume less energy and are more convenient to carry around.

2. Industrial application of embedded devices

Researchers were able to apply handheld devices such as smart phones and personal digital assistants to networked control system (NCS) [14]. Görlich et al. [15] developed an NCS where a Profibus-to-Bluetooth connection is used to control and parameterize a flow unit via Nokia 6280 mobile phone. Nonetheless, this system suffers from incompatibility, limited communication range and low bandwidth. Huang et al. [6] developed an oil and gas storage and transportation simulation system (OGSTSS). The proposed OGSTSS system uses wireless local area network (WLAN) and an android tablet PC in order to cover longer range and obtain high bandwidth. The tablet PC communicates with programmable logic controllers (PLCs) via the PC station. This approach has some problems: first, the workstation between the tablet PC and the PLCs exerts delay overhead even though WLAN is used to speed up communication between the tablet PC and the PLCs and second, albeit tablet PCs are hand-held PCs, their size may discomfort and/or distract the user in some industrial applications.

Nowadays, the developments in embedded systems allow researchers to replace tablets with smaller footprint but equally powerful processing devices. Yang et al. [16] developed a device that traces the wearer's body position and his movements using PIC microcontroller, a tri-axial accelerometer and an RF transmitter. This device is intended for medical applications. It comes in handy for people with the tendency to syncope while they are alone. However, the system lacks enough computation power. Hence it cannot cope with the rigorous data processing of industrial applications. Single board computers (SBCs) like Beagle Bone, Intel Galileo and Raspberry Pi have small footprint yet they have large processing power [17, 18, 19]. This enabled the researchers to replace phones, tablets and computers with SBC in control systems and WeSN [20]. Wang et al. [21] developed a wearable sensor network using Raspberry Pi. The system is used for crowd-sensing. It uses transcoding algorithm to reduce resolution of video files. Khelil et al. [22] developed a body sensor network (BSN) using Raspberry Pi. The system uses cooking hacks e-health shield, which contains an airflow sensor for measuring wearer's respiratory rate, a pulse oximeter for measuring oxygen saturation of hemoglobin and position sensor for motion tracking. Jutila et al. [23] used Intel Galileo as a sink node for a wearable system that helps track children at school.

The aforementioned SBC systems revealed promising results. However, there exists incompatibility between the different types of SBCs. Therefore, a WeSN system can only contain one type of SBC. This problem is a serious setback in industrial applications of WSN and/or WeSN. This incompatibility forces engineers to stick to one type of device, which exposes them to the tendency of discontinued support by the producing company. In order to tackle this problem, [24, 25] proposed the use of middleware. A middleware is a software capable of hiding the complexity and heterogeneity between system components. It increases the ease of managing systems' resources and improves the predictability of the system [26].

Fok et al. [27] developed a fire tracking system which uses mobile agents based on Agilla middleware. This system allows a fire tuple to be injected into the network, where they clone themselves and move from one node to another until they form a perimeter bounding the fire. The system recorded huge success and also proves that a multipurpose and multitask WSN can be developed. Zhai et al. [28] developed a publish/subscribe middleware called wireless message-oriented system (WMOS) on top of TinyOS. The authors were able to reduce overall energy consumption of the network through the use of self-adaptive quality of service (QoS) services and content/topic model. Zug et al. [29] successfully applied a publish/subscribe middleware Family of Adaptive Middleware for autonomOUs Sentient Objects (FAMOUSO) to an industrial automation system using Katana robot. The robot publishes its coordinates and angles and it subscribes to movement, speed and emergency stop commands. This middleware can only work optimally in certain fixed conditions as it does not have quality of service (QoS) parameters. These parameters can be set to fine tune the performance of the middleware thereby improving the performance of the system at large. Furthermore, the aforementioned middlewares are

WSN middlewares. Hence, they cannot be used in industrial environment, where a heterogeneous system consisting of embedded devices and computers is readily found.

The proposed system presented in this chapter is a smart hand glove that accepts hand gesture as input, processes it using an ARM microprocessor (Raspberry Pi) and then publishes it using the RTI connext professional (publish/subscribe middleware) via WiFi. With the proposed architecture, we were able to resolve the data processing constraints, the large hardware footprint problem and the communication delays encountered by the industrial-based wearable systems.

3. System architecture

Figure 1 shows the proposed system architecture. The system architecture consists of the following four main components: (1) a wearable sensor named Al-yad. The main function of this component is to convert the physical movements into electric signals. The electric signal is then conditioned so that it can be processed by the 3.3 v ARM-6 microprocessor, Raspberry Pi in the case of this research; (2) a real-time innovations connext professional is selected as a middleware to mediate between the wearable sensor and other computing devices of the system. This middleware is selected because of its ability to work on both embedded devices and workstations. It also has a wealth of QoS that can be used to optimize communication as well as energy consumption; (3) a computing device can be any device capable of running

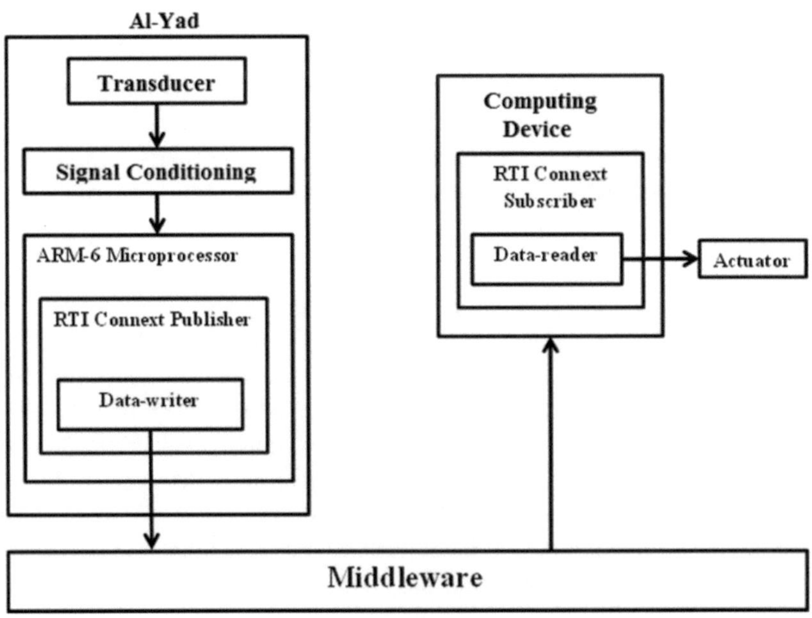

Figure 1. System structure.

connext professional (e.g., PC, Beagle Bone, Intel Galileo or Raspberry Pi); and (4) an actuator represents any industrial machinery and that is required to be controlled remotely.

3.1. Al-yad

Al-Yad is a smart hand glove made up of a combination of light sources (light emitting diodes (LEDs)) and light sinks (light dependent resistors (LDRs)). The 5 mm round red LED is chosen because the LDR is very sensitive to it and it consumes a maximum of 78 mw. A 50 mW, N5 AC-501085-type LDR is chosen for the proposed system because it has small-form factor and consumes small amount of energy.

The LDRs and the LEDs are in line of sight when fingers of the wearer are stretched straight. This gives a high signal (see **Figure 2**). As the user bends his fingers the light stops reaching the LDRs. This raises the resistance of the LDRs, thus, giving low output signal. In order to ensure no ambient light contaminates the system, the LDR is placed in a hollow cylinder of 1.45 cm in depth. Furthermore, the light source is pulsed at a frequency of 5 kHz, which is the highest frequency of PWM based on "wiringpi.h" implementation [30]. The highest frequency available is chosen because the higher the frequency, the smaller the capacitor required, therefore, the lighter the hand glove. Lemma 3.2 shows how the size of the capacitor C2 is derived. The pulsing light and the ambient light add up to form the simulated waveform, as shown in **Figure 3**. The direct current (DC) component of the wave

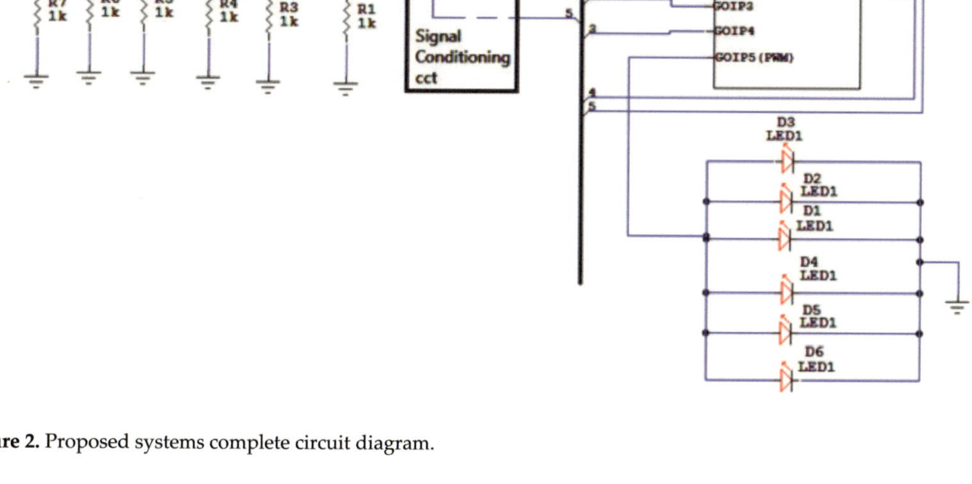

Figure 2. Proposed systems complete circuit diagram.

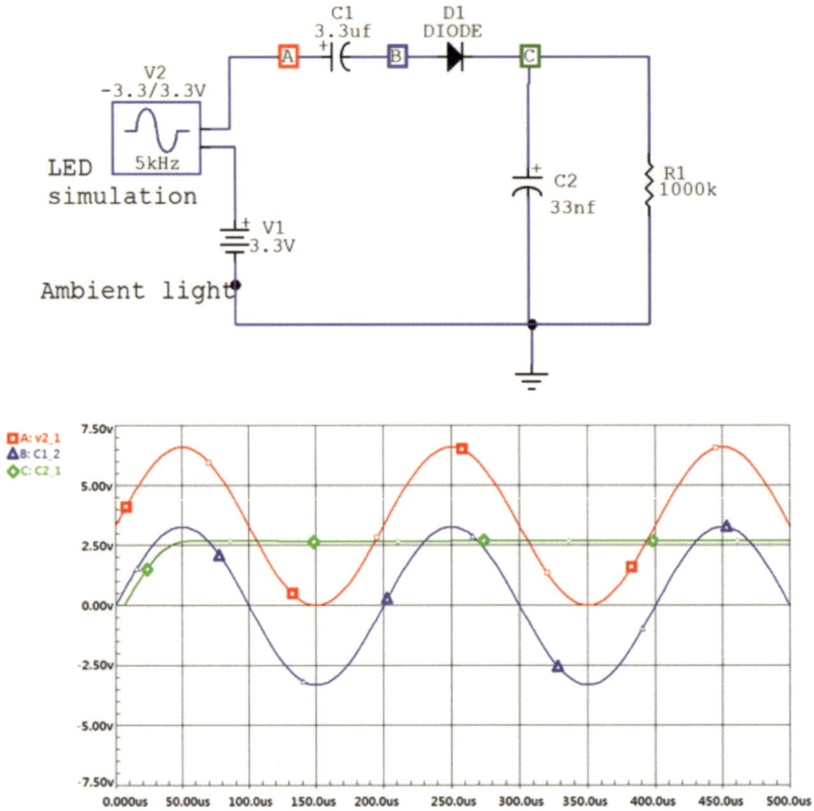

Figure 3. Transducer circuit design.

is formed as a result of the ambient light. A coupling capacitor (C1) is added to the circuit in order to remove the DC component. Lemma 3.2 shows how the value of the capacitor is derived.

Lemma 3.1: Let γ = 0.1% be the ripple factor of the half wave DC and f = 5 kHz be the frequency of the pulsing light. Since the input impedance (R_L) of a 7414 Schmitt trigger is very high, let R_L = 1 MΩ. Hence, the capacitance of C2 capacitor can be calculated as:

$$\begin{aligned} \gamma &= \frac{1}{2\pi \times f \times C2 \times R_L} \\ 0.1\% &= \frac{1}{2\pi \times 5000 \times C2 \times 1 \times 10^6} \\ C2 &= 31.8310 \times 10^{-9} \\ C2 &\approx 33nf \end{aligned} \tag{1}$$

Lemma 3.2: Let X_{C1} be the capacitive reactance of the capacitor C1 and the pulsing light frequency (f) is 40 kHz. Since the input impedance of a 7414 Schmitt trigger is very high, then any value can be chosen for X_{C1}. For simplicity, we choose X_{C1} = 1. Hence, capacitance of C1 capacitor can be calculated as follows:

$$\begin{aligned} X_{C1} &= \frac{1}{2\pi \times f \times C1} \\ 1 &= \frac{1}{2\pi \times 5000 \times C1} \\ C1 &= \frac{1}{2\pi \times 5000 \times 1} \\ C1 &= 3.1831 \times 10^{-6} \\ C1 &\approx 3.3\mu f \end{aligned} \tag{2}$$

In addition, a rectifying diode (1 N4001) is added to the system in order to prevent negative voltage from reaching the Schmitt trigger (74LS14). 1 N4001 is used because it is the smallest diode that can be used without compromising the reliability of the system. A C2 capacitor is added to the system in order to reduce the ripples down to 0.1% so that the Schmitt trigger receives an almost constant direct current (DC) voltage. Finally, the Schmitt trigger (LM7414) smoothens the signal, making it a perfect DC, which is then passed to the microprocessor without any debouncing effect [31]. In order to ensure accurate data acquisition, the sampling frequency for the hand movement is chosen to be 5 Hz which is a little more than what is recommended by Yuan et al. [32]. Six sensors are placed in three fingers, thereby producing six bits. Two zeros are padded to the upper two bits, as shown in line 3–10 of Algorithm 1. The resulting 8-bit signal is then converted into an integer, which is then published as a command.

The processing unit is made up of a single board computer (SBC) [33]. Raspberry Pi is chosen because it has abundant libraries in Python, C, C++ and Java. The SBC is programmed to scan the glove and convert the gloves input to an integer and publish it through the middleware. Algorithm 1 illustrates the behavior of the SBC in the system.

Quality of service	Value
Durability	kind = VOLATILE
Reliability	kind = RELIABLE max blocking time = 100 ms
Liveliness	kind = AUTOMATIC lease duration = 1 s
History	kind = KEEP LAST depth = 5
Presentation	Access_scope = TOPIC Ordered_access = True Coherent_access = False
Owner	kind = EXCLUSIVE
Owner_Strength	5
Time_Based_Filter	Minimum separation = 0:2 s

Table 1. QoS used in the proposed systems' middleware.

3.2. Middleware

Research has shown that DDS middleware can achieve up to 75% success rate in mobile node applications [34]. This and the reason that it hides heterogeneity of a system encourages the use of middleware in our proposed solution. The middleware chosen for this task is RTI connext middleware. This middleware is chosen due to its strict compliance with the object management group (OMG) specification for data distribution service middleware [35]. The middleware is implemented according to Algorithm 1 using C++ programming language.

Algorithm 1: Pseudo-code for Publisher side

```
gesture[ ]=0
while (True) do
    gesture[0] = GPIO0
    gesture[1] = GPIO1
    gesture[2] = GPIO2
    gesture[3] = GPIO3
    gesture[4] = GPIO4
    gesture[5] = GPIO5
    gesture[6] = 0
    gesture[7] = 0
    int gestureCommand = 0
    for (int i = 0; i < 8; i + +) do
    gestureCommand=gestureCommand + gesture[i] X 2^i
    end for
    Publisher.datawriter.write(gestureCommand)
    wait for 0.2s
end while
```

The QoS is configured in both the hand glove and the actuator in order to ensure compatibility and efficiency. **Table 1** shows the quality of services used in the proposed system. The first four QoSs in **Table 1** are the recommended QoSs for alarm/event systems by [36]. The remaining QoSs are selected by the authors in order to ensure that the system behaves in a controller-actuator manner. The following is an explanation for the used QoSs:

1. *Durability* is set to *"VOLATILE"* so that data is not stored for future datareaders (i.e., targeted machine/actuator). This will prevent the targeted machine from executing any command that was issued before it was turned on, hence reducing the possibility of accidents.

2. *Reliability* of communication is set to *"RELIABLE"* and the maximum blocking time is left at the default value, which is 100 ms. The setting ensures that all commands published reach the targeted subscriber.

3. *Liveliness* allows datareader to know when the datawriter (i.e., the wearable sensor) is dead or disconnected. It is set to *AUTOMATIC* and the lease time is set to 1 s so that the datareader automatically checks the presence of the datawriter every second.

4. *History* QoS allows the datawriter to cache some samples. Here, the datawriter is allowed to queue only five samples.
5. *Presentation* is set such that the middleware will ensure that the samples are accessed in the order they were sent by the operator.
6. *Owner* is set to *EXCLUSIVE* so that only the hand glove with the highest owner strength can control the target machine.
7. *Owner_Strength* is set to five. This is the maximum number of gloves that can be used in the system. Therefore, only when a glove with the strength of five is down the glove with strength four takes over and so on.
8. *Time_Based_Filter's* minimum separation is the minimum time period before a new command is sent to the datareader. Since the sampling rate is 5 Hz, then the minimum separation = 1/(5 Hz) = 0.2 s.

3.3. Processing device

When a user wants to control a machine, he/she makes a gesture equivalent to the required command. The gesture is converted into a byte of command and is transmitted to a processing device. The processing device is connected to the machine to be controlled. It is responsible for subscribing and converting the sent signal into commands usable by the target machine. Any device capable of using RTI middleware can be used as the processing device. In the case of our experiment a PC workstation with Intel Core i5 microprocessor and Atheros AR8161/8165 Gigabit Ethernet Controller is used.

4. Experiments

The sensor casings are designed using Blender 2.6 3D modeling software and they are printed with Replicator2 3D printer. LEDs and LDRs are placed in the 3D printed casings. They are connected in a manner, as shown in **Figure 4**. The conditioning circuit is then connected to the SBC/Raspberry Pi, where the data is processed and transmitted to the actuator. In our experiments, a PC is used as the actuator. A RTI subscriber is installed on a PC and publisher on the Raspberry Pi. The gesture commands sent by the Raspberry Pi are received by the PC and displayed on the monitor. It is with this setup that all experiments are carried out.

In this research throughput, latency, power consumption and comfort of the system are evaluated. The setup in **Figure 5** is used to measure both latency and throughput. RTI's performance toolkit, "Perftest," is used [37]. To test latency of the system, packets are sequenced and time stamped before they are transmitted to the datareader (receiver). The packets are immediately re-routed (i.e., bounced) back to the transmitter where the round trip time (RTT) is calculated. Similarly, the receiver counts the number of packets received per second in order to calculate the throughput. Finally, the transmitter reports the latency of the system while the receiver reports its throughput. As illustrated in the setup in **Figure 5**, Perftest was executed as a publisher on the Raspberry Pi and as a subscriber on the PC. The latency and

Figure 4. Proposed system.

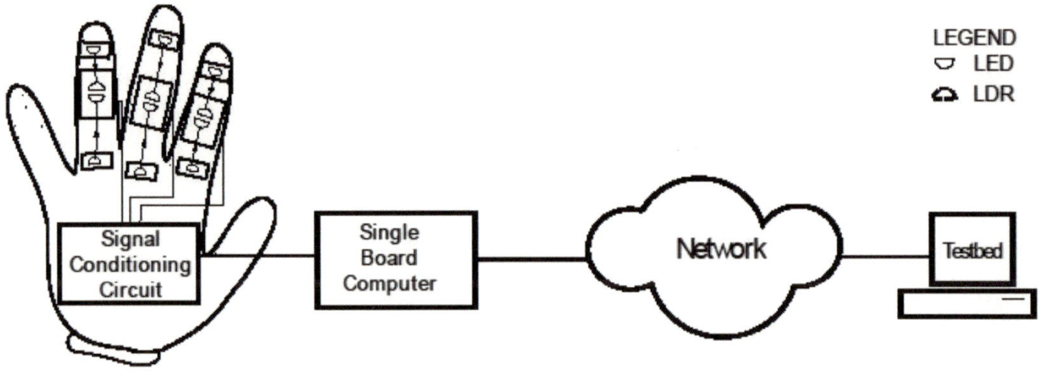

Figure 5. Experiment setup for measuring latency and throughput.

throughput of the system is then measured. The payload size for the experiment is 28 bytes. This packet size is chosen because the proposed system only sends integer values between 0 and 63 and the minimum allowable payload size in RTI connext is 28 bytes. Furthermore, the experiment is carried out using QoS settings and without QoS, so as to investigate the cost of applying the quality of service. **Figures 6** and **7** show the results obtained for latency and throughput of the system, respectively.

We performed an experiment to ascertain the response time of the transducer. This is necessary because LDRs naturally have a very slow response. Therefore, it is important to know what quota of the total delay is contributed to it. In this experiment, the hand glove is connected to an oscilloscope and the output voltage (Vo) is fed to the Raspberry Pi and to the oscilloscope. The light is blocked and the response time is obtained through the oscilloscope, as shown in **Figure 8**.

The setup in **Figure 6** is used to measure the energy consumption of the whole system. A 1 resistor is connected in series with the proposed system so that the current passing through the system and the resistor is the same. As such, the voltage drop across the resistor follows

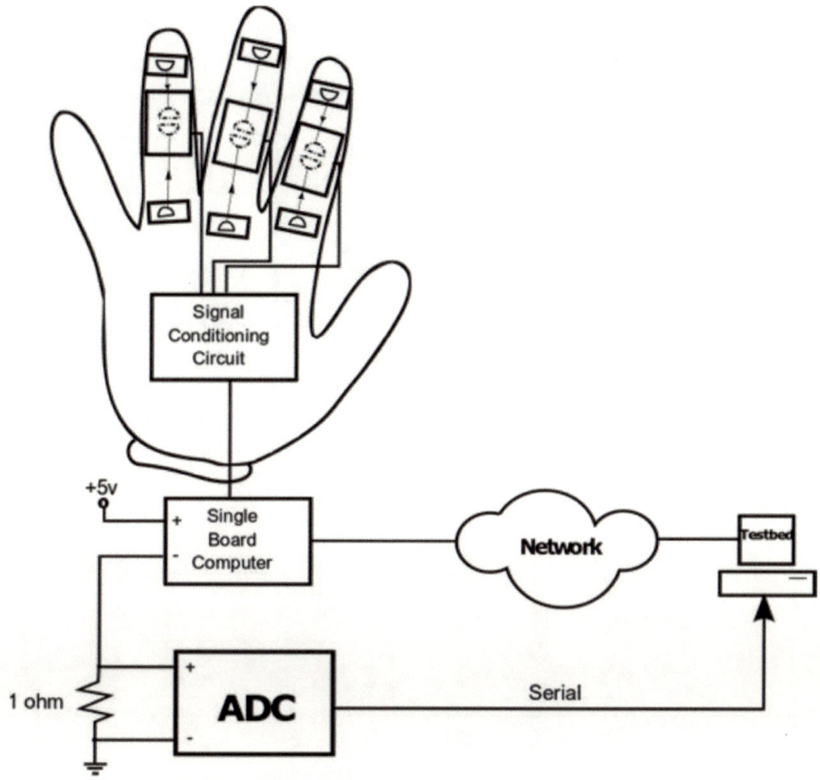

Figure 6. Experiment setup for measuring energy consumed by the system.

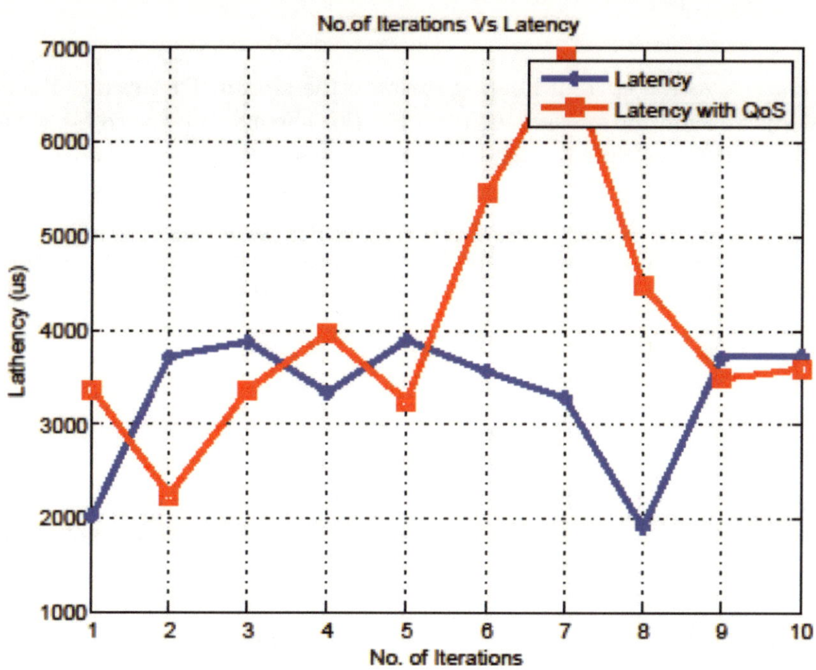

Figure 7. Latency of the proposed system.

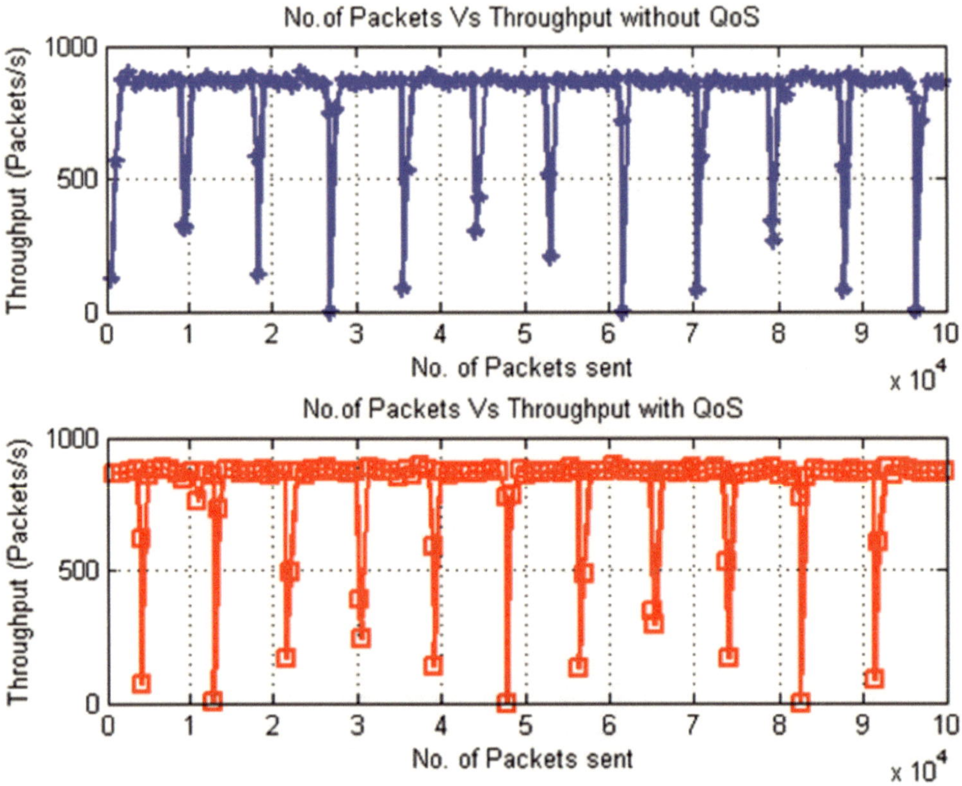

Figure 8. Throughput of the proposed system.

the same behavior as the current flowing through the circuit. To measure the actual current (I) flowing through the system, we used Eq. (3). The assumption in Eq. (3) is safe due to the fact that the Thevenin's resistance of the system is far greater than 1. An analogue to digital converter (ADC) is connected across the 1 resistor. It samples the voltage drop across the resistor every 200 ms. The acquired data is then transmitted to a computer via USB, where it is stored. The current flow through the system is recorded for two transmissions after which the WiFi dongle is immediately removed so as to record the overhead of the dongle itself.

$$
\begin{aligned}
I &= \frac{V}{R} \\
I &= \frac{V}{1} \therefore I = V
\end{aligned} \tag{3}
$$

Finally, a survey is contacted to investigate the ergonomics of the proposed system. Questionnaires are given to eight people. The questionnaire investigated the six attributes of comfort proposed by James and Chris [38]. Their comfort scale consists of the following attributes: emotion, attachment, harm, perceived change, movement and anxiety. The participants are asked to use the hand glove and then they are asked to answer 12 questions. The questions require them to score the attributes between 0 and 10. The score 0 represents the least while 10 represents most severe discomfort on the user.

5. Discussion of results

The latency of the system with and without QoS also showed a minor difference of 1 ms. This can be seen from **Figure 7**. The performance test toolkit calculates an approximation of the latency by dividing round trip time (RTT) by 2. Since this is not a reliable answer, the software also reports the average and the standard deviation, as shown in **Table 2**. The transmission's throughput and latency results obtained show that QoS has little effect on the performance of the proposed system. This is attributed to the small size of the packets transmitted.

Figure 8 shows that there is a little difference in the throughput of the system with or without using QoS. It is found that the average throughput of the system with QoS is 769 and 778 Mbps without QoS. In both cases no packets are lost during the 100,000 packet transmission. Given the fact that we are only sending commands in the form of integers over the network, this result is highly acceptable.

Moreover, the total latency of the system is the sum of the latency due to transmission and the inherent delay of the electric circuit that reads the hand gesture and places it on a bus for the SBC to read. **Figure 9** shows us that both the rising time of the signal and its falling time are approximately 20 ms. Therefore, the total delay of the system is 23 ms, as calculated using Eq. (4).

$$t_{delay} = t_{glove} + t_{trans}\, t_{delay} = 20 + 3.297\, t_{delay} = 23.3 \text{ ms} \tag{4}$$

Figure 10 shows the current (top) and the moving average of the current (bottom) flowing through the system during and after data transmission. Immediately after the last transmission, the WiFi dongle is removed, hence, the spike at 375th s. However, no change in the

Parameter	Value obtained
Ave latency	3.297 ms
Std. deviation	0.6983 ms
Min. latency	1.905 ms
Max. latency	3.881 ms
50th percentile	3.707 ms
90th percentile	3.881 ms
99th percentile	3.881 ms
99.99th percentile	3.881 ms

Table 2. Latency results.

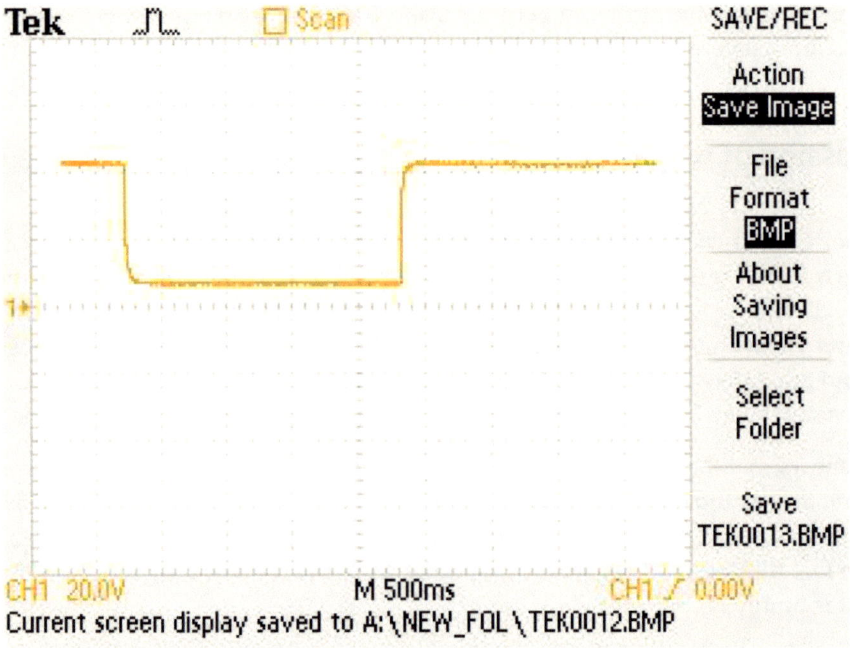

Figure 9. Time response of Al-Yad.

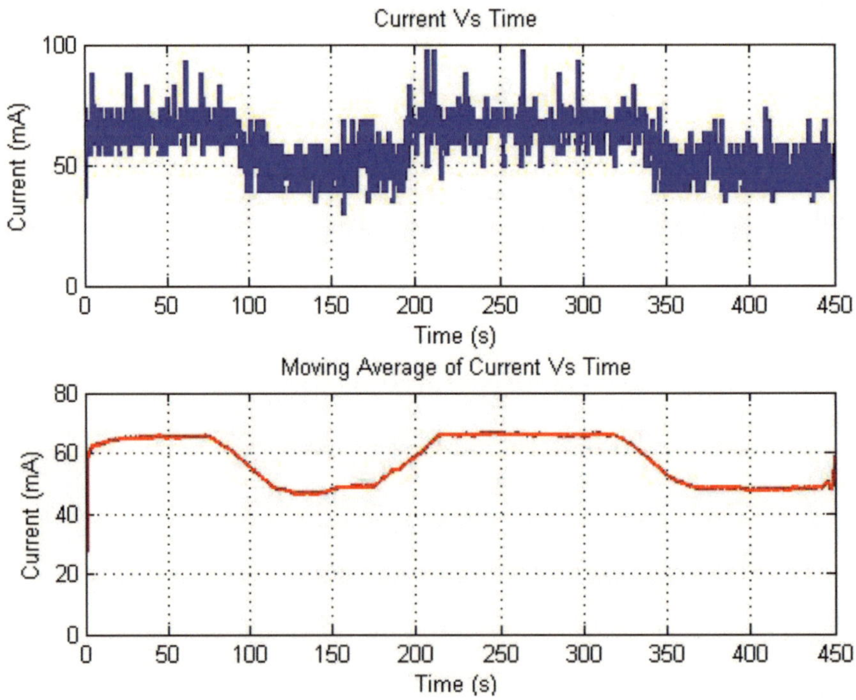

Figure 10. Current consumed by system while running.

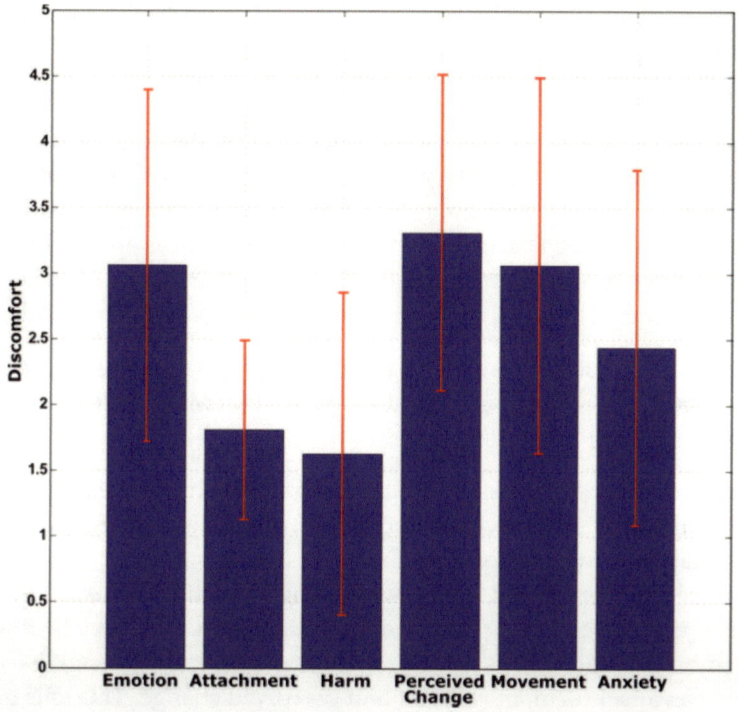

Figure 11. Survey result for comfort-ability of the hand glove.

	Emotion	Attachment	Harm	Perceived change	Movement	Anxiety
Emotion	1.0000	0.6436	0.8205	0.6207	0.5691	0.6397
Attachment		1.0000	0.6517	0.7298	0.7685	0.5202
Harm			1.0000	0.6726	0.7477	0.5358
Perceived change				1.0000	0.7273	0.0309
Movement					1.0000	0.4969
Anxiety						1.0000

Table 3. Correlation-coefficient matrix for comfort-ability of the hand glove.

flow of current is recorded after the dongle is removed. This shows that the energy overhead comes from data transmission itself.

Power dissipated by the system can be calculated by multiplying the current in **Figure 10** by the voltage across the system (i.e., 5 v). This gives an average of 310 mW during transmission and 220 mW when the system is not transmitting.

Finally, a survey was carried out to access the comfort of the hand glove. **Figure 11** shows the result of the survey and **Table 3** shows the correlation between the different attributes investigated. Eight participants were asked about the comfort of the hand glove. The participants were asked about their experience based on: emotion, attachment, harm, perceived change, movement and anxiety [38].

Emotion is whether the user is worried about his looks or on the edge because of wearing the glove. Attachment is the feeling the user gets when the glove is moving around—that constant feeling the user gets of wearing something. Harm is when the user fears being harmed by the glove. Perceived change is when the user feels physically different as opposed to emotion which is a psychological difference. Movement attribute is the measure of the glove which affects the user's movement. Anxiety is the worry a user gets about the glove's safety and reliability.

The attributes were tested on a scale of 0–10, with 0 being least negative and 10 being the most negative experience. The results show that the overall discomfort of the glove is well below the scale of five (see **Figure 11**).

However, perceived change, movement and emotion show higher readings. To investigate further, correlation coefficient between the different attributes is computed. **Table 3** shows the correlation matrix between all the attributes. It can be seen that there is a very strong relationship between emotion and harm, which suggests that the user's emotion toward the device is due to the user feeling some kind of mild harm. This may be due to the flying wires and the printed circuit boards (PCBs) that were glued to the glove. Also there is strong correlation between attachment, movement and perceived change. This tells us that the user could not move well or he is afraid of moving because he feels that the glove is loose, which leads to the user feeling physically different. Harm is strongly correlated with movement restriction because the user feels mild pain, in addition to moving parts mentioned earlier.

In conclusion, the users show emotion, perceived change and movement restriction the most because of the way the glove was packaged. To improve the glove, Raspberry Pi may be replaced with a smaller SBC like Beagle Bone. PCBs should be covered with slim plastic casing before attaching it firmly to the glove. This will help prevent parts from falling over and reduce harm due to protruding electronic components and heat. Flying wires can be covered by a layer of fabric. This will also reduce the movement of the components on the glove.

6. Summary

This chapter discussed our proposed intelligent glove designed to control actuators in an industrial environment setting. The glove is constructed using Raspberry Pi, a WiFi dongle and an assembly of light sources and light sensors. It was found that the system has a delay of approximately 23 ms, which is fast enough for controlling motors, pneumatic actuators and other mechanical actuators. Furthermore, the maximum power dissipated by the system is 310 mW, which showed that the system could be powered by a mobile phone's Lithium ion battery (3100 mAh) for up to 10 h of continuous data transmission. Moreover, a survey was carried out to investigate the comfort ability of the glove: It was found that the glove was acceptable but users have safety concerns and movement restrictions. This problem can be solved by repackaging the device.

Author details

Farouq Muhammad Aliyu* and Basem Almadani

*Address all correspondence to: g201303650@kfupm.edu.sa

Department of Computer Engineering, King Fahd University of Petroleum and Minerals, Dhahran, Saudi Arabia

References

[1] Waris MM, Sanin C, Szczerbicki E. Smart innovation engineering: Toward intelligent industries of the future. Cybernetics and Systems. [Online]. 2018:1-16. Available from: https://doi.org/10.1080/01969722.2017.1418708

[2] Gungor V, Hancke G. Industrial wireless sensor networks: Challenges, design principles, and technical approaches. IEEE Transactions on Industrial Electronics. 2009;**56**(10): 4258-4265

[3] Aliyu F, Sheltami T. Development of an energy-harvesting toxic and combustible gas sensor for oil and gas industries. Sensors and Actuators B: Chemical. 2016;**231**:265-275

[4] Elsarnagawy T, Farrag M, Haueisen J, Abulaal M, Mahmoud K, Fouad H, Ansari S. A wearable wireless respiration rate monitoring system based on fiber optic sensors. Sensor Letters. 2014;**12**(9):1331-1336

[5] Billinghurst M, Starner T. Wearable devices: New ways to manage information. Computer. 1999;**32**(1):57-64

[6] Huang Z, Lu Y. Wireless monitoring and control system via android tablet pc. In: International Symposium on Computer, Communication, Control and Automation (3CA 2013). Atlantis Press; 2013. pp. 449-452

[7] Aleksy M, Rissanen MJ. Utilizing wearable computing in industrial service applications. Journal of Ambient Intelligence and Humanized Computing [Online]. 2012;**5**(4):443-454 Available from: http://dx.doi.org/10.1007/s12652-012-0114-2

[8] Benavente-Peces C, Ahrens A, Filipe J. Advances in technologies and techniques for ambient intelligence. Journal of Ambient Intelligence and Humanized Computing. 2014;**5**(5):621-622

[9] Westelaken R, Hu J, Liu H, Rauterberg M. Embedding gesture recognition into airplane seats for in-flight entertainment. Journal of Ambient Intelligence and Humanized Computing [Online]. 2010;**2**(2):103-112 Available from: http://dx.doi.org/10.1007/s12652-010-0032-0

[10] Kikhia B, Stavropoulos TG, Meditskos G, Kompatsiaris I, Hallberg J, Savenstedt S, Melander C. Utilizing ambient and wearable sensors to monitor sleep and stress for people with bpsd in nursing homes. Journal of Ambient Intelligence and Humanized Computing [Online]. 2015:1-13 Available from: http://dx.doi.org/10.1007/s12652-015-0331-6

[11] Rachael R. Wearables vs. smartphone apps: Which are better to count steps? LiveScience [Online]. 2015. Web 15 May 2018. Available from: http://www.livescience.com/49756-smartphone-apps-wearables-step-counts.html

[12] Anliker U, Beutel J, Dyer M, Enzler R, Lukowicz P, Thiele L, Troster G. A systematic approach to the design of distributed wearable systems. IEEE Transactions on Computers. 2004;**53**(8):1017-1033

[13] Zhao Y-B, Liu G-P, Kang Y, Yu L. A brief tutorial of networked control systems. In: Packet-Based Control for Networked Control Systems. Singapore: Springer; 2018. pp. 1-11

[14] Jianjun Z, Min'Gang C, Man Z, Su L. Remote monitoring and automatic navigation system for agricultural vehicles based on wlan. In: 4th International Conference on Wireless Communications, Networking and Mobile Computing, 2008. WiCOM '08; October 2008. pp. 1-4

[15] Gorlich D, Stephan P, Quadflieg J. Demonstrating remote operation of industrial devices using mobile phones. In: Proceedings of the 4th International Conference on Mobile Technology, Applications, and Systems and the 1st International Symposium on Computer Human Interaction in Mobile Technology. ACM; 2007. pp. 474-477

[16] Yang C-C, Hsu Y-L. Development of a portable system for physical activity assessment in a home environment. In: Proceedings of International Computer Symposium; Taipei Taiwan; 2006. pp. 1339-1344

[17] Coley G. Beaglebone Black System Reference Manual; Beaglebone.org [Online]. 2013. Available from: https://cdn-shop.adafruit.com/datasheets/BBB_SRM.pdf

[18] Ramon MC. Intel galileo and intel galileo gen 2. In: Intel R Galileo and Intel R Galileo Gen 2. Springer [Online]; 2014. pp. 1-33. Available from: https://link.springer.com/book/10.1007/978-1-4302-6838-3

[19] Pi R. Raspberry pi 3 model b (8 April 2017), Retrieved April. Vol. 9; 2017. p. 2017

[20] Shah J, Patel VP. Real time interfacing & control techniques using an open source. International Journal of Science, Engineering and Technology Research (IJSETR). 2014;**3**(3)

[21] Wang S-T, Fan C-L, Huang Y-Y, Hsu C-H. Transcoding on wearable devices: Technical report; 2015

[22] Khelil A, Shaikh F, Sheikh A, Felemban E, Bojan H. Digiaid: A wearable health platform for automated self-tagging in emergency cases. In: 2014 EAI 4th International Conference on Wireless Mobile Communication and Healthcare (Mobihealth)[Online]; November 2014. pp. 296-299. Available from: http://ieeexplore.ieee.org/abstract/document/7015969/

[23] Jutila M, Strmmer E, Ervasti M, Hillukkala M, Karhula P, Laitakari J. Safety services for children: a wearable sensor vest with wireless charging. Personal and Ubiquitous Computing [Online]. 2015;**19**(1):1-13 Available from: http://dx.doi.org/10.1007/s00779-015-0838-z

[24] Park S, Hosen AS, Cho G. An dds based architecture in supporting of data centric wireless sensor network environments. International Journal of Control and Automation. 2014;**7**(1):251-258

[25] Boonma P, Suzuki J. Tinydds: An interoperable and configurable publish/subscribe middleware for wireless sensor networks. Principles and Applications of Distributed Event-Based Systems [Online]; 2009. pp. 206-231. Available from: http://www.cs.umb.edu/~jxs/pub/pernets09.pdf

[26] Mascolo C, Hailes S, Lymberopoulos L, Picco GP, Costa P, Blair G, Okanda P, Sivaharan T, Fritsche W, Karl M, et al. Survey of middleware for networked embedded systems. Project Report: http://www.ist-runes.org/docs/deliverables/D5 01.pdf; 2005

[27] Fok C-L, Roman G, Lu C. Mobile agent middleware for sensor networks: An application case study. In: Fourth International Symposium on Information Processing in Sensor Networks, 2005. IPSN 2005 [Online]; April 2005. pp. 382-387. Available from: https://dl.acm.org/citation.cfm?id=1147747

[28] Zhai L, Li C, Sun L. Research on the message-oriented middleware for wireless sensor networks. Journal of computers. 2011;**6**(5):1040-1046

[29] Zug S, Schulze M, Dietrich A, Kaiser J. Programming abstractions and middleware for building control systems as networks of smart sensors and actuators. In: 2010 IEEE Conference on Emerging Technologies and Factory Automation (ETFA)[Online]; September 2010. pp. 1-8. Available from: http://ieeexplore.ieee.org/stamp/stamp.jsp?tp=&arnumber=5641341&isnumber=5640954

[30] Weaver WP. Wiringpi. Software tone library [Online]. 2018. Available from: http://wiringpi.com/reference/software-tone-library/

[31] Instrument T. Hex schmitt-trigger inverters datasheet; 1988. pp. 1-11

[32] Yuan H, Perdoni C, He B. Relationship between speed and eeg activity during imagined and executed hand movements. Journal of Neural Engineering. 2010;**7**(2):026001

[33] Langdon GG. Single board computer. Computer Design. San Jose: Computeach Press; 1982

[34] Shen H, Li T. Data management of wireless sensor networks. In: Consumer Communications and Networking Conference, 2009. CCNC 2009. 6th IEEE [Online]; January 2009. pp. 1-2. Avalable from: https://ieeexplore.ieee.org/abstract/document/4784851/

[35] OMG. Data distribution service for real-time systems version 1.2. Object Management Group [Online]; 2015 . Available from: https://www.omg.org/spec/DDS/1.4/

[36] Hunt G. Is your data secured? RTI [Online]. July 2014. Available from: http://www.slideshare.net/RealTimeInnovations/ is-your-datasecure. pp. 20-21

[37] RTI. RTI Connext DDS Performance Test. Data distribution service (DDS) community RTI connext users. RTI [Online]; 2018. Available from: https://community.rti.com/kb/example-performance-test-rti-message-service

[38] James FK, Chris B. A tool to assess the comfort of wearable computers. Human Factors. 2005;**47**(1):77-91